Advances in Intelligent and Soft Computing

144

Editor-in-Chief

Prof. Janusz Kacprzyk
Systems Research Institute
Polish Academy of Sciences
ul. Newelska 6
01-447 Warsaw
Poland
E-mail: kacprzyk@ibspan.waw.pl

For further volumes:
http://www.springer.com/series/4240

Ford Lumban Gaol and Quang Vinh Nguyen (Eds.)

Proceedings of the 2011 2nd International Congress on Computer Applications and Computational Science

Volume 1

 Springer

Editors
Ford Lumban Gaol
Bina Nusantara University
Perumahan Menteng
Caking
Indonesia

Quang Vinh Nguyen
School of Computing and Mathematics
University of Western Sydney
Penrith
New South Wales
Australia

ISSN 1867-5662   e-ISSN 1867-5670
ISBN 978-3-642-28313-0   e-ISBN 978-3-642-28314-7
DOI 10.1007/978-3-642-28314-7
Springer Heidelberg New York Dordrecht London

Library of Congress Control Number: 2012932484

Printed on acid-free paper

Springer is part of Springer Science+Business Media (www.springer.com)

Foreword

The Proceedings of the 2011 2nd International Congress on Computer Applications and Computational Science (CACS 2011) are a compilation of current research findings in computer application and computational science. CACS 2011 comprises a keynote speech, plenary speech, and parallel sessions where all of the papers have been reviewed thoroughly by at least three reviewers and the program chairs.

This book provides state-of-the-art research papers in Computer Control and Robotics, Computers in Education and Learning Technologies, Computer Networks and Data Communications, Data Mining and Data Engineering, Energy and Power Systems, Intelligent Systems and Autonomous Agents, Internet and Web Systems, Scientific Computing and Modeling, Signal, Image and Multimedia Processing, and Software Engineering.

The book provides an opportunity for researchers to present an extended exposition of such new works in all aspects of industrial and academic development for wide and rapid dissemination. At all times, the authors have shown an impressive range of knowledge on the subject and an ability to communicate this clearly.

This book explores the rapid development of computer technology that has an impact on all areas of this human-oriented discipline. In the recent years, we are faced with new theories, new technologies and computer methods, new applications, new philosophies, and even new challenges. All these works reside in industrial best practices, research papers and reports on advanced inter-discipline projects.

The book makes fascinating reading and should be of great interest to researchers involved in all aspects of computational science and computer applications, and will help attract more scholars to study in these areas.

Ford Lumban Gaol
Quang Vinh Nguyen

Contents

A Microrobotic Gripper and Force Sensor 1
Ron Lumia

On-Line Path Planning for UAV in Dynamic Environment 9
Xiao Liang, Honglun Wang, Menglei Cao, Tengfei Guo

An Automatic Portal Radiograph Verification System for Proton Therapy 21
Neil Muller, Cobus Caarstens, Leendert van der Bijl

Compensation of Hydraulic Drag for an Underwater Manipulator Using a Real-Time SPH Fluid Simulator: Application in a Master-Slave Tele-operation 29
Haoyi Zhao, Masayuki Kawai

Design of Bipedal Bionic Electric Mobile Platform Based on MATLAB 39
Liu Donghui, Sun Xiaoyun, Yang Lili

On the Design and Implementation of a Wearable Hybrid Assisted Lower Limb Orthosis 43
Cheng-Chih Hsu, Jian-Shiang Chen

3D Point Sets Matching Method Based on Moravec Vertical Interest Operator 53
Linying Jiang, Jingming Liu, Dancheng Li, Zhiliang Zhu

Improvement and Simulation of Contract-Net-Based Task Allocation for Multi-robot System 61
Hao Lili, Yang Huizhen

A Hierarchical Reinforcement Learning Based Approach for
Multi-robot Cooperation in Unknown Environments 69
Yifan Cai, Simon X. Yang, Xin Xu, Gauri S. Mittal

Modeling Specific Energy for Shield Machine by Non-linear Multiple
Regression Method and Mechanical Analysis 75
*Qian Zhang, Chuanyong Qu, Zongxi Cai, Tian Huang, Yilan Kang,
Ming Hu, Bin Dai, Jianzhong Leng*

Design and Realization of College Finance OLAP Analyzer Based on
MDX 81
Peng Cheng, Tong Qiuli

Data Mining: Study on Intelligence-Led Counterterrorism 87
Wang Shacheng

Keywords Weights Improvement and Application of Information
Extraction 95
Yang Junhui, Huang Chan

Scientometric Study of the IEEE Transactions on Software
Engineering 1980-2010 101
Brahim Hamadicharef

Query-Based Automatic Multi-document Summarization Extraction
Method for Web Pages 107
Qi He, Hong-Wei Hao, Xu-Cheng Yin

Top-k Algorithm Based on Extraction 113
Lingjuan Li, Xue Zeng, Guoyu Lu

Keyword Extraction Based on Multi-feature Fusion for Chinese Web
Pages 119
Qi He, Hong-Wei Hao, Xu-Cheng Yin

Word Semantic Orientation Calculation Algorithm Based on Dynamic
Standard Word Set 125
Yun Sha, Ming Xia, Huina Jiang, Xiaohua Wang

A KNN Query Processing Algorithm over High-Dimensional Data
Objects in P2P Systems 133
Baiyou Qiao, Linlin Ding, Yong Wei, Xiaoyang Wang

Encouragement of Defining Moderate Semantics for Artifacts of
Enterprise Architecture 141
Shoichi Morimoto

Development of PC-Based Radar Signal Simulation Test System 151
Li Xiaochun, Ding Qingxin, Tian Fei

Decision Support System for Agribusiness Investment as
e-Government Service Using Computable General Equilibrium
Model 157
Arif Imam Suroso, Arief Ramadhan

Web Security Testing Approaches: Comparison Framework 163
Fakhreldin T. Alssir, Moataz Ahmed

An Evolutionary Model of Enterprise Involvement in OSS:
Understanding the Dynamism in the Emerging Strategic Engineering
Dimension of OSS 171
Toshihiko Yamakami

MI: An Information Support System for Decision Makers 181
Ohan R. Oumoudian, Ramzi A. Haraty

Cost Optimization in Wide Area Network Design: An Evaluation of
Created Algorithms 191
*Tomasz Miksa, Leszek Koszalka, Iwona Pozniak-Koszalka,
Andrzej Kasprzak*

Application of Domain Knowledge in Relational Schema Integration
with Uncertainty 199
Wen Bin Hu, Hong Zhang, Si Di Zhang

Multi-pose Face Recognition Using Fusion of Scale Invariant
Features 207
I Gede Pasek Suta Wijaya, Keiichi Uchimura, Gou Koutaki

Development of a Cost Predicting Model for Maintenance of
University Buildings 215
Chang-Sian Li, Sy-Jye Guo

Functional Treatment of Bilingual Alignment and Its Application to
Semantic Processing 223
Yoshihiko Nitta

Digital Image Stabilization Using a Functional Neural Fuzzy
Network 231
Chi-Feng Wu, Cheng-Jian Lin, Yu-Jia Shiue, Chi-Yung Lee

A Combined Assessment Method on the Credit Risk of Enterprise
Group Based on the Logistic Model and Neural Networks 239
Xiao Min, Liu Wenrui, Xu Chao, Zhou Zongfang

Research on Factors Influencing European Option Price by Using
Hybrid Neural Network 247
Zhang Hong-yan

Tibetan Processing Key Technology Research for Intelligent Mobile
Phone Based on Android 253
Nyima Trashi, Yong Tso, Qun Nuo, Tsi Qu, Duojie Renqian

Self-adaptive Clustering-Based Differential Evolution with New
Composite Trial Vector Generation Strategies 261
Xiaoyan Yang, Gang Liu

Research on Structure Design of Health Information Management
System for Women and Children 269
Hangkun Ling, Ning Ling

How Hybrid Learning Works in the Workplace 277
Li Zhang

Research on the Application of Twitter-Like Platform Based on
Web 2.0 Technology in College English Extracurricular Learning:
From the Perspective of Interactive Teaching 283
Xiao-Ling Zou, Ting Luo

A New Type of Voltage Transducer (EFVVT) 291
Qi Zengying, Xu Qifeng

Modelisation of Atmospheric Pollutant Emissions over the French
Northern Region Using Database Management System 297
P. Lebègue, V. Fèvre Nollet, M. Mendez, L. Declerck

Preliminary Study on Reliability Analysis of Safety I&C System in
NPP 303
Chao Guo, Duo Li, Huasheng Xiong

A Traceability Model Based on Lot and Process State Description 311
Xuewen Huang, Xueli Ma, Xiaobing Liu

Research on Three-Dimensional Sketch Modeling System for
Conceptual Design 321
Jingqiu Wang, Xiaolei Wang

MC9S12XDP512-Based Hardware Circuit Design and Simulation for
Commercial Vehicle's Electric Power Steering System 329
Liu Jingyu, Wang Tao, Hu Zhong

A Research on the Urban Landscape Value Evaluation Based on AHP
Algorithm 335
Jun Shao, Junqing Zhou, Junlei Yang, Hanxi Liu

Regional Degraded Trend through Assessing of Steppe NPP on
Remotely-Sensed Images in China 341
Suying Li, Wenquan Zhu, XiaoBing Li, Junjiang Bao

Examining the Effects of and Students' Perception toward the
Simulation-Based Learning 349
Peng-chun Lin, Shu-ming Wang, Hsin-ke Lu

Automated User Analysis with User Input Log 355
Jae Min Kim, Sung Woo Chung

Developing an Online Publication Collaborating among Students in
Different Disciplines 361
Kirsi Silius, Anne-Maritta Tervakari, Meri Kailanto, Jukka Huhtamäki,
Jarno Marttila, Teemo Tebest, Thumas Miilumäki

Programming of Hypermedia: Course Implementation in Social
Media 369
Kirsi Silius, Anne-Maritta Tervakari, Jukka Huhtamäki, Teemo Tebest,
Jarno Marttila, Meri Kailanto, Thumas Miilumäki

Cloud Computing for Genome-Wide Association Analysis 377
James W. Baurley, Christopher K. Edlund, Bens Pardamean

Finding Protein Binding Sites Using Volunteer Computing Grids 385
Travis Desell, Lee A. Newberg, Malik Magdon-Ismail,
Boleslaw K. Szymanski, William Thompson

Wood of Near Infrared Spectral Information Extraction Method
Research 395
Xue-shun Wang, Ting Yang, Zhong-jie Lin, Fengyan Yang, Kaixin Lu,
Songyue Qiao

Electromagnetic Energy Harvesting by Spatially Varying the
Magnetic Field 403
Rathishchandra R. Gatti, Ian M. Howard

The Inrush Current Eliminator of Transformer 411
Li-Cheng Wu, Chih-Wen Liu

Promoting Professional IT Training Environment Construction with
Virtual Machine Technology 421
Liu Qingyu, Zheng Mengze

The Study of Parametric Modeling and Finite Element Analysis
System of Skirt Support 429
Su Liu, Changming Yang

A Quantitative Research into the Usability Evaluation of China's
E-Government Websites 437
Zhou Guimei, Yan Taowei

**Experimental Power Harvesting from a Pipe Using a Macro Fiber
Composite (MFC)** 443
Eziwarman, G.L. Forbes, I.M. Howard

**Fast Access Security on Cloud Computing: Ubuntu Enterprise Server
and Cloud with Face and Fingerprint Identification** 451
*Bao Rong Chang, Hsiu Fen Tsai, Chien-Feng Huang, Zih-Yao Lin,
Chi-Ming Chen*

**Reform of and Research Based on the Higher Vocational Education of
"Database Principles" Teaching** 459
Liu Shan, Cao Lijun

**Application of Virtual Instrument Technique in Data Acquisition of
Gas-Water Pulse Pipe Cleaning Experiment** 465
Kun Yang, Chenguang Wu, Yixing Yuan, Jingyang Yu

**Quick Formulas Based on DDA Outputs for Calculating Raindrop
Scattering Properties at Common Weather Radar Wavelengths** 471
Wang Zhenhui, Guo Lijun, Dong Huijie, Zhang Peichang

Optimization Analysis for Louver Fin Heat Exchangers 477
Ying-Chi Tsai, Jiin-Yuh Jang

**Numerical Investigation of Nanofluids Laminar Convective Heat
Transfer through Staggered and In-Lined Tube Banks** 483
Jun-Bo Huang, Jiin-Yuh Jang

A Kind of High Reliability On-Board Computer 491
Pei Luo, Guofeng Xue, Jian Zhang, Xunfeng Zhao

**The Spatial Differentiation and Classification of the Economic
Strength of Counties along the Lower Yellow River** 497
Zhang Jinping, Qin Yaochen

**Developing the Software Toolkit on 3DS Max for 3D Modeling of
Heritage** 503
Min-Bin Chen, Ya-Ning Yen, Wun-Bin Yang, Hung-Ming Cheng

Author Index 509

A Microrobotic Gripper and Force Sensor

Ron Lumia

Abstract. This paper describes a microrobotic gripper made from an ionic polymer metal composite (IPMC) material. The grasping capabilities of this device are
described theoretically, predicting a 1.22 mN force when grasping a rigid object.
Experimental results of this grasping are shown, corroborating the expected force
with a measurement of 1.2 mN. In addition, a small modification to the finger, i.e.,
cutting one of the finger surfaces into two separate pieces allows the device to
function as both a sensor and actuator. The microfinger's sensing capability is described through its application as a force sensor.

1 Introduction

This paper describes an approach to micromanipulation using an ionic polymer
metal composite (IPMC) material. IPMC microgrippers show great potential in
applications related to the assembly of microsystems and bio-micromanipulation.
Assembly and testing of microsystems requires handling of MEMS components
while bio-micromanipulation requires handling flexible and fragile biological micro-objects such as cells and bacteria. Various microgripping solutions have been
presented for assembly of MEMS components, though few microgripping technologies are suited for bio-micromanipulation.

The ability to manipulate biological samples, especially small cells, has become a
great technological challenge for the future of bioengineering, microbiology and ge-
nomics. Operations such as positioning, grasping, and injecting materials into a cell
are expected to become increasingly important. Existing bio-micromanipulation techniques are mostly of non-contact type such as laser trapping [1], electro-rotation [2]
and dielectrophoresis [3]. The limitations of non-contact type micromanipulation
make mechanical contact type micromanipulation desirable. Micropipettes have been
used for microinjection of a transgene into a mouse embryo [4]. The use of microgrippers for mechanically gripping and manipulating micro-objects to a desired position represents a promising solution to the above limitations. Bio-micromanipulation
devices must be able to tolerate a wet environment. Cells, bacteria and embryos are
flexible and fragile objects and hence the microgrippers should not exert large forces
while handling them.

This paper describes the design and test of a microgripper using an ionic polymer metal composite (IPMC) as an actuator to grasp and manipulate micro-sized

Ron Lumia
Fellow, IEEE

F.L. Gaol et al. (Eds.): Proc. of the 2011 2nd International Congress on CACS, AISC 144, pp. 1–8.
springerlink.com   © Springer-Verlag Berlin Heidelberg 2012

objects. IPMC is a compliant material and can work in both wet and dry environments. In addition, a novel modification allows the IPMC microfinger to act as a sensor and actuator simultaneously.

2 Other Approaches for Micromanipulation

While IPMC as a material has advantages, other approaches for actuation and micromanipulation have been proposed. Zhang et al. [5] developed a shape memory alloy (SMA) microgripper used for tissue engineering. Goldfarb and Celenovic [6] developed a microgripper using a piezo-actuator. However, piezo devices require high voltage for actuation and also result in a stiff microgripper. Sun et al. [7] used an electrostatic actuator that is stiff and also cannot be actuated in an aqueous medium. Menciassi et al. [8] proposed a microgripper using an electromagnetic moving coil actuator yielding a stiff microgripper. Kim et al. [9] used Lorentz force-type actuators that cannot be used in wet environments. Another approach, proposed by Zesch et al., used a glass pipette with controllable vacuum tool is also used to grasp and release micro-objects [10]. Arai and Fukuda [11] proposed a method for micromanipulation that utilizes pressure change based on temperature change inside the micro-holes made on the end effector surface.

Shen et al. [12] used PVDF film as sensor for assembly of micro-mirrors. Keller [13] developed micro-tweezers manufactured by the hexsil process. He used piezoresistive strain gauges to obtain tactile feedback during grasping. Unfortunately these tweezers are particularly fragile. Carrozza et al. [14] present a LIGA fabricated force controlled microgripper by mounting semiconductor strain gauge sensors at flexure joints of the gripper. However, the strain gauge sensors cannot tolerate a wet environment.

The vast majority of the micromanipulation techniques described above is either stiff, fragile or cannot tolerate a wet environment. Consequently, these techniques have limitations in handling delicate micro-objects. Stiff microgrippers damage fragile and flexible micro-objects during handling and fragile microgrippers lack longevity. Deole et al. [15] previously developed a prototype IPMC microgripper and manipulation of flexible objects was demonstrated.

This paper describes the design and fabrication of a microgripper, proposes a theoretical force model for the microgripper, and provides experimental verification of the approach. In addition an approach to simultaneously actuate and sense on the same IPMC finger is proposed and verified experimentally.

3 IPMC Actuation and Sensing

IPMC has traditionally been used either as a sensor or an actuator.

3.1 Actuation

Ionic polymeric metal composites (IPMCs) are active devices, i.e., they deform significantly when excited by a relatively low voltage. Grodzinsky [16] and

Yannas et al. [17] were the first to present a continuum model for electrochemistry of the deformation of charged polyelectrolyte membranes. The application of these active polymers to artificial muscles can be traced to Shahinpoor et al. [18]. A variety of other researchers, e.g., Osada [19], Brock [20], and Bar-Cohen [21], have looked at various applications of artificial muscles.

A patented chemical process in which a noble metal, such as platinum or gold, is deposited within the molecular network of the base ionic polymer is the basis for this technology. IPMC is a sandwich of platinum-nafion-platinum, where the platinum is often only a few microns thick. One can visualize IPMC as a sheet of paper, where nafion is the paper. Platinum is plated onto the front and back of the sheet effectively forming a capacitor. However, unlike a capacitor, a voltage across the two platinum layers produces a deflection of the sheet.

At the macro-scale, the most significant disadvantage of using IPMC as an actuator is that it does not produce large forces. At the micro-scale, however, this property actually becomes an advantage because the IPMC microgripper will not damage fragile objects, such as cells. Hydration was a concern in the past, but recently a parylene film coating has been developed that prevents the IPMC from desiccation when exposed to air [22].

Lumia and Shahinpoor [23], [24] observed empirically that the IPMC material retains its electroactive characteristics even as it is cut smaller and smaller. This became the rationale justifying the use of IPMC for the fingers of a microgripper.

3.2 Sensing

IPMC can also be used as a sensor to measure deflection. Rather than exciting the IPMC strip with voltage for use as an actuator, the strip is physically moved, and the resultant voltage across the electrodes is measured. The phenomenon is explained as follows: when the composite is bent, a stress gradient builds on the outer fibers relative to the neutral axis. The mobile ions then shift toward the favored region where opposite charges are available. The deficit in one charge and excess in the other translates into a voltage gradient across the platinum that is easily sensed by an instrumentation amplifier. Sadeghipour et al. [25] used ion-exchange-membrane materials to sense pressure in an accelerometer. Shahinpoor [26] discussed the phenomenon of the "flexogelectric" effect in connection with dynamic sensing of ionic polymeric gels, where manually moving the device resulted in a measurable electric field.

As the size of microgripper fingers becomes smaller and smaller, the approach to sensing described above becomes increasingly problematic. There is simply not enough charge in the finger to create a voltage large enough to be extracted from noise. For example, it is common to measure microvolts for microfingers when moving them mechanically. An alternative approach to sensing position is described in the next section.

4 Design of Microrobotic Gripper

4.1 Design of Microgripper

Figure 1 shows the concept of a microrobotic manipulator using the "pincher" design. The green IPMC microfingers, which can be cut in any arbitrary shape, are sandwiched together.

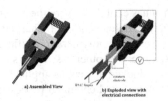

Fig. 1 Pincher design

The actuation is organized such that the fingers move toward each other when a voltage is provided.

A microgripper force model is developed to estimate the force exerted by the IPMC fingers when grasping an object. The input to the model is the weight of the object being lifted and the output is the amount of force that needs to be exerted by the IPMC fingers to grasp that object securely.

Assumptions: The object is spherical in shape; The object and the IPMC fingers contact at the tip of the IPMC fingers.

Figure 2 shows the free body diagram of a spherical test object where F = Frictional force; N = Normal reaction; μ = Coefficient of friction between IPMC and object; θ = Angle of contact between IPMC finger and object; m = mass of the micro-object; g = acceleration due to gravity. By using the static equilibrium condition for the free body diagram shown, we obtain

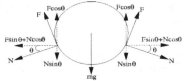

Fig. 2 Free body diagram of object that is manipulated

$$2F\cos\theta = 2N\sin\theta + mg \tag{1}$$

Since the friction is proportional to the normal force, i.e., F = μN, we can derive the normal force that the microgripper finger must exert.

$$N = \frac{mg}{2\mu\cos\theta - 2\sin\theta} \tag{2}$$

N is the normal reaction exerted by the object and is equal and opposite to the force exerted by each IPMC finger. Therefore, equation (2) computes the force that an IPMC finger exerts to grasp a sphere of known weight.

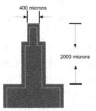

Fig. 3 IPMC microfinger with a 2 micron-wide cut in the platinum of one surface of the IPMC.

4.2 Simultaneous Actuator and Sensor

As stated earlier, the ability to sense deflection becomes problematic as the size of the IPMC shrinks. Therefore, a new and unique approach has been developed. The approach is to cut a path through one layer of the IPMC surface, as shown in Fig. 3. The 2-micron channel is cut using a Signatone laser probe station. This process creates two separate surfaces on one side of the IPMC. The larger, central surface is used for actuation, as previously described. The smaller surface is used to sense deflection. Rather than use the stored charge concept, as described in Section 3.1, the approach is to measure the change of resistance to measure deflection. When moved in one direction, the platinum molecules, which are about 2 microns thick on each side of the IPMC, stretch apart. This increases the resistance. Conversely, when the finger is moved in the other direction, the molecules are forced closer together, which decreases the resistance. It is common to see a nominal 9 Ω resistance (no deflection) vary by 0.10 Ω as the finger is moved, i.e., roughly 0.05 Ω in each direction. Of course, the exact amount of resistance change is related to the size and shape of the sensor strip created by the channel.

5 Experimental Verification

5.1 Resistance Calibration

Fig. 4 shows the resistance measured as a function of displacement for a typical IPMC sample. Physically moving the tip of the microfinger and measuring the resulting resistance obtained this graph. The graph shows a reasonably linear relationship between resistance and deflection for the range of 0-4 mm.

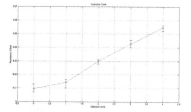

Fig. 4 Resistance as a function of displacement.

Fig. 5 shows the actual measurement apparatus as a single microgripper finger is moved against a load cell.

The finger starts with some space between it and the force sensor. As the excitation voltage is increased, the IPMC finger moves toward the sensor, ultimately colliding with it. As the actuation voltage continues to increase, the load cell measures the force, as shown in Fig. 6. The figure demonstrates a linear relationship between input voltage and force generated. Note that no force is generated until the finger finishes moving through free space

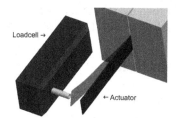

Fig. 5 Microfinger pushing against a force sensor.

and then contacts the sensor. With this information, one can compute the force generated after contact.

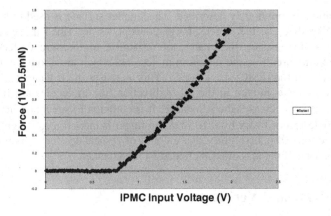

Fig. 6

5.2 Actuation

Fig. 7 illustrates the "pincher" microgripper grasping a (roughly) spherical solder ball. The coefficient of friction between the platinum (surface of the IPMC microfinger) and the object (solder ball) is 0.30. The mass of the solder ball was measured on a scale to be 15 μg. The angle of contact was measured via camera image to be approximately 20°. Therefore, using equation (2), the expected force will be 1.22 mN. The actuation voltage of a 1 V delta after contacting the object from Fig. 6 shows roughly 1.2 mN of force, which is very close to the predicted value. (1.2 V * 0.5mN/V * 2 fingers)

Fig. 7 Microgripper holding a (roughly) spherical solder ball.

6 Conclusion

This paper described a microrobotic gripper that has the capability to actuate and sense deflection simultaneously. By cutting a 2-micron channel in the surface of one side of the IPMC microfinger, each surface can be used independently. One part can be used as the standard actuator, where deflection is proportional to input voltage. The actuator is slightly weaker because the area of the side is smaller than it was before the 2-micron channel cutting process. The tradeoff for the slightly weaker actuator is the ability to sense deflection by measuring the change in resistance of the platinum on the surface of the microfinger.

Future work will be to use the deflection measurement in a closed loop feedback system for a microgripper controller. Also, there are plans to model the microfinger using Comsol Multiphysics to predict the interaction between the actuator and sensing components of the microfinger.

Acknowledgment. This material is based upon work supported by the National Science Foundation under Grant No. IIS-0911133.

References

[1] Ashkin, A.: Optical trapping and manipulation of neutral particles using lasers. Proceedings of National Academy of Sciences 94, 4853–4860 (1997)

[2] Masuda, S., Washizu, M., Kawabata, I.: Movement of blood cells in liquid by non-uniform travelling field. IEEE Transactions on Industrial Applications 24, 217–222 (1988)

[3] Lee, G.B., Fu, L.M.: Platform Technology for manipulation of Cells, Protiens and DNA. In: Proceedings of International Conference on Robotics and Automation, Taipei, Taiwan, September 14-19, pp. 3636–3641 (2003)

[4] Kumar, R., Kapoor, A.A., Taylor, R.H.: Preliminary experiments in robot/human cooperative microinjection. In: Proceedings of International Conference on Intelligent Robots and Systems, Las Vegas, Nevada, October 27-31, pp. 3186–3191 (2003)

[5] Zhang, H., Bellouard, Y., Sidler, T., Burdet, E., Poo, A.N., Clavel, R.: A monolithic Shape Memory Alloy Microgripper for 3-D Assembly of Tissue Engineering Scafolds. In: Brugeuteds, J.-M., Nelson, B.J. (eds.) Proceedings of SPIE-Microrobotics and Microassembly III, Boston, MA, October 29-30, vol. 4568, pp. 50–60 (2001)

[6] Goldfarb, M., Celanovic, N.: A flexure-based gripper for small scale manipulation. Robotica 17(2), 181–188 (1999)

[7] Sun, Y., Piyabongkarn, D., Sezen, A., Nelson, B.J., Rajamani, R., Schoch, R., Potasek, D.P.: A Novel Dual-Axis Electrostatic Micro-Actuation System for Micromanipulation. In: Proceedings of International Conference on Intelligent Robots and Systems, EPFL, Switzerland, October 2-4, pp. 1796–1801 (2002)

[8] Menciassi, A., Hannaford, B., Carrozza, M.C.: 4-Axis Electromagnetic Microgripper. In: Proceedings of International Conference on Robotics and Automation, Detroit, Michigan, May 10-15, pp. 2899–2904 (1999)

[9] Kim, S.M., Kim, K., Shim, J.H., Kim, B., Kim, D.H., Chung, C.C.: Position and force control of a sensorized microgripper. In: Proceedings of International Conference on Control, Automation and System, Muju Resort, Jeonbuk, Korea, October 16-19, pp. 319–322 (2002)

[10] Zesch, W., Brunner, M., Weber, A.: Vacuum Tool for Handling Microobjects with a Nanorobot. In: Proceedings of International Conference on Robotics and Automation, Detroit, Michigan, May 10-15, pp. 2899–2904 (1999)

[11] Arai, F., Fukuda, T.: Adhesion-type Micro endeffector for Micromanipulation. In: Proceedings of the International Conference on Robotics and Automation, Albuquerque, NM, pp. 1472–1477 (April 1997)

[12] Shen, Y., Xi, N., Li, W.J.: Force-Guided Assembly of Micro Mirrors. In: Proceedings of International Conference on Intelligent Robots and Systems, Las Vegas, Nevada, October 27-31, pp. 2149–2154 (2003)

[13] Keller, C.G.: Microgrippers with Integrated Actuator and Force sensors. In: Proceedings of the International Symposium on Robotics and Automation, World Automation Conference, Anchorage, AK (May 1998)

[14] Carrozza, M.C., Dario, P., Menciassi, A., Fenu, A.: Manipulating Biological and Mechanical Micro-objects with a LIGA-Microfabricated End Effector. In: Proceedings of the International Conference on Robotics and Automation, Leuven, Belgium, May 16-20, pp. 1811–1816 (1998)

[15] Deole, U., Lumia, R., Shahinpoor, M.: Grasping flexible objects using artificial muscle microgrippers. In: World Automation Congress: International Symposium on Manufacturing and Applications (ISOMA), Seville, Spain, June 28-July 2 (2004)

[16] Grodzinsky, A.J.: Electromechanics of Deformable Polyelectrolyte Membranes, in Dept. of Elec. Eng. vol. Sc.D. Dissertation: MIT (1974)

[17] Yannis, I.V., Grodzinsky, A.J.: Electromechanical Energy Conversion with Collagen Fibers in an Aqueous Medium. Journal of Mechanochemical Cell Motility 2, 113–125 (1973)

[18] Shahinpoor, M.: Continuum Electromechanics of Ionic Polymeric Gels as Artificial Muscles for Robotic Applications. Int. Journal of Smart Material and Structures 3, 367–372 (1994)

[19] Osada, Y., Hasebe, M.: Electrically Activated Mechanochemical Devices Using Polyelectrolyte Gels. Chemistry Letters, 1285–1288 (1985)

[20] Brock, D., Lee, W., Segalman, D., Witkowski, W.: A Dynamic Model of a Linear Actuator Based on Polymer Hydrogel. In: International Conference on Intelligent Materials, pp. 210–222 (1994)

[21] Bar-Cohen, Y., Xue, T., Joffe, B., Lih, S.-S., Shahinpoor, M., Simpson, J., Smith, J., Willis, P.: Electroactive polymers (IPMC) low mass muscle actuators. In: SPIE Conference on Smart Materials and Structures, San Diego, California (1997)

[22] Kim, S.J., Lee, I.T., Lee, H.Y., Kim, Y.H.: Performance improvement of an ionic polymer-metal composite actuator by parylene thin film coating. Smart Materials and Structures 15, 1540–1546 (2006)

[23] Lumia, R., Shahinpoor, M.: Design of a Microgripper Using Artificial Muscles. In: World Automation Conference, Anchorage, AK (1998)

[24] Lumia, R., Shahinpoor, M.: Microgripper design using electro-active polymers. In: Smart Structures and Materials 1999: Electroactive Polymer Actuators and Devices, Newport Beach, CA, USA. SPIE-Int. Soc. Opt. Eng, vol. 3669, pp. 322–329 (1999)

[25] Sadeghipour, K., Salomon, R., Neogi, S.: Development of a novel electrochemically active membrane and 'smart' material based vibration sensor/damper. Smart Materials and Structures 1, 172–179 (1992)

[26] Shahinpoor, M.: A New Effect in Ionic Polymeric Gels: The Ionic Flexogelectric Effect. In: North American Conference on Smart Structures and Materials, San Diego, California, vol. 2441 (1995)

On-Line Path Planning for UAV in Dynamic Environment

Xiao Liang, Honglun Wang, Menglei Cao, and Tengfei Guo

Abstract. Under the premise of predicting the dynamic obstacles, a UAV path planning strategy of variable rolling window combined with potential flows is proposed. By using an autoregressive model, the expression of prediction for dynamic obstacles between discrete sampling points is given. The rolling window is designed to a triangle, also with adaptive function based on speed and angle of incidence of the dynamic obstacles. Potential flows method is used in the rolling windows to plan the obstacle avoidance route. At last the whole route is smoothed to satisfy the UAV constraints of maximum turning angle. The problem solution implementation is described along with several simulation results demonstrating the effectiveness of the method.

Keywords: UAV, dynamic environment, autoregressive prediction, on-line path planning.

1 Introduction

The active area of UAV (Unmanned Aerial Vehicle) is often a dynamic uncertain environment, so it is necessary for UAV to avoid some dynamic obstacles such as ground radar vehicles and mobile artillery threats. In this circumstance, how to generate a new flight path quickly becomes one of the prerequisites for UAV to complete mission successfully. Since the environment of battlefield is constantly changing,

Xiao Liang · Honglun Wang · Menglei Cao · Tengfei Guo
Science and Technology on Aircraft Control Laboratory,
Beijing University of Aeronautics and Astronautics, Beijing, China
and

Research Institute of Unmanned Aerial Vehicle,
Beijing University of Aeronautics and Astronautics, Beijing, China
e-mail: connyzone@yahoo.com.cn, hl_wang_2002@yahoo.com.cn,
    llcml@126.com, gtfei28@sina.com

F.L. Gaol et al. (Eds.): Proc. of the 2011 2nd International Congress on CACS, AISC 144, pp. 9–19.
springerlink.com   © Springer-Verlag Berlin Heidelberg 2012

UAV can not be well informed of all the information within the planning area, so the path planning method of such cases should have local on-line capability. Most of current path planning methods such as A* [1] and PRM [2] are suitable for static global planning, so they have to make local amendments to solve the problem of dynamic obstacles for the cost of re-planning is huge. Therefore, hybrid path planning algorithm has been studied. The global optimum of the ant colon algorithm can overcome the local minimum problem of the artificial potential field, thus ant colony algorithm can be employed for static global planning while artificial potential field for avoiding obstacles [3]. Taking the advantage of parallelism of genetic algorithm, static planning and dynamic planning can be combined, but the planning results depend on the selection of the fitness function [4].

To the real-time capability, as the representative of the artificial potential field, Virtual Forces [5] and Potential Flows (Stream Function) [6] are more computationally efficient. Stream Function had been proven be able to prevent artificial potential field from getting local minimum [7]. Also its path is smoother and more suitable for UAV after several local optimizations. Rolling window was proposed in robot real-time path planning for the first time [8]. By using local optimum instead of global optimum, it improved real-time capability under the premise of ensuring accessibility. This method was utilized for robot path planning in unknown environment with dynamic obstacles, and achieved satisfactory results [9].

Some prediction algorithm was introduced to path planning in order to solve the problem of dynamic obstaclesand it was proved to be more effective[10-12]. The common methods include time series prediction, grey prediction and so on. According to the information of last moment, CAS (Crash Avoidance System) algorithm can directly make prediction. However, the accuracy of prediction is difficult to ensure, for the prior information has not been fully utilized [13]. The neural network prediction requires high on the initial samples of dynamic obstacles, and the routes are not suitable for UAV [14].

In this paper, we use autoregressive prediction for dynamic obstacles, and make path planning based on the predicted position. When planning, the dynamic obstacles are treated as instant and static constraints. A hybrid path planning algorithm for UAV is proposed, which combines the rolling window and potential flows. Rolling window is used to guide UAV to target. In the window, we utilize the characteristics of potential flows with high real-time capability and smoother trajectory to avoid obstacles. Through taking the constraints of UAV into account, the method makes the planning path more reasonable. The simulation results show that this method can effectively solve the problem of UAV path planning in uncertain environment with dynamic obstacles.

2   Environment Modeling and Prediction of Dynamic Obstacles

To begin path planning, fly domain and obstacles need to be described by mathematical formulas firstly. The article assumes that UAV is in the two-dimensional movement, and can be expressed by the center point.

2.1 Environment Modeling

For terrain information is needed to provide the movement and change of rolling window with coordinates and measurement, grid modeling of terrain is adopted. Grid map method records terrain information in each grid as a unit, and the environment is quantified into a mesh of grids with certain resolution. Though there is a conflict between grid density and environment information [15], the method we used can avoid this problem. Because the rolling window just provides the heuristic direction, and the exact path planning is made in potential flows by stream function.

Grid map of the flight area can be divided into two categories: No-Fly Domain (NFD) and Fly Domain (FD). Then the problem of UAV path planning can be described as follows: Make sure the UAV avoid obstacles (NFD) safely, and find a path from the starting point (P_{begin}) to the finishing point (P_{goal}), $P_{begin} \neq P_{goal}$, $P_{begin}, P_{goal} \in FD$.

2.2 Prediction of Dynamic Obstacles

According to the information of dynamic obstacles sampled by sensor, statistical modeling method can be utilized to predict the sampling information of next moment. Since the speed of obstacles in the flight area is relatively low, we use autoregressive model to predict dynamic obstacles, for this method requires a few initial information and can ensure real-time capability. Suppose the location of the k th dynamic obstacle at the t moment is $p_k(t)$, and apply n-order autoregressive to model the predicted location $p_k(t+1)$ at $t+1$ moment. Then the position sequence $p_k(t)$ can be described as follows:

$$p_k(t) = \sum_{i=1}^{n} \alpha_i p_k(t-i) + e(t),\qquad(1)$$

here $p = [x,y]^T$, t is sampling moment, $e(t)$ is the error of prediction and coefficient α_i is a matrix of 2×2 dimension in two-dimensional planning. As the velocity of dynamic obstacle is low, acceleration $a_k(t)$ of O_k can be modeled by 1-order of Eq. 1:

$$a_k(t) = \beta_{k,t} a_k(t-1) + w(t),\qquad(2)$$

here $w(t)$ is the error of prediction and coefficient $\beta_{k,t}$ means the movement of obstacle along x,y direction is related to sampling time. Use $v_k(t)$ to represent the speed of dynamic obstacle and we can derive the followings from Eq. 2:

$$
\begin{aligned}
a_k(t) &= \frac{v_k(t) - v_k(t-1)}{\Delta t} \\
&= \frac{[p_k(t) - p_k(t-1)] - [p_k(t-1) - p_k(t-2)]}{\Delta t^2} \\
&= \frac{p_k(t) - 2p_k(t-1) + p_k(t-2)}{\Delta t^2}.
\end{aligned}
\qquad(3)
$$

Make $\Delta t = 1$, from Eq. 2 and Eq. 3 we have:

$$p_k(t) = (2 + \beta_{k,t})p_k(t-1) - (2\beta_{k,t} + 1)p_k(t-2) + \beta_{k,t}p_k(t-3) + w(t), \qquad (4)$$

$p_k(t)$, $p_k(t-1)$, $p_k(t-2)$ and $p_k(t-3)$ can all be obtained from sensor but $w(t)$ is unknown, so we can not use Eq. 4 to calculate the value of $\beta_{k,t}$ directly. Here we use the method in [16, 17] to estimate $\beta_{k,t}$ and obtain the estimated value of $\beta_{k,t}$:

$$\hat{\beta}_{k,t} = arg\min_{\beta_{k,t}} \sum_{i=4}^{t} \lambda^{t-i}[a_k(i) - \beta_{k,t}a_k(i-1)]'[a_k(i) - \beta_{k,t}a_k(i-1)]. \qquad (5)$$

When a_k is small, weight factor $\lambda \in (0, 1]$ is close to 1. Here we use $\lambda = 0.9$ after experiment and then the solution of Eq. 5 is:

$$\hat{\beta}_{k,t} = \frac{\Delta_{k,t}}{\eta_{k,t}}, \qquad (6)$$

where $\Delta_{k,t} = \lambda \Delta_{k,t-1} + a_k'(t)a_k(t-1), \eta_{k,t} = \lambda \eta_{k,t-1} + a_k'(t-1)a_k(t-1)$.

After obtaining the value of $\hat{\beta}_{k,t}$, the location of O_k at $t+1$ moment can be estimated by Eq. 7:

$$\hat{p}_k(t+1) = p_k(t) + v_k(t)\Delta t + \hat{\beta}_{k,t}a_k(t)\Delta t^2. \qquad (7)$$

To ensure the safety of UAV, we expand the $\hat{p}_k(t+1)$ location with safe distance based on the error of prediction σ, then treat the expanded location of obstacles as instant and static constraints to make path planning.

3 Hybrid Path Planning of Rolling Window and Potential Flows for UAV

Rolling Window can decompose the complex planning space. On the other hand, potential flows have better real-time capability and its routes are more smooth. Therefore, the hybird path planning algorithm combined by the two methods is more suitable to solve the problem of path planning in dynamic environment.

3.1 The Rolling Window for UAV

Rolling planning originated from robot path planning [8]. It uses the local optimum of every window instead of global optimum to reduce the scale of problem, which makes it more suitable for path planning of uncertain environment with dynamic obstacles. Robot's detection range and the characteristic of stop and turn anytime, determine most of robot rolling windows are rectangular and circular [18]. For UAV, the detection range may be round, but the terrain information having been flying over is not very important, and UAV flight is limited by maximum turning angle,

Fig. 1 Rolling window of
UAV

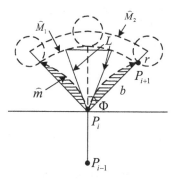

which makes its flexibility less than robot. In view of this, we design the UAV rolling window as shown in Fig. 1.

Φ is maximum turning angle of UAV. Suppose the location of UAV in previous time is $P_{i-1}(x_{i-1}, y_{i-1})$, the location in current time is $P_i(x_i, y_i)$, and the location of maximum deflection in next time is $P_{i+1}(x_{i+1}, y_{i+1})$. Define $\mathbf{a}_i = (x_i - x_{i-1}, y_i - y_{i-1})$ as the leg vector from P_{i-1} to P_i, and similarly $\mathbf{a}_{i+1} = (x_{i+1} - x_i, y_{i+1} - y_i)$ is the leg vector from P_i to P_{i+1}. Then the maximum turning angle of UAV Φ can be calculated by Eq. 8.

$$\Phi = \cos^{-1}\left(\frac{\mathbf{a}_i^T \bullet \mathbf{a}_{i+1}}{|\mathbf{a}_i||\mathbf{a}_{i+1}|}\right), \quad here \ i = 2, \cdots, n. \tag{8}$$

b is the distance from P_i to P_{i+1}, and it has relations with the speed of UAV. L is the tangent segment of arc $\widehat{m}$ and its length is also b. From Fig. 1 we know that shaded area is the inaccessible part of UAV. Therefore, UAV rolling window is the area surrounded by segment L and arc $\widehat{m}$. Since the calculation of arc is complicated, the triangle area is used as the rolling window of UAV to approximate the region surrounded by arc $\widehat{m}$ and arc $\widehat{M}_1$. To ensure continuity of the routes, the symmetric center of rolling window is the tangential direction of previous route segment. For the situation that the rolling window influenced by some obstacles whose center is not in it, we design the expanded window. Suppose the dashed circle in Fig. 1 is the largest obstacle in the environment and its radius is r. The expanded window is the area surrounded by segment $b + r$ and arc $\widehat{M}_2$, hence it can ensure the safety of rolling window, while minimizing the amount of computation. The sub goal can only exist in the rolling window, and the stream function is used in the expanded window.

The rolling window and the expanded window are all designed basing on the behavior of the stretch function, so that they can adjust the size according to the density of obstacles. Through the introduction of L_{in} and L_{out} (Fig. 2), the velocity and incident angle of dynamic obstacles are taken as the basis of adjusting window.

Fig. 2 The definition of L_{in} and L_{out}

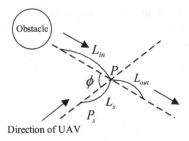

The definition of L_{in} and L_{out} [19] is described in Eq. 9, where v_u and v_o is the speed of UAV and obstacle respectively, and L_s is safe distance.

$$L_{in} = \frac{L_s}{\sin\phi}\left[\left(\frac{v_u^2 + v_o^2 - v_u v_o \cos\phi}{v_u^2}\right)^{\frac{1}{2}} + \frac{v_o}{v_u}\right]$$

$$L_{out} = \frac{L_s}{\sin\phi}\left[\left(\frac{v_u^2 + v_o^2 - v_u v_o \cos\phi}{v_u^2}\right)^{\frac{1}{2}} - \frac{v_o}{v_u}\right].\qquad(9)$$

When UAV reaches position P_s, L_{in} and L_{out} will be calculated. If the predicted position of obstacles at next moment is in the range of L_{in} and L_{out}, the chord length of arc $\widehat{M_2}$ in Fig. 1 should be expanded to $L_{in} + L_{out}$, and then select the sub goal in the window.

Learning form the heuristic direction thought of A* algorithm, the contact between final goal and sub goal can be created to guide UAV always flying toward the finishing point. We improved the selection of the sub goal in [18]: Note $Win(t_1 + i)$ is the current window of t_i moment, if $P_{goal} \in Win(t_1 + i)$, then select sub goal $P_{sub} = P_{goal}$, otherwise establish heuristic function $F(P)$, and find a point P which makes the value of $F(P)$ minimum as P_{sub}. $F(P)$ is shown in Eq. 10, where $G(P)$ is the cost of current location to P, $H(P)$ is the cost of P to final goal P_{goal}.

$$F(P) = G(P) + H(P),\qquad(10)$$

here if $P \in FD(t)$, $G(P) = 0$, otherwise $G(P) = +\infty$. And $H(P) = MD(P, P_{goal})$, where MD means Manhattan Distance: $MD((x_1, y_1), (x_2, y_2)) = |x_1 - x_2| + |y_1 - y_2|$.

Heuristic function constructed by Eq. 10 may result in the situation that sub goal is the same as the current location. Therefore, it is needed to determine whether P is the starting point of current window after calculating. If such case exists, consider the stretch function of the window, and change the size of current rolling window to recalculate until sub goal is different from the starting point. So far, the selection of sub goal is deduced to the following optimization problem:

$$\min J = \min_P MD(P, P_{goal}), \quad P \in Win(t) \cap FD(t).\qquad(11)$$

3.2 The Improvement of Potential Flows for UAV

The potential flows method are implemented in [6, 7], and the background of stream functions is employed in these papers. Consider an uniform flow with strength into which a single in 2D, stationary obstacle with radius a is placed at (b_x, b_y), let $b = b_x + ib_y$, the complex potential ω would become [20, 21]:

$$\omega = Uz + U\left(\frac{a^2}{z-b} + \bar{b}\right), \qquad (12)$$

here $z = x + iy$, the complex velocity can then be calculated using:

$$\omega'(z) \equiv u - iv, \qquad (13)$$

here $u = \dot{x}$ and $v = \dot{y}$. Suppose the current coordinates of UAV is (x_1, x_2) and the obstacle is located in the flight area, i.e., $(b_x, b_y) = (0,0)$ for simplicity, then we have:

$$\dot{x}_1 = U\left[1 - \frac{a^2}{r^4}\left(x_1^2 - x_2^2\right)\right], \quad \dot{x}_2 = U\frac{a^2}{r^4}2x_1x_2, \qquad (14)$$

where $r = \left(x_1^2 + x_2^2\right)^{\frac{1}{2}}$, and $\dot{x} = (\dot{x}_1, \dot{x}_2)$ describes the UAV movement along the flow lines. By imitating the motion of fluid flow, potential flows method achieves obstacle avoidance. The technique for single obstacle can be extended for cases with multiple obstacles by using a method called 'addition and thresholding' [7].

Although the routes by the stream function are relatively smooth, they still need improvement to satisfy the constraints of UAV. Design the waypoint to segment the route in stream function as Fig. 3, and then calculate the tangent of each waypoint in the route. We can get tangent angle between tangents of the neighbor waypoints such as α, β in Fig. 3. Take tangent angle to compare with maximum turning angle Φ, if $\alpha > \Phi$ then the routes need smooth. If the new tangent angle $\beta < \Phi$, the process of smooth is finished, otherwise the routes need second smooth until suitable for UAV flight.

The method of smooth is: remove the unsuitable waypoints, then under the premise that the tangent angle is less than Φ, make interpolation and fitting between the rests of the waypoints. In Fig. 3, original route (dotted line) contains three

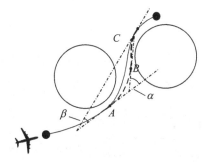

Fig. 3 The schematic diagram of improving potential flows

waypoints A, B, C. If $\alpha > \Phi(\alpha$ is the tangent angle of $AB)$, remove B and then consider the relationship between Φ and β(tangent angle of AC). Fig. 3 is a schematic diagram, and the actual waypoints are more intensive in general, so removing part of the waypoints will not cause great changes of the whole route.

4 Simulations and Analysis

This section will first verify the accuracy of autoregressive model, and then based on the error of prediction, our method was applied in a complex terrain with two kinds of trajectory of obstacles.

4.1 Trajectory Prediction of Autoregressive Model

The value satisfying normal distribution in interval $[-10\%a_k(t-1), 10\%a_k(t-1)]$ is taken as the error of prediction in Eq. 2. We have designed two kinds of trajectories for obstacles: one is S-shaped movement which is similar to sine; the other is approximate to parabolic curve. As the calculation of $\beta_{k,t}$, the movement of obstacles along x, y direction is designed to relate previous moment. Then we can verify the accuracy of prediction by autoregressive model. For ease of display, Fig. 4 shows the results of predicted position every two sampling time. In Fig. 4, the circle points are trajectories of obstacles, while '+' points and '*' points represent the predicted position calculated at the discrete sampling points respectively. The error of prediction σ is shown in Table 1. According to the value of σ, the predicted position of obstacles at next moment will be expanded to 1.5 grid length.

4.2 On-Line Path Planning for UAV in Dynamic Environment

The UAV starting point is [0,0], and finishing point is [30,30]. The simulation combines the prediction of dynamic obstacles with the path planning method proposed in the paper. Fig. 5 shows the results of predicted position every two sampling time. The flight area has 38 static obstacles and two dynamic obstacles, while the ac-

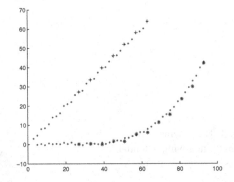

Fig. 4 The prediction of dynamic obstacle in autoregressive model
tab:1

Table 1 The accuracy of prediction in autoregressive model

Trajectory	σ_x	σ_y	σ^*
S-shaped movement	0.1142	0.0828	0.1410
Parabolic movement	0.1199	1.3542	1.3595

* means it is calculated in **RMSE**

tual trajectories of dynamic obstacles are marked by '+' points and '*' points separately. After expanded, the predicted positions of dynamic obstacles are expressed by solid round. Fig. 5 divides the process of path planning into 4 sub-graphs. From the figures, the method can satisfy requirements of UAV path planning in dynamic environment, and avoid dynamic obstacles with smooth routes.

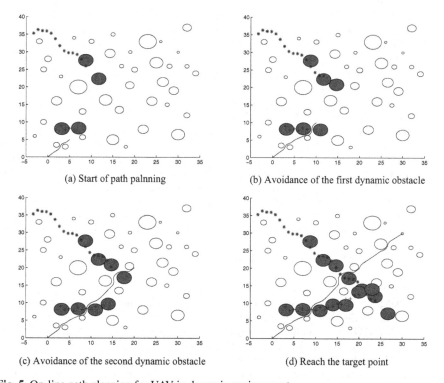

(a) Start of path palnning

(b) Avoidance of the first dynamic obstacle

(c) Avoidance of the second dynamic obstacle

(d) Reach the target point

Fig. 5 On-line path planning for UAV in dynamic environment

5 Conclusions

In this paper, according to the characteristic of low velocity, we used autoregressive model to predict the trajectories of the dynamic obstacles. By combining the rolling window and potential flows, we proposed a method of path planning for UAV in dynamic environment. Through the improvement of rolling window and the smooth of routes, the method was made more practical. A set of analytical results, extensive simulation studies as well as preliminary experimental demonstrations illustrate the viability of the proposed method.

Acknowledgements. This research has been funded by the Fundamental Research Funds for the Central Universities NO. YWF-10-01-B015 which is supported by Ministry of Finance: 2010.1-2011.12.

References

1. Koenig, S., Likhachev, M., Furcy, D.: Lifelong planning A*. Artificial Intelligence (2004), doi:10.1016/j.artint.2003.12.001
2. Wang, N., Gu, X., Chen, J., Shen, L., Ren, M.: A Hybrid Neural Network Method for UAV Attack Route Integrated Planning. In: Yu, W., He, H., Zhang, N. (eds.) ISNN 2009, Part III. LNCS, vol. 5553, pp. 226–235. Springer, Heidelberg (2009), doi: 10.1007/978-3-642-01513-725
3. Mei, H., Tian, Y.T., Zu, L.N.: A hybrid ant colony optimization algorithm for path planning of robot in dynamic environment. International Journal of Information Technology 12, 78–88 (2006)
4. Parry, A.C., Ordonez, P.: Intelligent path planning with evolutionary computation. In: AIAA Guidance, Navigation, and Control Conference, Toronto, Ontario Canada (2010)
5. Bortoff, S.A.: Path planning for UAVs. In: American Control Conference (2000), doi:10.1109/ACC.2000.878915
6. Sullivan, J., Waydo, S., Campbell, M.: Using stream sunctions for complex behavior and path generation. In: AIAA Guidance, Navigation and Control Conference, Austin, USA (2003)
7. Waydo, S., Murray, R.M.: Vehicle motion planning using stream functions. In: Proc. IEEE International Conference on Robotics and Automation (ICRA 2003) (2003), doi:10.1109/ROBOT.2003.1241966
8. Zhang, C.G., Xi, Y.G.: Sub-optimality analysis of mobile robot rolling path planning. Science in China(Series F) (2003), doi:10.1360/03yf9010
9. Zhang, C.G., Xi, Y.G.: Robot rolling path planning based on locally detected information. Acta Automatica Sinica 29, 38–44 (2003)
10. Lambert, A., Gruyer, D., Mangeas, M., et al.: Safe path planning and replanning with unmapped objects detection. In: Intelligent Vehicle Symposium (2002), doi:10.1109/IVS.2002.1187946
11. Wong, C.C., Lin, B.C., Cheng C.T.: Safe path planning and replanning with unmapped objects detection. Fuzzy Systems (2001), doi: 10.1109/FUZZ.2001.1007257
12. Shi, H.Y., Sun, C.Z.: Motion planning for mobile robot under unstructured environment. Robot 26, 27–31 (2004)
13. Jeong, C.D.: Testing neural network crash avoiding system in mobile robot. Case Western Reserve University (2001)

14. Dung, L.T., Komeda, T., Takagi, M.: Reinforcement learning in non-markovian environments using automatic discovery of subgoals. In: SICE 2007 Annual Conference (2007), doi:10.1109/SICE.2007.4421430
15. AI-Hasan, S., Vachtsevanos, G.: Intelligent route planning for fast autonomous vehicles operating in a large natural terrain. Robotics and Autonomous Systems (2002), doi: 10.1016/S0921-8890(02)00208-7
16. Lee, D., Morf, M., Friedlander, B.: Recursive least squares ladder estimation algorithms. IEEE Transactions on Acoustics, Speech and Signal Processing (1981), doi:10.1109/TASSP.1981.1163587
17. Shensa, M.J.: Recursive least squares lattice algorithms-a geometrical approach. IEEE Transaction on Automatic Control (1981), doi:10.1109/TAC.1981.1102682
18. Zhang, C.G., Xi, Y.G.: Robot path planning in globally unknown environments based on rolling windows. Science in China Series E: Technological Sciences (2001), doi:10.1007/BF03014623
19. Miura, J., Shirai, Y.: Modeling motion uncertainty of moving obstacles for robot motion planning. In: Proc. IEEE, Robotics and Automation (ICRA 2000) (2000), doi:10.1109/ROBOT.2000.846363
20. Thomson, L.M.: Theoretical Hydrodynamics. Macmillan, London (1968)
21. Currie, I.G.: Fundamental Mechanics of Fluids. MC Graw-Hill, USA (1993)

An Automatic Portal Radiograph Verification System for Proton Therapy

Neil Muller, Cobus Caarstens, and Leendert van der Bijl

Abstract. In conformal radiotherapy, the final position should be confirmed with a portal radiograph showing the relationship of the patient to the beam line. Historically, this radiograph is verified by manual comparison with a reconstructed radiograph of the expected patient position. In this paper, we discuss a new automatic verification system developed for the proton therapy facility at iThemba LABS[1].

1 Introduction

Proton therapy is a useful treatment method for a number of lesions. The dose distribution properties of the proton beam allow higher doses to be delivered to the target volume with much lower doses to the surrounding tissue. Due to the high cost of treatment, it is often reserved for lesions that are difficult to treat with conventional radiotherapy, such as inter-cranial lesions or lesions close to critical structures such as the spine. For more information see e.g. [12]. iThemba LABS has been involved with proton therapy for over ten years. Due to cost restrictions, iThemba LABS uses a fixed beam-line and a motorised chair to position the patient (see [7]).

The patient is fitted with a mask carrying retro-reflective markers, whose relationship with the target volume is known. The position of the patient during setup and treatment is monitored by several cameras and stereo techniques are used to calculate the patient's position at any time. The retro-reflective markers are easily observed by the vision system, and so the patient position can be tracked accurately

Neil Muller
Department of Mathematical Sciences, University of Stellenbosch,
Stellenbosch, South Africa
e-mail: `neil@dip.sun.ac.za`

Cobus Caarstens · Leendert van der Bijl
Department of Medical Radiation, iThemba LABS, Cape Town, South Africa

[1] Laboratory for Accelerator Based Science.

F.L. Gaol et al. (Eds.): Proc. of the 2011 2nd International Congress on CACS, AISC 144, pp. 21–27.
springerlink.com   © Springer-Verlag Berlin Heidelberg 2012

in real time, and can also be monitoring throughout treatment to guard against unexpected patient motion. For previous work on the vision aspects see [8, 10].

One issue with the positioning system is that the relationship between the mask and the patient is established before treatment when the patient is scanned for planning. This relationship can change over time, due to inaccuracies in the positioning of the mask relative to the patient. To address this, iThemba LABS uses a portal radiograph (an X-ray view of the patient taken along the beam axis). This radiograph is compared to a synthetic digital reconstructed radiograph (DRR), constructed using the expected view of the patient. Treatment proceeds only if these are sufficiently similar. The existing system relies on manual verification by a physician. A new system is being developed to replace this with a automatic system, using a Varian PaxScan 4030R flat panel detector to acquire the portal radiographs.

2 Verifying the Portal Radiograph

2.1 The Verification Algorithm

To automate the verification step, Van der Bijl [2] proposed a 2D-3D image registration system that aims to be accurate and robust. 2D-3D image registration is a process where 3D CT data acquired pre-operatively is registered to a 2D PR image obtained intra-operatively [11]. From the patient positioning system, we know the patient position under ideal circumstances. We assume that the actual patent position is perturbed by a small rigid transformation

$$\mathbf{T}_{DRR} = \begin{bmatrix} \delta_x & \delta_y & \delta_z & \delta_\theta & \delta_\phi & \delta_\rho \end{bmatrix} \tag{1}$$

where $\delta_x, \delta_y, \delta_z$ are the translational errors and $\delta_\theta, \delta_\phi$ and δ_ρ are the rotational errors. The expected treatment position is $\mathbf{T}_{DRR} = 0$. The actual transformation is $\mathbf{T}_{cur}$. When $\mathbf{T}_{DRR} \approx \mathbf{T}_{cur}$, we expect the DRR to resemble the radiograph and the image similarity score to be high. The best estimate for $\mathbf{T}_{cur}$ is found by maximising

$$O(\mathbf{T}_{DRR}) = M(PR, DRR(\mathbf{T}_{DRR})) \tag{2}$$

where PR is the acquired portal radiograph, DRR is the calculated DRR using the appropriate transformation and M is a suitable image comparison function. Due to the design of the mask, the range of possible misalignments is constrained, so we can restrict the search to a small interval around $\mathbf{T}_{DRR} = 0$.

Van der Bijl [2] used Powell's minimiser and the Mutual Information similarity measure to perform registration using DRRs generated with the ray cast algorithm. A schematic representation of the process is shown in figure 1.

2.2   Fast DRR Generation

The algorithm described requires repeated generation of a number of DRRs, so this DRR generation step is a significant potential bottle neck. Consequently, we need to ensure that DRR generation is both fast and accurate.

Let $\rho(i,j,k)$ denote the voxel density or attenuation coefficient in a 3-dimensional CT volume and $l(i,j,k)$ the length of the intersection of an X-ray with that voxel, then the radiological path length is defined as

$$d = \sum_i \sum_j \sum_k l(i,j,k)\rho(i,j,k) \tag{3}$$

The radiological path length is an approximation of the physics involved when an X-ray image is generated. Computing DRRs using the radiological path is very inefficient $(O(n^3))$ as only a few voxels actually contribute to a path, since most $l(i,j,k)$ values are zero. Figure 2 shows how a DRR is constructed using this approach.

Caarstens [4] considered several different DRR generation techniques. He concluded that light fields, proposed by Levoy and Hanahan [9], were the best choice.

A light field can be described as a way of parameterising the set of all rays that emanate from a static scene. Each ray is identified by its intersection with two arbitrary planes. By convention, the first plane has coordinates (u,v) and is called the *focal plane*. The second plane has coordinates (s,t) and is called the *image plane*. So each ray in the scene is represented as a point $\mathbf{p}_i = (u_i, v_i, s_i, t_i)$.

A *light slab* is the shape that is created when the focal plane and the image plane are connected. This represents all the light that enters the restricted focal plane and exits the restricted image plane.

If we can generate infinitely many rays inside a light slab, we can recreate almost any image with a focal point inside the light slab, by finding the correct rays and their corresponding pixel values (figure 3). Since we cannot generate infinitely many rays, we generate a large number and compute the missing rays using interpolation.

Light fields are a simple method to construct novel views from arbitrary camera positions. This is achieved by re-sampling a set of existing images and is therefore called *image-based rendering* [9]. This parameterisation does not include, for instance, rays parallel to the two planes. Since the projection space for our DRRs is constrained, this limitation does not concern us.

Light fields attractive for the DRR generation as most computation can be done pre-operatively. During patient treatment, when computation time must be minimised, images can be quickly generated from the pre-computed data. Images are

Fig. 1 Overview of the 2D-3D registration process.

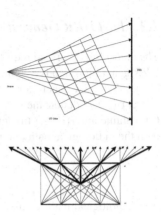

Fig. 2 A 2D view of DRR
generation. The DRR is
the set of values for the
radiological paths from the
source to the pixels on the
image plane.

Fig. 3 A 2D view of a light
slab illustrating (in bold)
the view generated for an
arbitrary focal point.

generated by interpolation of the pre-computed data. This can be done in constant
time, since the computation time is not dependent on the complexity of the image.

A pixel value in a general light field is the light reflected off the first surface a
ray intersects with. For DRRs, however, the pixel values are the radiological path
lengths (eq. 3) that the rays encounter from the projection point to the image plane.

To generate DRRs, we associate each point $p_i = (u_i, v_i, s_i, t_i)$ with a scalar function $p_i \mapsto q(p_i)$ which maps a point to the radiological path length of the ray R_{p_i}.

In order to trace a ray through the CT data and maintain the same parameterisation of rays in space as traditional light fields we must cast the rays beyond a *virtual
image plane* onto an *effective image plane*. The values on the effective image plane
is used for the light field generation.

In traditional light field rendering as well as light field DRR generation, the generated image is a skewed perspective image. However, where in traditional light field
rendering the image plane remains fixed between the scene and the focal plane, in
DRR generation the virtual image plane remains fixed while the effective image
plane can move and lies on the other side of the scene from the focal plane.

Furthermore, since each path traced through the light slab is independent, the
process can be trivially parallelised to take advantage of multi-core machines.

2.3   The System in Operation

Once the patient has been positioned, the flat panel is lowered into position, as seen
in figure 4. The radiograph is then verified and, if the discrepancy is too large, the
patient is repositioned using the appropriate correction factor.

The left of figure 5 shows the comparison between the a section of the predicted
radiograph (in blue) and the observed x-ray (in red) for the mask being deliberately
misaligned, while the right shows the same section after the correction has been
applied and the patient repositioned. As can be seen, the correspondence between
the x-ray and predicted DRR improves significantly.

Fig. 4 Head phantom in the beam line with flat panel in position.

Fig. 5 Effect of the calculated correction factor.

3 Calibrating the Portal Radiograph System

For the automatic verification of the portal radiograph to succeed, the calculated DRR must be of high accuracy. The geometry of the portal radiograph setup must be correctly modelled for the DRR calculation, and the position of the patient relative to the portal radiograph system must be known to a high degree of accuracy.

While the mechanical aspects of the portal radiograph system can position the x-ray tube and flat panel detector to within 0.1 mm over repeated trials, routine maintenance tasks can change the relative alignment of these, so a calibration procedure is required to establish the current geometry of the system.

The position of the patient mask relative to the actual beam-line is known to a high degree of accuracy from the the patient positioning system, so calibration of the portal radiograph system needs to establish the geometric relationship between the beam line and the portal radiograph system.

3.1 Initial Calibration Step

Following Brack [3], we model the X-ray imaging system as a pinhole camera, as illustrated in figure 6. From the geometry, we can see that

$$\frac{X}{d} = \frac{x}{z} \Rightarrow X = d\frac{x}{z} \tag{4}$$

which is the classic formula for a pinhole camera. Since we are using a flat-panel detector, there is no radial distortion due to lens artifacts, and we can, for the initial calibration stages, simplify the problem by ignoring distortion.

Since the system is modelled as a pin-hole camera, traditional approaches, such as the direct linear transform [1], can be used without any modification.

For the DLT, we need to be able to accurately detect pints in the portal radiograph whose real-world positions are well known. For calibration, we use a hollow cube with four rings of small steel ball-bearings embedded on the inside of two opposite faces of the cube, as illustrated in figure 7. There is a wire arrow embedded in the centre of one face of the cube, so that the orientation of the cube can be determined on the x-ray image, which provides robustness against mis-positioning of the cube.

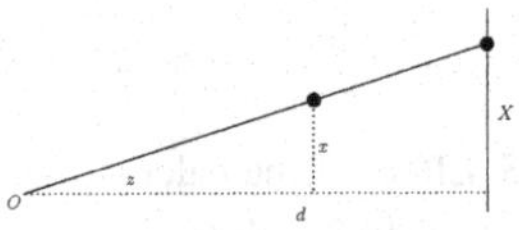

Fig. 6 Portal radiograph
system model.

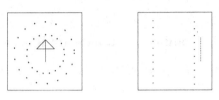

Fig. 7 Calibration object.

An actual radiograph view is shown in figure 8. We calculate the initial set of calibration parameters using the standard DLT approach.

Fig. 8 Portal X-Ray view
of the calibration object,
showing the ball bearings
and the directional arrow

3.2 Calibration Refinement

To improve robustness against image artifacts that can reduce the accuracy of the DLT calibration, we refine the results using the approach presented by Eisert [5].

Since we have accurate knowledge of the composition of the calibration object, and the approximate geometric setup of the system, we refine the geometric properties of the setup, by minimising the difference between the observed portal radiograph and rendered DRRs of the calibration object using small perturbations of the values obtained using the DLT calibration. The initial estimate to the geometric position of the cube given by the DLT ensures that we are close to the final solution, so the computational cost of this optimisation step is reasonable, and we need not be concerned about local minima leading to incorrect solutions.

4 Conclusion

In this paper, we describe some aspects of the new portal radiograph verification system developed at iThemba LABS. The new system removes a lot of the uncertainty around the current manual verification process, and allows us to correct for errors in the alignment of the patient using the existing patient positioning system, which significantly improves the speed and quality of treatments.

References

1. Abdel-Aziz, Y.I., Karara, H.M.: Direct linear transformation from comparator coordinates into object space coordinates in close-range photogrammetry. In: Proceedings of the Symposium on Close Range Photogrammetry, pp. 1–18 (1971)
2. van der Bijl, L.: Verification of patient position for proton therapy using portal X-rays and digitally reconstructed radiographs. Master's thesis, University of Stellenbosch (2006)
3. Brack, C., Götte, H., Grosse, F., Mocrezuma, J., Roth, M., Schweikard, A.: Towards accurate x-ray-camera calibration in computer-assisted robotic surgery. In: Proceedings of the International Symposium on Computer Assisted Radiology, pp. 721–728 (1996)
4. Carstens, J.E.: Fast generation of digitally reconstructed radiographs for use in 2D-3D image registration. Master's thesis, University of Stellenbosch (2008)
5. Eisert, P.: Model-based camera calibration using analysis by synthesis techniques. In: Greiner et al. [6], pp. 307–314
6. Greiner, G., Niemann, H., Ertl, T., Girod, B., Seidel, H.P. (eds.): Vision, Modeling, and Visualization 2002 (2002)
7. Jones, D.T.L., Schreuder, A.N., Symons, J.E., Rüther, H., van der Vlugt, G., Bennet, K.F., Yates, A.D.B.: Use of stereo-photgrammetry in proton radiotherapy. In: Rüther, H. (ed.) Proceedings of the International FIG Symposium on Photogrammetry in Enigineering Surveying, pp. 138–152 (1995)
8. de Kock, E., O'Kennedy, B., Muller, N.: Calibrating a stereo rig and CT scanner with a single calibration object. In: Greiner et al. [6]
9. Levoy, M., Hanrahan, P.: Light field rendering. In: SIGGRAPH 1996: Proceedings of the 23rd Annual Conference on Computer Graphics and Interactive Techniques, pp. 31–42. ACM Press, New York (1996), http://doi.acm.org/10.1145/237170.237199
10. Muller, N., de Kock, E., van Rooyen, R., Trauernicht, C.: A stereophotogrammic system to position patients for proton therapy. In: Ranchordas, A., Araújo, H., Vitriá, J. (eds.) VISAPP 2007, Second (2007)
11. Russakoff, D.B., Rohlfing, T., Rueckert, D., Shahidi, R., Kim, D., Maurer Jr., C.R.: Fast calculation of digitally reconstructed radiographs using light fields. In: Sonka, M., Fitzpatrick, J.M. (eds.) Medical Imaging 2003: Image Processing. Proceedings of the SPIE, vol. 5032, pp. 684–695 (2003), doi:10.1117/12.481888
12. Webb, S.: The physics of three-dimensional radiotherapy: Confromal radiotherapy, radiosurgery and treatment planning. Insitute of Physics Publishing, Bristol and Philadelphia (1993)

Compensation of Hydraulic Drag for an Underwater Manipulator Using a Real-Time SPH Fluid Simulator: Application in a Master-Slave Tele-operation

Haoyi Zhao and Masayuki Kawai

Abstract. This paper discusses a control method of a tele-operation system for an underwater manipulator robot. Since fluid drag affects the dynamics of the underwater manipulator and degrades the accuracy of the control, it is necessary to decrease the influence of water. Generally, in order to compensate the fluid drag, theoretical approaches are used, but the applications are restricted by the simple modeling of the robot. Instead, we adopt a method to compensate the drag using a real-time fluid simulation. By using the simulation, the method can be applied to a robot with a complicated shape. For the method, we develop a real-time fluid simulator using smoothed particle hydrodynamics. The simulator calculates the fluid drag with the position of the underwater manipulator, and the drag is fed back to the controller to eliminate the real fluid drag. Finally, experiments have been performed to evaluate the effectiveness of the proposed method.

1 Introduction

This paper researches a tele-operation system with an underwater manipulator robot, which is controlled by a symmetric master-slave; MS system. In the MS system, since the slave robot is designed only to perform the same motion as the master robot does, an operator can manipulate an object at the fingertip of the slave robot easily and intuitively by operating the master robot. However, when the slave robot is located in water, its motion can be greatly affected by the hydraulic drag. Therefore, the feeling of the operation and the transparency of the system is also affected by the drag. Some methods to eliminate the influence of the drag have been studied in researches of underwater robots. In these researches, theoretical approaches with a simple modeling of the robot and a basic theory of hydraulic dynamics are adopted [1]-[3]. Since a robot usually has a complicated shape, thus the applications seem to be restricted.

In this paper, we propose a new method to compensate the hydraulic drag using a real-time fluid simulation instead of the theoretical approaches. First, we construct a

Haoyi Zhao · Masayuki Kawai
University of Fukui, 3-9-1 Bunkyo Fukui, Japan
email: m_kawai@u-fukui.ac.jp

F.L. Gaol et al. (Eds.): Proc. of the 2011 2nd International Congress on CACS, AISC 144, pp. 29–38.
springerlink.com   © Springer-Verlag Berlin Heidelberg 2012

real-time fluid simulator using smoothed particle hydrodynamics; SPH, which is a kind of particle methods in computational fluid dynamics; CFD. As the simulator calculates motion of virtual water using information of the slave robot, and can obtain force to the slave robot from the virtual water, so the system can uses the force to the control of the underwater tele-operation system to cancel the real hydraulic drag.

In this paper, firstly we show the overview of a MS system. Secondly, we introduce the proposed method, and the real-time fluid simulator which we have constructed. Finally, experiments have been performed to shows the effectiveness of the proposed method.

2   Background

2.1   *Dynamics of an Underwater Tele-Operation System*

Dynamics of a usual MS system is described as

$$\tau_m = M_m(\theta)\ddot{\theta} + h_m(\theta, \dot{\theta}) + J_m{}^T f_m \tag{1}$$

$$\tau_s = M_s(\theta)\ddot{\theta} + h_s(\theta, \dot{\theta}) + J_s{}^T f_s \tag{2}$$

where τ_* are the joint torques, θ_* are the joint angles, M_* are the inertia matrices, h_* are the nonlinear terms, J_* are the Jacobian matrices, and f_* are the external forces. The suffix m and s indicate the terms of the master and slave robots, respectively. Fig. 1 shows an illustration of an underwater tele-operation system.

Fig. 1 Underwater Tele-operation System

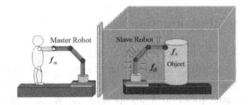

In the underwater tele-operation system, hydraulic drag such as fluid resistance and viscosity exerts on the whole surface of the robot. As the dynamics of the slave robot, (2) is replaced by

$$\tau_s = M_s(\theta)\ddot{\theta} + h_s(\theta, \dot{\theta}) + J_s{}^T f_s + f_{di} \tag{3}$$

where f_{di} is the hydraulic drag.

2.2   *Previous Researches*

Many researchers have been researching about underwater robots including manipulators especially for the undersea investigating robots and the robots for the undersea exploration which can help human to explore or work in the depth of sea [2],[3],[4]. In these methods, using the motion equation to the underwater robot with external force, the dynamic property of manipulator can be obtained through the inertia force, which is proportional with the acceleration, and the fluid force which is square to the velocity of the object. But in these methods, fluid is assumed to be static, or modeling of robots is simple. Therefore, the applications seem to be restricted. Considering the disadvantage of these methods, the paper proposes a new control method for complex shaped manipulators.

2.3 Symmetric Control of MS System

In this research, a symmetric control is utilized to control the MS system. Fig. 2 shows the block diagram of the symmetric control. x_m and x_s denote the fingertip position of master and slave robots, respectively.

Fig. 2 Block Diagram of a Symmetric Control

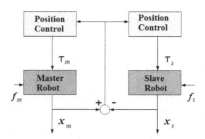

The symmetric control uses a feedback control to decrease the displacement between the position of master and slave robots. As the control method requires only the position and the velocity, and does not need any force sensors or similar devices, so it is highly stable and easy to construct the system.

3 Compensation of Hydraulic Drag Using a Real-Time Fluid Simulator

3.1 Overview of the Proposed Method

The purpose of the research is to enable that the operator feels as if the slave robot was located on the ground. For the purpose, the measurement of the drag f_{di} in (3) is necessary. But it is not easy to measure the real drag occurred on the surface of robot using sensors. Instead of the measurement, the drag is calculated using CFD. In the CFD simulation, it is easy to obtain the value of drag f_v on a surface of a virtual robot, and the value can be utilized to compensate the real drag. Using a real-time simulation, we propose a symmetric control as,

$$f_m = K(x_m - x_s) + D(\dot{x}_m - \dot{x}_s) \tag{4}$$

$$f_s = K(x_s - x_m) + D(\dot{x}_s - \dot{x}_m) - f_v \tag{5}$$

where K is a proportional gain and D is a differential gain, f_v is simulated drag in the simulation. If the simulated drag f_v corresponds to the real drag f_{di}, the real drag can be compensated and will not be felt by the operator. Fig. 3 shows the illustration of this method, which is called 'Slave Compensation' (SC method) in the paper.

Fig. 3 Block Diagram of the SC Method

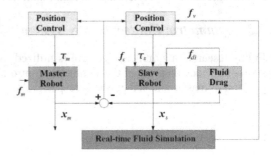

3.2   A Real-Time Fluid Simulator

In the proposed method, a real-time fluid simulator is necessary, and we have been developing a real-time fluid simulator using SPH [8], [9]. SPH was originally developed to simulate the divisions and collisions of planets in cosmophysics [5]. Subsequently, the method was used to study incompressible fluid dynamics. In the method, fluid is divided to particles and the fluid dynamics is calculated at each position of particles. The method can simulate fluid quickly because fluid can be basically considered compressible, and the constraint of incompressibility is not imposed. Instead of the constraint, pressure is used as a feedback to control the density at the desired value. Some quick methods to calculate SPH have been researched for constructing real-time virtual environments [6], [7].

3.3   Equations of SPH

In SPH, fluid is divided into a set of small volume elements named particles. Fig. 4 schematically illustrates SPH.

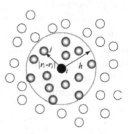

Fig. 4 Particles in SPH

Each particle has mass m, and the position of the i th particle is expressed by r. The state variables of the i th particle are calculated by interpolation of the state variables of peripheral particles in order to produce a smooth distribution of state variables. A state A_i is calculated by

$$A_i = \sum_j m \frac{A_j}{\rho_j} w_{ij} \tag{6}$$

where the w_{ij} is a function serving as a weight in the interpolation and is called the kernel function. The summation index j refers to the particles that are located within an efficient radius of h around the i th particle, as shown in Fig. 5. The kernel function and the radius are chosen considering the stability of motion and computation time.

With the interpolation of (6), the density of i th particle ρ_i can be obtained by

$$\rho_i = \sum_j m w_{ij} \tag{7}$$

In SPH, the gradient and laplacian of state variables are also described by a summation of the product of the gradient and laplacian of the kernel function. In general, the Navier- Stokes (NS) equation and the equation of continuity are used for incompressible fluids. In the SPH formula, the NS equation is known to be given by

$$\ddot{r}_i = m \sum_j \left\{ -\left(\frac{P_i}{\rho_i^2} + \frac{P_j}{\rho_j^2} \right) \nabla w_{ij} \right\} + a_i \tag{8}$$

where P_i is the pressure of the i th particle and a_i is a term that includes viscosity, gravity, and external forces, except the pressure. The differential of the pressure in a general NS equation is replaced with that of the kernel function. In a real-time SPH, the density should be controlled to the desired density ρ_0 instead of the equation of continuity. In order to control the density, the pressure P_i is used as feedback considering the stability and computation time. Most of the control laws of P_i in the previous researches can be expressed by

$$P_i = \beta \left\{ \left(\frac{\rho_i}{\rho_0} \right)^\gamma - 1 \right\} \tag{9}$$

where constant values β and γ are chosen depending on the stability of motion and the dimension of simulation.

We have been constructing a real-time simulator using SPH, in which the number of particles and the values of m, h, and β are automatically adjusted for the real-time process. In this paper, we use the simulator to control an underwater teleoperation system.

4   Experiments

In order to evaluate the effectiveness of the proposed method, we perform an experiment with two 2-DOF manipulators and a 3-dimensional real-time fluid simulator.

4.1   Experimental System

Fig. 5 shows the experimental system in this research.

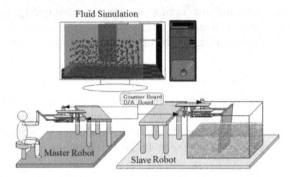

Fig. 5 Overview of Experimental System

This system uses master and slave robots. A part of slave robot soaks in water. The computer plays two roles; control of the master and slave robots, and the calculation of SPH. The specification of the computer is CPU: Core2Quad 2.66GHz, Memory: 4GByte, OS: Windows XP, Visual C++. Generally, it is hard to calculate these tasks in the same computer, but the automatic adjustment of parameters, which we have developed, makes it possible in the system. In the experiment, the slave robot does not perfectly sink in water because of the hardware issues.

4.2   Conditions in Experiment

Fig. 6 shows the overview of master and slave robots. The master and slave robots in the experiment are 2-DOF parallel link mechanisms with the same length of links. There is a handle at the finger tip of the master robot for the operation. The slave robot has an aluminum; Al board at the fingertip instead of the handle of the master robot. The robots use DC serve motors and encoders.

In the experimental system, a part of the Al board at the fingertip of the slave arm soaks in water, which is stored in a plastic container. The volume of water is 390mm× 230mm× 210 mm, as shown in Fig. 7. The size of the Al board is 100mm× 210 mm× 1mm thickness. 100 mm of the board soaks in water. These lengths are also used in the simulation of SPH.

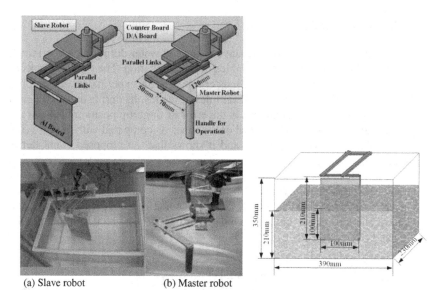

| (a) Slave robot | (b) Master robot |

Fig. 6 Overview of Master and Slave Robots   **Fig. 7** Fluid Environment

4.3   Introduction of Experiments

In the experiments, operators move the master robot with the same trajectory for 3 kinds of control methods. The shape of the trajectory is set to circle, as shown in the Fig. 8. The cycle of motion is about 2.5s.

Fig. 8 Trajectory in the Experiment

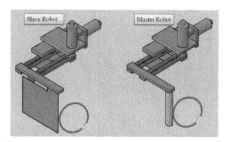

We measure error between the fingertip position of master and slave robots in one cycle, and also record the velocity of the slave and master robots.

In the experiment, 3 kinds of conditions are used and compared. In the case (I), water is not stored in the container, and the slave robot is not affected by fluid drag. The system is controlled by a normal symmetric PD control, which corresponds to $f_v = 0$ in (5). In the case (II), water is stored, and water drag occurs to the surface of the slave robot, but the controller is a normal symmetric PD controller. In the case (III), water is also stored, but the system uses the SC method. In the experiments, 3 subjects try the experiments. Each subject tries 8 cycles for each condition.

4.4   Experimental Results

Fig. 9 (a) shows the overview of the slave robot and water during the experiment, and Fig. 9 (b) shows the computer graphics of the simulation. The top right links in the figure represent the position of the master robot. In the experiments, the number of particles is 400. Fig. 10 shows the examples of the results during a cycle in each case. In the figure, the horizontal axis is the error of fingertip position between the master and slave robots and the vertical one is time. As compared with the case (II), the results of the case (III) decrease, and it shows that the proposed controller is effective.

Next, Fig. 11 shows the average value of the error between the fingertip of master and slave robots for each condition. The blue left bar is the result of the case (I). The error occurs even when the slave robot is not affected by water, because the robots have frictions in the joints. The red center bar and the green right bar are the case (II) and (III), respectively.

Fig. 9 Overview of Experiment
and Simulation

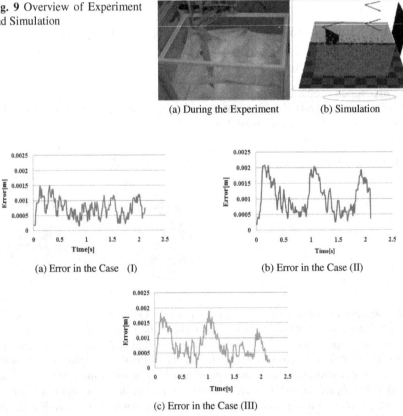

(a) During the Experiment   (b) Simulation

(a) Error in the Case   (I)   (b) Error in the Case (II)

(c) Error in the Case (III)

Fig. 10 Examples of the Experimental Results

Fig. 11 Average Error of the Experiments

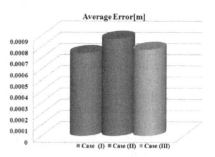

4.5 Discussions

When the slave robot is in water in the case (II), the average error is the worst. By using the SC method, the error is decreased by about 76.52%. So it shows that the proposed SC method using the real-time fluid simulator is effective to cancel the real fluid drag and make the system more accurate. All the operators feel the light operation when using the SC method. When the velocity of the operation becomes larger, the error in the case (II) increases conspicuously, because the fluid drag also becomes larger. In the proposed method, even with a large velocity, the accuracy can also be ensured. However, these quantitative and qualitative results of the SC method do not reach to the target; the result of the case (I). This will be because the accuracy of the fluid simulator is not enough. The accuracy of the fluid simulator depends on the algorithm of SPH and the sampling periods of the system. So we need to improve the real-time fluid simulator.

In the experiment, we use a simple plane board and just a part of it soaks in water. A merit of the proposed method is that it has possibility to apply to robots with more complicated shape when comparing to the previous theoretical methods. To apply it to such robots, we need to increase the number of particles in the fluid simulation. We also need to improve the fluid simulator more accurate and faster for this point of view.

5 Conclusion

In this paper, we proposed a new method to compensate the hydraulic drag in an underwater teleoperation system using a real-time fluid simulator. In the paper, we introduced a real-time fluid simulator that we have developed, and proposed to use the simulator for a symmetric control of a master-slave system to compensate the real drag and to achieve a light and easy operation. Finally, In order to show the effectiveness of the proposed method, we performed experiments using two 2-DOF robots and a 3-dimensional fluid simulator. The results of the experiments showed that the proposed method is effective to compensate the real drag and to realize a light operation, but it also showed that it is needed to improve a more accurate and fast simulator.

References

1. Leabourne, K.N., Rock, S.M.: Model Development of An Underwater Manipulator for Coordinated Arm-Vehicle Control. In: OCEANS 1998, vol. 2, pp. 941–946 (1998)
2. de Wit, C.C., Daíz, E.O., Perrier, M.: Nonlinear Control of an Underwater Vehicle/Manipulator with Composite Dynamics. IEEE Transactions on Control Systems Technology 8(6), 948–960 (2000)
3. Antonelli, G., Chiaverini, S., Sarkar, N.: External Force Control for Underwater Vehicle-Manipulator Systems. IEEE Transactions on Robotics and Automation 17(6), 931–938 (2001)
4. Ishitsuka, M., Ishii, K.: Control of an underwater manipulator mounted for an AUV considering dynamic manipulability. International Congress Series, vol. 1291, pp. 269–272 (2006)
5. Lucy, L.B.: A Numerical Approach to the Testing of the Fission Hypothesis. Astronomical Journal 82, 1013–1024 (1977)
6. Monaghan, J.J.: Smoothed Particle Hydrodynamics. Annual Review of Astronomy and Astrophysics 30, 543–556 (1992)
7. Müller, M., Charypar, D., Gross, M.: Particle-Based Fluid Simulation for Interactive Applications. In: Proc. of 2003 ACM SIGGRAPH Symp. on Computer Animation, pp. 154–159.
8. Kawai, M., Hirota, K., Kuroyanagi, S.: Development of a Real-time Fluid Simulator for an Interactive Virtual Environment: Improvement of Density Feedback in SPH. In: IEEE/ASME Conference on Advanced Intelligent Mechatronics, pp. 1701–1706 (2009)
9. Kawai, M., Suzuki, Y., Zhao, H.: An Automatic Adjustment of Parameters in a Real-time SPH Fluid Simulation for Interactive Virtual Environments. In: 2010 IRAST International Congress on Computer Applications and Computational Science, pp. 355–359 (2010)

Design of Bipedal Bionic Electric Mobile Platform Based on MATLAB

Liu Donghui, Sun Xiaoyun, and Yang Lili

Abstract. This paper designs a new type of transport-double foot bionic electric mobile platform based on MATLAB. With the simulation software of virtual reality toolbox, video acquisition Toolbox, GUI and SIMULINK to complete the whole simulation, a 3D lower limb model to complete the controlling simulation is built and driven; By GUI, the movement of lower limbs through a camera is got, the changed angle data of hip joints, knee joints and ankle joints in movement is extracted, the storage of the lower limbs move data is completed.

1 Introduction

A common wheelchair can surely bring convenience to the life of the old and disabled in a certain extent, but it has many obvious limitations. For example, it will be difficult without other people's help when they met barriers such as ditch, upstairs or downstairs. For these questions, this paper proposes a new style of foot-type chair, which is simulated based on MATLAB. By using the principle of bionics to simulate the alternate movement of double foot, foot-type chair could operate together and complete the whole walk movements with the user's control, synergy of hip joint and knee joint.

In GUI (Graphical User Interface), the callback function of acquisition button uses video acquisition toolbox of MATLAB to capture the movement of lower limb, and then compiles the program function to complete the function of control. At last, we store the data of joints in the pictures with an algorithm, put the data into simulation block, connect with the 3D model in virtual world, and make the model moves according to our command.

Donghui Liu
Hebei University of Science & Technology,
Shijiazhuang, China
e-mail: liu_donghui@126.com

Xiaoyun Sun
Shijiazhuang Tiedao University
Shijiazhuang, China
e-mail: sunxy@hebust.edu.cn

Lili Yang
Hebei University of Science & Technology,
Shijiazhuang, China
e-mail: iloveyou_yangli.com

F.L. Gaol et al. (Eds.): Proc. of the 2011 2nd International Congress on CACS, AISC 144, pp. 39–42.
springerlink.com   © Springer-Verlag Berlin Heidelberg 2012

2 Establish GUI and Take Video Acquisition with the Callback Function

Edit the GUI and set up callback functions for different purpose [1]. Collect the lower limb's movement image and storage the data for backup, so the human lower limb motion parameters of human body are got. Fig. 1 shows the interface.

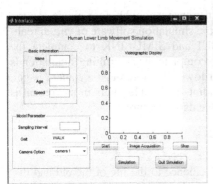

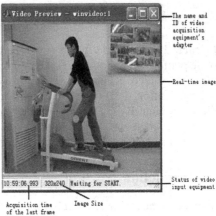

Fig. 1 The overall effect diagram   **Fig. 2** Video preview window

Image Acquisition Toolbox which is one of the most important software boxes in the MATLAB toolbox provides user with a kind of object-oriented image acquisition method. We make the image acquisition device connect to MATLAB and collect the image with the function of toolbox. Meanwhile, if we adjust the controlled objects' attribution, we can decide when to begin collecting and collecting how many data [2].

Connected the external video equipment with MATLAB, using the callback function 'imaqhwinfo' in MATLAB command window, then we can choose the adapter in MATLAB. The returned value of a function includes the adapter and other related information chooses in Installed Adaptors filed. For example, input 'imaqhwinfo' in the MATLAB common window, the optional adapter for this system named 'winvideo' is got, so, a video input object myvid= videoinput ('winvideo') is created. If we input 'preview(myvid)' in command window, which is shown in Fig. 2, a video review window is got. We can call the return value 'Supported Formats' of function 'imaqhwinfo' to check the formats that the video equipment supports. For example, the video input object is 'my_vid= imaqhwinfo ('winvideo',1)', check video format with 'my_vid . Supported Formats'.

Use the command 'closepreview(myvid)'to close the previewed window. There are other kinds of adapters except for 'winvideo', like 'coreco' (Mainly applies to video equipment which is produced by the Coreco) , 'dcam'(mainly used in image acquisition equipment of IEEE1394) , 'dt'(used in video equipment of changing data) and 'matrox'(used in video equipment of Matrox electronic system)[3].

Video acquisition toolbox provides plenty of function to set the related parameters. We can control the quantity of video data acquisition, the brightness of the collected image, the tonal and the saturation through setting. The collected image data can be stored to AVI files to backup. Image acquisition tools provide the corresponding function. First, execute the code 'aviobj=avifile('my_datalog.avi')' to create a AVI file. By amending the attribute value of the AVI file object's return value to set the quality of the stored data and the data compression, etc. For example, 'aviobj.Quality=50' means to change the image quality, the numerical smaller and then the file smaller.

By catching the movement of lower limb, extract the changed angle data of hip joints, knee joints and ankle joints in movement in video acquisition module .Then we store the data into the corresponding files named with the suffix of '.mat'.

After using three circular tags with red, blue and green color as sports mark of hip, knee and ankle respectively in the experiment, we could extract three colors in the acquisition image and assign a white value to the background. In addition, a video input object is established to start the data collection. The collected data is a 4D matrix including a time vector which is recorded as a time sequence of image acquisition. For example, if we know that 'vid(; , ; , ; ,1)'means the first collected image and 'vid(; , ; , ; ,n)'means the collected image at position 'n' , we can do the image processing .Read one picture in the right way to extract its color.

After completing the background removal job, it is needed to find the center of red, green, blue three circular tags. Image with removed background only have red, green, blue three labels and the background is white. Because the image pixels was idealized in the previous process, that is, the RGB component of white background in the picture is (255, 255, 255), red color is (255, 0, 0), green color is(0, 255, 0), blue color is (0, 0, 255). Finding the centre coordinates of the three colors substantially is looking for its row and column's value of center.

In order to control 3D model better, a further processing that means to change the coordinates in plane into rotating angle values of each joint is taken. So we simulate lower limbs' moving process through the control on angle rotation.

3   Simulation of Double Foot Bionic Electric Mobile Platform

According to the simulation design requirements of double foot bionic electric mobile platform, a structure of lower limb skeleton should be analyses first. Then set up the ball-strict model of human skeleton according to the skeleton movement association between lower limbs. Finally, use the V-Realm Builder to build a 3D lower limb model.

We read the collection data with the module named 'from file' in SIMULINK.

In VR (Virtual Reality), the rotation needs two numbers to control its rotation, one is the direction of the rotation and the other one is the angle of the rotation. The translation needs three-dimensional direction to control. Then connect with 3D model through virtual reality toolbox and drive it to complete the simulation of lower limb movement. The simulation model is shown in Fig.3

Use the VRML browser of MATLAB to view the 3D simulation results, which is shown in Fig.4. According to simulation results, we can undertake some corresponding experimental research, verify the accuracy of the models and affection of the control algorithm, and build a manned physical platform and do some engineering research on this basis.

Fig. 3 Block diagram of simulation

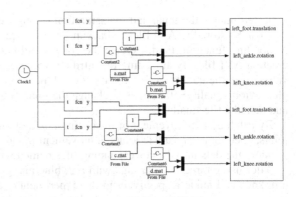

Fig. 4 Results of motion simulation

4 Conclusion

This paper realized the movement simulation of double foot bionic electric mobile platform's early research. Based on this paper, we can do more research on the platform, for example, we can write some program to control the platform, make it moving in a constant velocity and make the hip in Fig.4 more stable, make the disabled people who sit on this platform more comfortable.

Acknowledgments. The research presented in this paper is supported by Natural Science Foundation of China and Hebei Province (Grant No: 50874035, F2009000732, CPRC023).

References

1. Zhang, Q., Li, Y.: Creating Graphical User Interface with MATLAB. Computer Development & Applications 3, 10 (2003)
2. Texas Instruments, TMS320DM642 Video/Imaging Fixed-Point Digital Signal Processor (2005), http://www.ti.com
3. Wang, Z., Liu, M.: Proficient in MATLAB 7. Publishing house of Electronics Industry (2006)
4. Sims, E.M.: Reusable, lifelike virtual humans for mentoring and roleplaying. Compute and Education(s0360-1315) 49(1), 75–92 (2007)
5. Tong, J., Wang, L., Wu, Y.: Simulation of arm rehabilitation robot based on MATLAB virtual reality toolbox. Applied Science and Technology 33(10), 63–65 (2006)

On the Design and Implementation of a Wearable Hybrid Assisted Lower Limb Orthosis

Cheng-Chih Hsu and Jian-Shiang Chen

Abstract. The objective of this research is aimed to design and implement a Wearable Hybrid Assisted Lower Limb Orthosis (HALLO), which could provide up to 10% of knee torque support. The proposed device adopted a specially designed M-shaped flexible mechanism to store the potential energy due to different human body gestures, such as stand-to-sit, a DC motor would act as a variable damper to control the releasing rate of potential energy. To provide a proper trajectory tracking, a discrete-time output feedback sliding-mode controller design [7] is adopted to perform efficient tracking. Finally, wearing the device to perform a variety of basic movements is demonstrated through experiments.

1 Introduction

Knee joint degeneration is mainly due to the cartilage between joints worn-out, and the muscle power to the joint is not large enough to support. There are many researches on the design of wearable orthosis or exoskeleton, for example, the Hybrid Assisted Limb (HAL) system[3][4], BLEEX at U.C. Berkely[5], the MIT Exoskeleton[10] at MIT, and Quasi-Passive Knee Exoskeleton[1] are the most recent representative research. These applications are mainly focused on weight-carrying and military purposes, therefore the overall weight is not a major concern, however. This study is aimed to design a light weight wearable lower limb orthosis for the knees. To reduce the overall weight, here uses DC motors and a semi-active device as major components.

Cheng-Chih Hsu
Department of Power Mechanical Engineering, National Tsing Hua University, Hsinchu, Taiwan
e-mail: alvis0606@gmail.com

Jian-Shiang Chen
Department of Power Mechanical Engineering, National Tsing Hua University, Hsinchu, Taiwan
e-mail: jschen@pme.nthu.edu.tw

F.L. Gaol et al. (Eds.): Proc. of the 2011 2nd International Congress on CACS, AISC 144, pp. 43–51.
springerlink.com © Springer-Verlag Berlin Heidelberg 2012

The orthosis is mainly consisted of a specially designed M-shaped flexible mechanism [2] (Hereinafter referred to as torsion spring) and DC motors. The torsion spring would store potential energy due to different human body gestures, such as stand-to-sit, the DC motors would act as a variable damper, to control the releasing rate of potential energy from the torsion spring. Counteraction of these two forces can provide appropriate assisting-torque to the knee joint, subsequently, reducing the user`s burden, and providing the user a relief on knee discomfort.

2 System Description

Fig.1 shows a walking gait, it could be separated into Stance Phase and Swing Phase [9]. The leg which in stance phase bears body's weight, generates torque to the joint by muscles, and push the body forward; the leg which in swing phase does not have to bear the body weight, simply makes swing motion and into the next cycle. Based on that, while the leg is at stance phase, the orthosis will provide assisting torque to user`s knee, and is called the "stance leg mode"; and while the leg is at swing phase, no assisting is needed from the orthosis but to lift the leg, the body simply tries to balance the user`s motion, and is called the "free leg mode". It is noted that the torque command for stance and free leg mode would be different, and should be treated with different control modes.

Fig. 1 A normal human walking gait cycle [9]

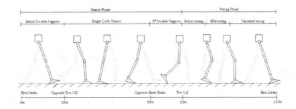

2.1 Stance Leg Mode

Fig.3 shows the working principle of the orthosis. Since the orthosis is mainly composed by torsion springs[2] and DC motors, counter-action of these two forces can provide propoer assisting-torque to the knee joint, and the assisting troque, τ_f, is the diffierence between τ_{spring} and τ_{motor}, and can be defined as,

$$\tau_f = SR \cdot \tau_1 = \tau_{spring} - \tau_{motor}$$
$$\Rightarrow T_{motor} \cdot L = T_{spring} \cdot L - SR \cdot \tau_1 \tag{1}$$

Where τ_1 is the loading torque of user`s knee while SR is the supporting ratio provided by the orthosis. From Eq. (1), τ_{spring} is proportional to the spring's deformation, instead of measuring torque, a tension sensor in series with the cable,

T_{spring} is measured to control the tension T_{motor}, indirectly. The force arm, L (see Fig.2), could be easily calculated by leg's geometry.

As shown in Fig.3 [8], in stance phase, the loading torque τ_1 can be determined by

$$\tau_1 = f(L_0 \cos q_0 - r) - m_0 g(L_0 - l_{c0}) \cos q_0 \tag{2}$$

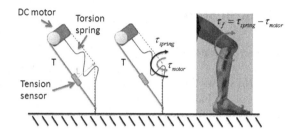

Fig. 2 Schematic diagram of assisting principle

In Eq. (2), on the right side of the equation, the first term is due to whole body's weight, and the second term is due to the weight below the knee, in comparison, the second term can be ignored. So that it could be simplified as [8]

$$\tau_1 = f(L_0 \cos q_0 - r) \tag{3}$$

The knee torque could be calculated by measuring the reacting force from ground and the ankle angle.

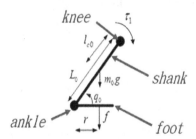

Fig. 3 Lower limb dynamics[8]

Here, τ_1 is the knee torque $(N \cdot m)$, τ_{spring} is the spring torque acting on the knee $(N \cdot m)$, τ_{motor} is the motor torque acting on the knee $(N \cdot m)$, m_0 is the weight of calf (kg), L_0 is the length of calf (m), l_{c0} is the distance between the center of gravity of calf and the knee (m), f is the reacting force from ground (N), and r is the distance between the center of gravity of whole body (m).

2.2 Free Leg Mode

From Eq.(3), the reacting force from ground will be zero as the foot lifted, then τ_1 will be zero, and τ_{spring} will equal to τ_{motor} . It means that while the foot leaves the ground, the motor is used to balance the spring force to allow the leg to move freely.

3 Control Strategy

3.1 Linearized Model

Eq. (4) [6] shows the dynamics of DC motor with reduction ratio N, since the time constant of the inductance is relatively small, it will be ignored for simplification. The units are all in MKS.

$$(I_a + N^2 I)\ddot{\theta}_m + N^2 (b + \frac{K_t^2}{R})\dot{\theta}_m = N\frac{K_t}{R}u - \tau_{load} \tag{4}$$

$$\tau_{load} = Mr_m^2 \ddot{\theta}_m + K_s r_m \theta_m \tag{5}$$

Here, u is the input voltage (V), τ_{load} is the loading torque $(N \cdot m)$, r_m is the radius of the roll wheel (m), θ_m is the rotation angle of the roll wheel (rad), R is the armature resistance (Ω), b is the rotating friction coefficient $(N \cdot m \cdot sec/ rad)$, K_t is the torque constant $(N \cdot m/ A)$, K_s is the torsion spring constant (N/ m), M is the load mass (kg) , I is the motor shaft inertia $(kg \cdot m^2)$, and I_a is the reflected inertia of load $(kg' \cdot m^2)$.

The transfer function from u and θ_m is

$$\left.\frac{\Theta_m(s)}{U(s)}\right|_{d=0} = \frac{N\frac{K_t}{R}}{(I_a + N^2 I + Mr_m^2)s^2 + N^2(b + \frac{K_t^2}{R})s + K_s r_m^2} \tag{6}$$

$$T = K_s \cdot \theta_m \tag{7}$$

A linear torsion spring constant is assumed to yield

$$\left.\frac{T(s)}{U(s)}\right|_{d=0} = \frac{N\frac{K_t}{R} \cdot K_s}{(I_a + N^2 I + Mr_m^2)s^2 + N^2(b + \frac{K_t^2}{R})s + K_s r_m^2} \tag{8}$$

Fig.4 shows the control block diagram, where T is the tension tracking output, while the tracking command T_d is pre-filtered with a LPF to avoid drastic action; the system input u is the modulation voltage to the DC motor,; and d is the force between user`s leg and the orthosis, it will be treated as output disturbance.

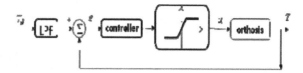

Fig. 4 Control block diagram

3.2 Controller Design

This system is an output feedback tracking control in nature, a robust controller that can tolerate system uncertainty controller is needed. Here, a discrete-time output feedback sliding-mode controllers [7] is chosen to complete the design. The following shows the design process, detailed derivatation can be found in[7].

For a nth order, m-input, and p-output linear system

$$x(k + 1) = Gx(k) + H[u(k) + d(k)] + H_r r(k)$$
$$y(k) = Cx(k)$$
(9)

The integral term $\mathbf{x}_r$ and τ, an integral constant, is added,

$$\mathbf{x}_r(k+1) = \mathbf{x}_r(k) + \tau[\mathbf{r}(k) - \mathbf{y}(k)]$$
(10)

Together with the estimated states

$$\mathbf{x}_c(k+1) = \mathbf{\Phi}\mathbf{x}_c(k) + \mathbf{\Gamma}_1\mathbf{y}(k) + \mathbf{\Gamma}_2\mathbf{x}_r(k) + \mathbf{\Gamma}_3\mathbf{r}(k)$$
(11)

Define $S = CG^{-1}$ and let S be an output matrix of the virtual system (G, H, S), and a linear transfer matrix could transfer (G, H, S) into a canonical form $(\bar{G}, \bar{H}, \bar{S})$ and has the following form

$$\bar{G} = \begin{bmatrix} \bar{G}_{11} & \bar{G}_{12} \\ \bar{G}_{21} & \bar{G}_{22} \end{bmatrix}, \bar{H} = \begin{bmatrix} 0 \\ H_2 \end{bmatrix}, H_2 \text{ is nonsingular.}$$
$$\bar{S} = \begin{bmatrix} 0_{p\times(n-p)} & \bar{T}_{p\times p} \end{bmatrix}, \bar{T} \text{ is orthonormal.}$$
(12)

The control law without any states could be formed as

$$\mathbf{u}(k) = -(\mathbf{FC}_{all}\mathbf{G}_{all}^{-1}\mathbf{H}_{all})^{-1}\mathbf{Fy}_{all}(k) + \mathbf{F}_r\mathbf{r}(k)$$
(13)

Where $(G_{all}, H_{all}, C_{all})$ is the extended matrices which contain $\mathbf{x}_r$ and $\mathbf{x}_c$. With this control law, the last row of the close-loop system matrix $\tilde{G}_c$ will all be zeros as Eq. (14) shows.

Fig. 5 Expansion of closed-loop
system matrix $\tilde{G}_c$

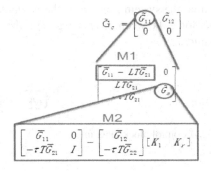

$$\tilde{G}_c = \tilde{G} - \tilde{H}(F\tilde{S}\tilde{H})^{-1}F\tilde{C} = \begin{bmatrix} \tilde{G}_{11} & \tilde{G}_{12} \\ 0 & 0 \end{bmatrix} \tag{14}$$

Therefore, as Fig.5 shows, the eigen values of M1 and M2 are closed-loop system poles.

In Fig.5, L、K_1、K_r can be solved by pole placement method to assign the eigen vaules of M1 and M2 respectively; in addition, S_r is related to the tracking speed only and can be adjusted accordingly. Φ、Γ、Γ_2、Γ_3 in Eq. (11) and F、F_r in Eq. (13) can be determined by L、K_1、K_r、S_r, their relationship can be referred to [7].

4 Experimental Results

4.1 Implementation of Controller

Since the bandwidth of this system is lower than 10HZ, the sampling frequency is chosen as 40Hz which is larger than twice of the Nyquist frequency, with proper discretization process.

M1、M2 can be found as

$$M1 = 0.9999 - L(-0.0079)$$

$$M2 = \begin{bmatrix} 0.9999 & 0 \\ 0.0001975 & 1 \end{bmatrix} - \begin{bmatrix} 0.0247 \\ -0.0244 \end{bmatrix} [K_1 \quad K_r] \tag{15}$$

The eigen values of M2 is set to provide a damping of 0.707 and rise time about 1.2 second; and the eigen value of M1 is set to be 0.6, which would be ten times faster than M2. Therefore, L、K_1、K_r are

$$L = 466.48, \ K_1 = -0.0027, \ K_r = 0.097 \tag{16}$$

Where S_r was related to tracking speed only, here is assigned as 0.02 to yield,

$$\Gamma_1 = 466.48, \ \Gamma_2 = 30.0518, \ \Gamma_3 = 6.1931, \text{and } \Phi = -0.2289. \qquad (17)$$

From Eq.(11), the observed state x_c would be

$$x_c(k + 1) = -0.2289x_c(k) + 466.48y(k) + 30.0518x_r(k) + 6.1931r(k) \qquad (18)$$

Then the control law would be

$$\mathbf{u}(k) = \begin{bmatrix} 3.5294 & 95.6865 & -1446.4 \end{bmatrix} \begin{bmatrix} \mathbf{x}_c(k) \\ \mathbf{x}_r(k) \\ \mathbf{y}(k) \end{bmatrix} + 33.6975 \cdot \mathbf{r}(k) \qquad (19)$$

4.2 Free Lag Mode – Swinging

As Fig.6 shows, while the feet is lifted, the tracking command tends to follow the torsion spring resilience, and the difference was due to the force of the leg to the orthosis.

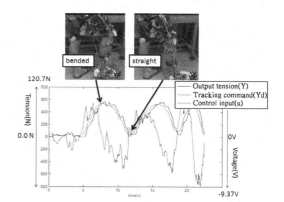

Fig. 6 Experiment results of free leg mode

4.3 Stance Leg Mode – Squatting and Standing

As Fig.7 shows, during 3s to 12s, the required assisting torque of the knee increased while the user squatted, and thus y_d decreased to make difference between torsion spring resilience and the cable tension, the difference is the assisting force by the orthosis. No assisting force to the knee before 3s and after 12s while the leg is stretched.

Fig. 7 Experiment result of squatting and standing

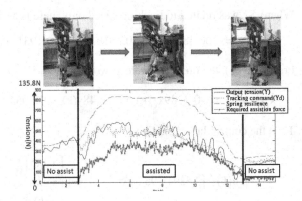

5 Concluding Remarks

The HALLO (Hybrid Assisted Lower Limb Orthosis) could provide 10% of knee torque from active device, reducing the loading of knee. This paper completed the swing motion by tracking control, and verifid its efficacy of the proposed method through experiments.

Acknowledgement. This work was supported by the National Science Council of Taiwan(ROC) under the contract number of NSC98-2221-E007-56- MY3.

References

[1] Dollar, A.M., Herr, H.: Design of a Quasi-Passive Knee Exoskeleton to Assist Running. In: Proc. IEEE/RSJ Int. Conf. Intell. Robots Syst. (IROS), pp. 747–754 (2008)

[2] Huang, L.-C.: Analysis and Design of the Flexible Component of a Lower Limb Orthosis, Master thesis, National Tsing Hua University, Taiwan (2007)

[3] Kawamoto, H., Sankai, Y.: Power Assist System HAL-3 for Gait Disorder Person. In: Miesenberger, K., Klaus, J., Zagler, W.L. (eds.) ICCHP 2002. LNCS, vol. 2398, pp. 196–203. Springer, Heidelberg (2002)

[4] Kawamoto, H., Lee, S., Kanbe, S., Sankai, Y.: Power Assist Method for HAL-3 Using EMG-based Feedback Controller. In: Proc. IEEE Int. Conf. Syst. Man, Cybern, vol. (2), pp. 1648–1653 (2003)

[5] Kazerooni, H., Steger, R.: The Berkeley Lower Extremity Exoskeletons. ASME Journal of Dynamics Systems, Measurements and Control (128), 14–25 (2006)

[6] Kuo, B.C., Jacob, T.: DC Motors in Incremental Motion Systems. SRL, Champaign (1978)

[7] Lai, N.O., Edwards, C., Spurgeon, S.K.: On Output Tracking Using Dynamic Output Feedback Discrete-Time Sliding-Mode Controllers. IEEE Transactions on Automatic Control (52), 1975–1981 (2007)

[8] Lu, T.-C.: Dynamic Control and Portability Achievement of Wearable Lower Limb Orthosis, Master thesis. National Tsing Hua University, Taiwan (2008)

[9] Rose, J., Gamble, J.G.: Human Walking, 2nd edn. Williams and Wilkins, Baltimore (1994)

[10] Walsh, C.J., Paluska, D., Pasch, K., Grand, W., Valiente, A., Herr, H.: Development of a Lightweight, Underactuated Exoskeleton for Load-carrying Augmentation. In: Proc. IEEE Int. Conf. Robot. Autom., pp. 3485–3491 (2006)

3D Point Sets Matching Method Based on Moravec Vertical Interest Operator

Linying Jiang, Jingming Liu, Dancheng Li, and Zhiliang Zhu

Abstract. The purpose of this paper is to solve the problem of matching two 3D point sets quickly in the field of robot vision. Moravec vertical interest operator is used to extract vertical edge feature of objects. The method of the sum squared difference (SSD) is used to match the feature points and obtain the 3D point sets which contain vertical line feature. Find the transformation relation of rotation and translation of corresponding straight lines in two 3D point sets according to the projection point of vertical line projected into x-y plane. Depending on the transformation relation, it can match two 3D point sets. The experiment results illustrate that this method has good exactness and robustness.

1   Introduction

The 3D scene is always represented by a three dimensional point set. Because of a huge amount of data of 3D point set and stochastic distribution of inner points, it takes lots of computing time to get a good matching result. Currently, researchers have resolved the matching problem of 2D scene information and the most widely applied ones are the matching of point to point and point to line, combined with the thought of evidence grid proposed by Hans P. Moravec and Albert Elfes, etc. The least squares and iterative closest point (ICP) are most widely used in matching 3D point sets, because it's quick and robust. This paper proposed an algorithm of fast matching two 3D point sets extracted from the actual scenes. This method first extracts vertical feature of objects in the scene, and obtain points' spatial location information by matching feature points based on SSD matching strategy. Then it counts the weight of each point in the 3D point set and extracts vertical line information.

In a 3D space, the points could constitute a vertical line, and the horizontal plane projection coordinates of these points should be the same, namely, the rate of occurrence of vertical and horizontal coordinates of these points are more than others. Setting a threshold can help extract straight line of different lengths. According to the different lengths of lines, appropriate weights are assigned as the

Linying Jiang · Jingming Liu · Dancheng Li · Zhiliang Zhu
Software College, Northeastern University, Shenyang, China
email: jiangly@swc.neu.edu.cn, jingming_de@126.com,
lidc@swc.neu.edu.cn, zhuzl@swc.neu.edu.cn

F.L. Gaol et al. (Eds.): Proc. of the 2011 2nd International Congress on CACS, AISC 144, pp. 53–59.
springerlink.com   © Springer-Verlag Berlin Heidelberg 2012

criteria for judging matching degree. Finally, corresponding relation of rotation and translation is calculated and the correct matching matrix is found by comparing the weight of matched vertical lines and the corresponding relation with other straight lines.

2   Theories and Experimental Methods

2.1   Vertical Feature Extracting and Feature Points Matching

Moravec feature extraction algorithm divides the image into several grids of non-overlapping windows, and on each computes the sum of squares of differences of pixels adjacent in each of four directions: horizontal, vertical and right and left diagonals. The minimum value of the sums is used to represent the value of the window. Then by setting the threshold it can determine whether the character of the region is obvious, which means whether they are feature points. In the indoor scene, objects usually have vertical edge features, so we can only extract the vertical feature of the object; and compute the horizontal direction value shown as (1).

$$Value = \sum_{i=1}^{n} (A_i - B_i)^2 \tag{1}$$

The sum squared difference approach computes the sum of squared difference of corresponding pixel points in the two windows. The original cost of the match is shown by (2), L and R represent the point sets of left image and right image respectively; i and j are the index of two sets; d is the offset of the responding point's abscissa. The best matching point location can be found by moving SSD window.

$$C(x, y, d) = \sum_{i=1}^{n} \sum_{j=1}^{n} [L(x + i, y + j) - R(x + d + i, y + j)]^2 \tag{2}$$

2.2   The Thought of 3D Point Set Matching Method Based on the Property of Vertical Line

Moravec operator is used to compute only the horizontal direction value, that is, just extract the vertical edge features of the object. Even if the vertical edges of object are not absolutely perpendicular to the x-y plane, because Moravec operator divides the image into several sub-alignment of square window, it can make the edges strictly perpendicular to the x-y plane as long as selecting the same pixel point location as feature point of the window in the column windows.

SSD is used to match the feature points and find corresponding points of two images. The vertical edges of object which shows linear relationship after extracted as feature points keep linear relationship in the 3D space between feature points after the depth information of these points is calculated. The longer straight line is given greater weight. After extracting the vertical line of great weight, it's projected into x-y plane. The projected points have the same weight value. Thus,

3D point set matching problem is simplified to 2D point set matching problem. Depending on different weight, different priorities are set when matching points which makes the greater weight have higher priorities. Generally, the probability of matching is higher if the point has greater weight, which reduces the time to match.

We can take any two points in a point set, and calculate Euclidean distance, while traversing all points of the other 2D point set until a close distance is found. After finding the relation of rotation and translation, this transformation relation is applied on the target point set. In the fusion point set, look for any two from different set of points corresponding to the other point, and record the number of corresponding points on. The correct relation of rotation and translation between two point sets can't be obtained until the numbers of corresponding points reach the largest one.

2.3   The Design of 3D Point Set Matching Method Based on the Property of Vertical Line

1. The extraction of vertical line in the space is the core step of 3D point matching method. And the concrete steps of extracting vertical lines are as follows:
   a. Count the number of occurrences of each point's abscissa.
   b. Set a threshold, select a higher frequency of the abscissa value x', and in accordance with the frequency giving appropriate weight to the point.
   c. Find the index in original 3D point set corresponding vector that the abscissa is equal to the value of x'.
   d. According to the index, respectively, identify the value of the corresponding coordinate in the direction of Y axis and Z axis.
   e. Get the value of coordinate in the direction of X axis, Y axis, and Z axis, that is, extract the coordinate of the vertical line.
2. After extracting the vertical line in the space, we can find the corresponding relation of rotation and translation in the two-dimensional plane according to value of the coordinate (x, y) of vertical line projected into x-y plane. The concrete steps are as flows:
   a. Arrange the points in the point set in descending order by weight.
   b. The point P_{11} and P_{12} in point set A constitute a line segment L_1, and The point P_{21} and P_{22} in point set B constitute a line segment L_2, until the two sets are traversed completely.
   c. Calculate the difference of lengths between L_1 and L_2 to determine whether it is within certain error range, if not, return a), and update P_{11}, P_{12}, P_{21}, P_{22}, to form a new segment.
   d. Calculate the matrixes of transformation between L_1 and L_2.
   e. Determine whether the matrix is correct, if not, return a), and update P_{11}, P_{12}, P_{21}, P_{22}, to form a new segment, until the two sets are traversed completely.
   f. Save the matrix.

Finally, find the matrixes of rotation and translation, and then apply this transform matrix on the target point set. The computing process ends.

3   Experiments

3.1   Experimental Data Collection and Extraction of Vertical Edge

The experimental images are acquired from a stereo camera platform which is built by the author. The platform consists of two Logitech C-310 cameras, and the camera focal length is 814.25381±2.59872 (pixels), the binocular distance is 71.29306±0.25871 (mm), the location of left head's main point is (324.23399, 229.86646) (pixels) through the camera calibration method based on Bouguet algorithm.

Fig. 1 The resolution of the original image pair collected by the stereo camera platform is 480×640. It can be seen that the rotation between the two scenes is relatively large, as Fig. 1 shows.

Fig. 2 In the experiment, the window size is 8×8, and the threshold is the largest eigenvalue/1000. 956 feature points are extracted from this image in this image, and the result is shown as Fig. 2.It is obvious that the extracted edges are vertical. Although the left edge of the bucket is inclined, as shown in pink rectangular box, it also shows vertical feature through vertical feature extracted by Moravec operator.

3.2   The Vertical Line and Projected Point of 3D Point Set

Count the frequency of point's coordinate (x, y) appears, and by setting threshold to save the point with higher frequency, then assigns appropriate weight. Table 1 shows the statistics of points of scene 1, and the 'Value' represents the value of abscissa, the 'Count' represents the number of occurrences, the 'Percent' represents the percentage of the number of occurrences and the overall points. The threshold is set to extract the point which frequency is greater than 2%.

Table 1 The statistics of points

Value	-70.37101	-36.5687	-21.73728	-21.11622	111.7455	115.1317	126.5769
Count	7	19	10	7	8	8	17
Percent	4.93%	13.38%	7.04%	4.93%	5.95%	5.63%	11.97%

Fig. 3 The vertical line in 3D point set is extracted after completing the statistics of frequency. The extracted lines are shown as Fig. 3.

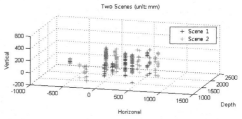

If the extracted vertical line is projected into x-y plane, the original 3D point sets could form two 2D point sets, and we can find the transformation relation between the two sets according to the points in 2D point set. The projected point sets through rotation transformation are shown as Fig. 4 shows.

Fig. 4 The two points in scene 1 are matched with the two points in scene 2 through the transformation of rotation and translation. And we can find other points in scene 1 which correspond with the ones in scene 2 after transformation. The judgment points circled with the box in Fig. 4 are such points. The couple of points' distance difference is less 50mm, so we can determine they are the same point in the space. It illustrates the transformation relation is correct.

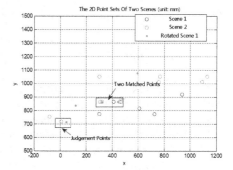

3.3   The Rotation of 3D Point Set

As step *B.* is shows, the extraction of vertical edge feature based on Moravec operator works well in extracting the vertical feature of object. There is little influence even though the camera has been inclined slightly. According to this characteristic, we can expand the transformation relation of two 2D point sets into 3D point sets as step *C.* obtained, namely, the 3D point just rotates around Z axis. Equation (3) is the rotation matrix of 2D point set. Equation (4) is the rotation matrix of 3D point set.

$$(x', y') = (x, y) \begin{bmatrix} \cos\theta & \sin\theta \\ -\sin\theta & \cos\theta \end{bmatrix} \tag{3}$$

$$(x', y', z') = (x, y, z) \begin{bmatrix} \cos\theta & \sin\theta & 0 \\ -\sin\theta & \cos\theta & 0 \\ 0 & 0 & 1 \end{bmatrix} \tag{4}$$

It can be seen from the above two equations that the transformation from 2D rotation matrix to 3D rotation matrix just adds one dimension [0 0 1]. The two raw 3D point sets are shown as Fig. 5(a), and the matched 3D point sets are shown as Fig. 5(b). The blue points in Fig. 5(a) are extracted from the scene 1 as the left image of Fig. 2; the pink points are extracted from scene 2 as the right image of Fig. 2. As the contours of the case is obvious, we can regard it as reference object, and

the blue and pink points in 3D point sets is close to each other in Fig. 5(b), namely, the edge feature of it is matched well.

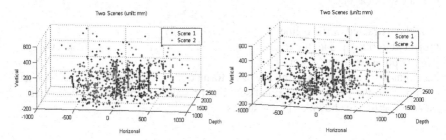

Fig. 5 (a) the original point sets of two scenes; (b) the matched point sets

4   Conclusion

The research background of this paper is the construction of environment model of
visual navigation robot. It is aimed to find a fast method to match the 3D point
sets information captured from stereo vision platform. By analysis and practice,
we find that extracting vertical edge feature based on Moravec operator works
well, and the extracted points could form perpendicular line in the space. The result is used to design a fast algorithm to match 3D point sets. The experiment results indicate that this method has good exactness and robustness.

References

[1] Yamauchi, B., Growley, J.: A comparison of position estimation techniques using oc-
    cupancy grids. Journal of Robotics and Autonomous Systems 12, 163–171 (1994)
[2] Moravec, H.P., Elfes, A.: High Resolution Maps from Wide Angle Sonar. In: IEEE
    Iternational Conference on Roboics and Automation, pp. 116–121 (March 1985)
[3] Haehnel, D., Schultz, D., Burgard, W.: Map Building with Mobile Robots in Popu-
    lated Environments. In: Proceedings of the International Conference on Intelligent
    Robots and Systems(IROS) (2002)
[4] Besl, P.J., McKay, N.D.: A method for registration of 3-d shapes. IEEE Transaction
    Pattern Analysis and Machine Intelligence 14(2), 239–256 (1992)
[5] Chetverikov, D., Svirko, D., Stepanov, D., Krsek, P.: The Trimmed Iterative Closest
    Point Algorithm. In: Proceedings of International Conference on Pattern Recognition,
    Quebec City, Canada, pp. 545–548 (August 2002)
[6] Qing, R.: The Research of Three-Dimensional Reconstruction Based On Multi-Depth
    Images. Zhejiang University Master's Degree thesis (2006)
[7] Se, S., Lowe, D.G., Little, J.J.: Vision-Based Global Localization and Mapping for
    Mobile Robots. IEEE Transactions on Robotics 21(3) (June 2005)

[8] Scharstein, D., Szeliski, R.: A Taxonomy and Evaluation of Dense Two-Frame Stereo Correspondence Algorithms. International Journal of Computer Vision 47(1/2/3), 7–42 (2002)

[9] Shupeng, W., Lili, F.: The Analysis of Using Moravec Operator to Extract Feature Points. Computer Knowledge and Technology (26), 125–126 (2006)

[10] Bradski, G., Kaebler, A.: Learning OpenCV, pp. 433–436. O'Reilly Media, Inc., Sebastopol (2008)

Improvement and Simulation of Contract-Net-Based Task Allocation for Multi-robot System

Hao Lili and Yang Huizhen

Abstract. Due to the shortage of ignoring the balance between the mission success rate and the executive ability of the whole system in the common contract net models, a threshold-limited load balance strategy for awarding is proposed. The mathematical model of negotiation process based on improved contract net is established. The conceptual model of the multi-robot system based on improved contract net is implemented by Colored Petri Nets (CPN), and its correctness and dynamic properties such as liveness property, home property and fairness property are validated by using CPN Tools. Simulation results show that this method can effectively improve the performance, and ensure the overall system load balance.

1   Introduction

Task allocation reflects the organization and operation mechanism in the high level of multi-robot cooperative system. It is the premise of multi-robot cooperation and determines the effectiveness of the system in large extent as well. Contract net is negotiation-oriented through the imitation of economic behavior in the process 'task announcement-bid-award' so as to allocate tasks, and has been widely applied in the formation cooperation, satellite, multi-UCAV system [1] and other fields.

The strategies of task announcement, bid, awarding are the key technologies of task allocation based on contract net. [2] pointed out that the goal of task allocation is lower cost, higher efficiency and limited the maximum number of the allowed allocated tasks to reduce the computational complexity. In [3, 4], the contract net model based on Petri Net was researched primarily. Trust and punishment mechanism was introduced to the traditional contract net protocol in [5], which reduced the traffic by restricting the scope of announcing tasks and controlling the number of task evaluated. [6] discussed the influence of a various mind parameters in deciding whom the task is announced to, such as familiar and active. The bid buffer pool was used to limit the bid number in order to achieve the purpose of reducing communication traffic. All of the above works reduce the traffic in the

Hao Lili · Yang Huizhen
School of Marine, Northwestern Polytechnical University, Xi'an, Shaanxi, China
email: onceinthemoon@163.com, rainsun@nwpu.edu.cn

F.L. Gaol et al. (Eds.): Proc. of the 2011 2nd International Congress on CACS, AISC 144, pp. 61–67.
springerlink.com   © Springer-Verlag Berlin Heidelberg 2012

negotiation for complex system, but it also bring some problems such as the bidders with higher credit degree always won the task, while the lower is in idle, finally the task load is not balance.

In order to ensure the stronger whole executive ability as well as achieve higher overall system performance of the designed multi-robot system, we proposes a load-balance-based awarding algorithm with limited threshold. The other sections are organized as follows: In section 2, the task allocation problem of multi-robot system is described formally. Section 3 establishes the mathematic model based on improved contract net. In order to simulate the studied task allocation method and analysis the models' security, the improved contract net models are formally established and validated by CPN and CPN Tools in Section 4 and 5 separately. A summary and discussion for future direction are presented in Section 6.

2   Formalize of Task Allocation Problem for Multi-robot System

Each robot executes task autonomously in a distributed structure. Task types are different, and so are the capacities of the robot. When the new task is discovered, it's necessary to allocate tasks dynamically, quickly, and logically according to the current situation. The designed multi-robot task allocation system should satisfy two targets, and it can be formalized as follow:

(1) Higher efficiency [7] If the number of tasks to be distributed exceeds the maximum task number that the multi-robot cooperative system could performed, the tasks are selected in accordance with the principle of the highest overall efficiency after completing the tasks.

(2) The system's overall executive capacity is improved. That is allocating tasks evenly and making the task load of each robot tend to balance. Suppose that the task load of robot R_i is $TL_i(S_i)$. It is the ratio of the current task number of robot R_i, and the task number it can execute, the average task load $\overline{TL}$, and then $\sum_{i=1}^{N_V} | TL_i(S_i) - \overline{TL} |$ must be minimum to ensure the task load balance.

The total efficiency attained by multi-robot system completing tasks is determined by the task efficiency and the robot's ability.

Definition 1. Task efficiency: The efficiency $U_i(T_j)$ attained by robot R_i executing task T_j reflects the rewards $Reward_i(T_j)$ and corresponding cost $Cost_i(T_j)$ .Its formula is given as follow:

$$U_i(T_j) = \operatorname{Re} ward_i(T_j) - Cost_i(T_j) \tag{1}$$

And the reward $\operatorname{Re} ward_i(T_j)$ is related to the type of task. As a result of the different implementation capacity of each robot, the cost $Cost_i(T_j)$ mainly includes distance costs and time costs, given as:

$$Cost_i(T_j) = \alpha_1 \times Cost_i(length(T_j, R_i)) + \alpha_2 \times Cost_i(Time(T_j, R_i)) \qquad (2)$$

Where, α_1 and α_2 is the weight of each cost, and $\alpha_1 + \alpha_2 = 1$.

3 The Improvement of Contract-Net-Based Task Allocation Algorithm

In order to ensure the strongest whole executive ability under the premise of achieving maximum efficiency, threshold-limited load balance strategy for awarding is proposed to improve the contract net proposed by Smith [8] in 1980. This method avoids large gap of the executive ability among the robot caused by trust. It prevents one auction optimum other than the overall task allocation optimum caused by the different task discovery and implementation order too. It can be denoted as $\{R_i, R_j, T_k, U_{i,j}^{sale}, TL_t(S_i)\}$, where T_k is the task R_i transited to R_j, $U_{i,j}^{sale}(T_k)$ is the changed efficiency of the overall system after T_k transited. The negotiation algorithm is described as follows:

Step1: Task announcement. If robot R_i finds a new task T_k but could not accomplish independently, it announces the task, and the changed efficiency $U_i^-(T_k)$ is denoted as:

$$U_i^-(T_k) = U_i(S_i \setminus \{T_k\}) - U_i(S_i) \qquad (3)$$

Where, $U_i(S_i)$ is the efficiency of R_i before task announcement, $U_i(S_i \setminus \{T_k\})$ is the efficiency of R_i executing other tasks except T_k, and $U_i^-(T_k) < 0$.

Step2: Bid. After receiving the announced task, robot R_j calculates changed efficiency $U_j^+(T_k)$ according to its current capabilities and situation, and decides to whether buy this task.

$$U_j^+(T_k) = U_j(S_j \cup \{T_k\}) - U_j(S_j) \qquad (4)$$

Where, $U_j(S_j \cup \{T_k\})$ is the efficiency of R_j executing all tasks involved T_k, and $U_j^+(T_k) > 0$. The changed efficiency of whole system $U_{i,j}^{sale}(T_k)$ is denoted as:

$$U_{i,j}^{sale}(T_k) = U_j^+(T_k) + U_i^-(T_k) \qquad (5)$$

If $U_{i,j}^{sale}(T_k) >= 0$, R_j sends the bid $Bid_j^{buy}(T_k, U_{i,j}^{sale}(T_k))$ to R_i.

Step3: Awarding. When bids for T_k are collected from R_j ($j \neq i$) or time has expired, the auctioneer determines winner for T_k through choosing the maximal bid, which is also formalized as:

$$U_{i,j}^{sale}(T_k) = \max_{q=1,\dots,N_v} U_{i,q}^{sale}(T_k) \qquad (6)$$

Step4: Improved awarding strategy based on load balance with limited threshold. Firstly, suppose that threshold of efficiency deviation $\xi_u > 0$. Secondly, if R_a is the bidder who gives the maximum bid, let's choose another bidder R_b who satisfied $U_{i,a}^{sale}(T_k) - U_{i,b}^{sale}(T_k) < \xi_u$ from the left robots. Finally, comparing the task load $TL_a(S_a)$ with $TL_b(S_b)$, we determine the robot with the smaller task load as the winner.

Step5: Award acceptance. Once R_j received the bid winner information, the task set is updated to $S_j \cup \{T_k\}$. Then, R_j perform the tasks.

4   Improved Contract Net Model for Multi-robot System Based on CPN

A.   Problem Description

Robot1, 2, and 3 execute the tasks T1~T6 cooperatively, where T1, T4, T5, T6 are Type2, and T2, T3 are Type1. Suppose Type2 tasks are more emergence than the Type1. The pure profit of completing one Type2 task is 500, which is 200 more than the Type1. The initial condition of the problem is that robot1's task is T1, robot2's task is T2, and robot3's task is T3. Robot1 find the following tasks T4-T6 in the task executed process, and the three robots allocate all tasks efficaciously through negotiation and competition based on improved contract net.

B.   The Improved Contract Net Model

Colored Petri Net [9] is a high level Petri Net with graphical form, hierarchy structure and well-known semantic, it is suitable to describe concurrent, discrete and complicated system. Furthermore, formal modeling, simulation and formal validation are integrated by applying CPN Tools. We establish and validate the improved contract models of multi-robot system by using CPN and CPN Tools.

By analyzing the task executed process of multi-robot system, the multi-robot system model is divided into three sub-models named robot1, robot2, robot3, and one top model named contract net model which is established as figure 1.

In figure 1, the place "num" is used to record the current task numbers of Type1 and Type2. When the robot1 discovers a new task, the guard function of "invite biding" is defined as *noty1+noty2>0* to judge whether the current task could be handled, and its code segment is used to calculate the changed profit according to formula (1, 2). If robot1 is not capable to handle the current task, it broadcasts the bidding, and robot2 and 3 evaluate it. Take the robot2 as an example. In order to avoid deadlock caused by task conflict, we suppose robot2 must execute the next task after the current task completed. Therefore, it's necessary for robot2 to evaluate the bidding "receive biding2" according to the forecasted position

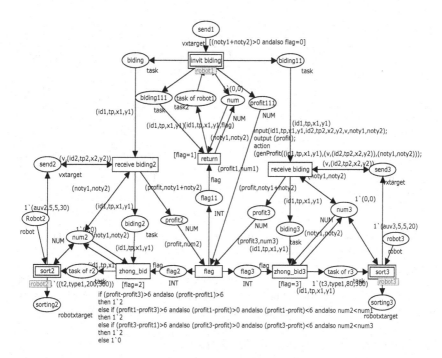

Fig. 1 Top level model of multi-robot system

"send2" at the moment of completing current task, the current task type number "num2" and the task announced. Robot1 selects the winner based on the proposed threshold limited load balance strategy for awarding, which involves parameters such as each robot's evaluated profit "profit1""profit2", "profit3", and "num1, num2, num3". Their awarding flags are 1, 2, 3 respectively, and their losing flag is 0. In other words, if robot2 is winner, the transition "flag" sends 2 to robot2 submodel. Meanwhile, its tasks are updated by guard function of "zhong_bid". Additionally, in order to avoid one robot bidding repetitively, we add the "FLAG" to the task information which is initially marked as 0, and guard function of "invite biding" is changed to $flag = 0$. If inviting bidding is unsuccessful, the "FLAG" in task information changes to 1.

5   Simulation and Validation

A.   Analysis of Simulation Results

In order to analyze the performances of improved contract-net-based multi-robot system proposed in this paper, we take the basic contract net model in contrast. The comparison of simulation results is shown in Table 1.

Table 1 Comparison of simulation results

	Improved Contract Net	The Basic Contract Net
finished tasks	robot1: T1, T5, T6; robot2: T2, T4; robot 3: T3	robot 1: T1; robot2: T2, T4, T5, T6; robot 3: T3
Time to complete tasks (Min)	robot 1:32; robot 2:22; robot 3:15	robot 1:14; robot 2:35; robot 3:15
Total efficiency	1664	1673
The relative workload of overall system	0.2808	0.3808

The simulation results indicate that if we don't consider the load balance, the three tasks found in task execution process by robot1 allocate to robot2 totally. But based on threshold-limited load balance Strategy for awarding, only T4 is allocated to robot2. Although the total efficiency of improved contract nets model is slightly lower, the model takes a shorter time to complete the tasks, and the overall relatively workload is smaller than that without load balance.

B. *Dynamic Performance Validation*

As shown in figure 2, we have the following validation results:

(1) There are total 146 states and 374 arcs in the state space of improved contract-net-based multi-robot system as many as those in the SCC-graph. Therefore there are no infinite occurrence sequences when the model is executed. And the fairness properties report proves it again.

(2) Home properties report shows that the initial marking is not a home marking, which means the initial marking can't initialize itself.

(3) Liveness properties reflect that the 146th node is a dead marking, which is the terminal state as well. Therefore, we can say that the model has liveness property since there is no dead transition from the initial states to the terminal state.

Fig. 2 By running "calculating the state space"and "computing strongly connected components map", and "saving standard state space report", CPN Tools generate the validation report for improved contract-net-based multi-robot system model. It includes blondeness, home, liveness, and fairness properties

```
CPN Tools state space report for:
  Statistics              Home Properties
-------------------------------------------------------------------
  State Space             Home Markings    Initial Marking is not a home marking
    Nodes:  146           Liveness Properties
    Arcs:   374          ------------------------------------------------
    Secs:   0             Dead Markings          [146]
    Status: Full         Dead Transition Instances    None
  Scc Graph              Live Transition Instances    None
    Nodes:  146          Fairness Properties
    Arcs:   374         ------------------------------------------------
    Secs:   0             No infinite occurrence sequences.
```

6   Closing

In this paper, the task allocation method for multi-robot system based on contract net is studied, and the threshold-limited load balancing strategy for awarding is proposed. Conceptual model based on improved contract net is established by CPN. The correctness, liveness, home and fairness property of the designed model are validated by CPN Tools, which provides reference and application value for the study of multi-robot system's mission planning and coordination control methods. There are many influence factors of the robot system performance, spatial conflict influence on multi-robot system task allocation is the further research direction.

References

[1] Fan, T.S., Qin, Z.: Task Assignment for Multi-UAV Using Contract Net Based on Filter Model. In: 2010 6th International Conference on Wireless Communications, Networking and Mobile Computing, Chengdu, pp. 343–346 (September 2010)

[2] Tang, F., Parker, L.E.: A Complete Methodology for Generating Multi-Robot Task Solutions using ASyMTRe-D and Market-Based Task Allocation. In: 2007 IEEE International Conference on Robotics and Automation, Roma, Italy, vol. 10(14), pp. 3351–3358 (April 2007)

[3] Zhang, G.S., Jiang, C.J.: Research of Contract Net Model Based on Cost Timed Petri Net. Journal of System Simulation 20, 5438–5441,5445 (2008)

[4] Hsieh, F.: Modeling and Analysis of Contract Net Protocol. In: Proceedings of International Conference on Web Engineering, Berlin, Germangy, pp. 142–146 (2004)

[5] Singh, A.: Introducing Trust Establishment Protocol in Contract Net Protocol. In: 2010 International Conference on Advances in Computer Engineering, pp. 59–63 (2010)

[6] Liu, N., Gao, F.: Research on the Negotiation Strategy of Multi-Agent Based on Extened Contract Net. In: 2009 ETP International Conference on Future Computer and Communication, Wuhan, pp. 105–109 (2009)

[7] Lee, J., Lee, S.-J., Chen, H.-M.: Dynamic Role Binding with Agent-centric Contract Net Protocol in Agent Organizations. In: 2008 IEEE International Conference on Systems, Man and Cybernetics, pp. 636–643 (2008)

[8] Smith, R.G.: The Contract Net Protocol: High Level Communication and Control in a Distributed Problem Solver. IEEE Transactions on Computers C29, 1104–1111 (1980)

[9] Jensen, K., Kristensen, L.M.: Colored Petri Nets and CPN Tools for modeling and validation of concurrent system. International Journal on Software Tools for Technology Transfer 9, 213–254 (2007)

A Hierarchical Reinforcement Learning Based Approach for Multi-robot Cooperation in Unknown Environments

Yifan Cai, Simon X. Yang, Xin Xu, and Gauri S. Mittal

Abstract. Reinforcement learning is a good method for multi-robot systems to handle tasks in unknown environments or with obscure models. MAXQ is a hierarchical reinforcement learning algorithm, which is limited by some inherent problems. In addition, much research has focused on the completion of the task, rather than the ability to deal with new tasks. In this paper, an improved MAXQ approach is adopted to tune the parameters of the cooperation rules. The proposed scheme is applied to target searching tasks by multi-robots. The simulation results demonstrate the effectiveness and efficiency of the proposed scheme.

1 Introduction

Multi-robot cooperation is a popular topic in robotics and artificial intelligence. A group of robots can solve the complex task that is difficult for individual robot and noticeably improve the work efficiency. In unknown environments, the cooperative robots are capable of learning from the surroundings and their team mates. Reinforcement learning (RL) is a classical machine learning scheme, but it has difficulty in dealing with complicated tasks or large working space. To resolve it, hierarchical reinforcement learning (HRL) is introduced. HRL involves several levels of decision making that together achieves the final goal.

For HRL, there are some typical frameworks that have been proposed, including Options [1], HAMs [2] and MAXQ [3]. The traditional MAXQ runs with state abstraction to reduce the storage space. But the actual applications cannot always offer the clearly refined MAXQ states. In [4], an improved HRL approach is introduced by integrating Options into MAXQ. The method is applied to path planning.

Yifan Cai · Simon X. Yang · Gauri S. Mittal
School of Engineering, University of Guelph, Guelph, Canada
e-mail: {ycai,syang,gmittal}@uoguelph.ca

Xin Xu
College of Mechatronics and Automation, National University of Defense Technology, Changsha, China
e-mail: xinxu@nudt.edu.cn

F.L. Gaol et al. (Eds.): Proc. of the 2011 2nd International Congress on CACS, AISC 144, pp. 69–74.
springerlink.com   © Springer-Verlag Berlin Heidelberg 2012

Though the optimal path could be gained in the end, the method is not practical due to its learning process. Furthermore, most current research on RL only focuses on the results. In new environments, new learning processes are needed. To improve it, Juang and Hsu [5] proposed the method to tune the special parameters of the control strategy by reinforcement learning. This method is applied to wall-following problem. After learning, the robot can effectively follow the walls in new working spaces. However, the control rules in such scenarios are easy to make, and the explored surroundings are very simple.

In this paper, multi-robot cooperation for target searching is studied. Two cooperation rules are designed to handle new tasks in new environments. An improved MAXQ algorithm similar as the proposed approach in [4] is adopted here to tune the parameters of the rules. The experiment results show that the proposed scheme can effectively and efficiently lead the team of robots to complete new target searching tasks in unknown environments.

2   The Proposed Improved MAXQ Algorithm

The research in this paper tries to find proper methods to build the hierarchies and abstract the state in an effective way, similar as the HRL approach proposed in [4]. In the hierarchies, the Options are defined as the combinations of subtasks. In target searching tasks, the multi-robot system needs to choose subtasks for the robots. Some of the subtasks are cooperative, while in non-cooperative subtasks robots just pay attention to their own works. In reinforcement learning hierarchy, a level with cooperation subtasks is regarded as a cooperation level. The learning process with three robots is implemented, with different target points. The example of 7 target points is shown in Fig. 1. The overall task is composed of a series of subtasks, which are covering T 1, $\cdots$, T 7. A subtask here is defined as one robot finding one target point. To be exact, it contains two basic primitive subtasks: navigating to the target point and stopping for coverage.

In the design, each primitive subtask needs to find basic actions to employ. In this paper, there are three basic actions for the primitive subtask to choose: setting orientation angle, setting angular velocity and setting linear velocity. To build the proposed improved MAXQ framework, a primitive subtask m_i is defined as a four tuple (s_i, a_i, π_i, r_i). The first parameter s_i is the state. The set of s_i includes all the possible states in a Markov Decision Process (MDP). The second parameter a_i represents the action that the robot chooses to finish the subtask i, under the policy of π_i. The last parameter r_i is the pseudo-reward, which tells the reward of the transition from state s_i to a new state s_i'. The function $Q(i,s,a)$ denotes the action value of taking action a in state s in the context of parent task i. The function of $Q(i,s,a)$ composes of two part: one is the projected value function $V(i,s)$ and the other one is the completion function $C(i,s,a)$, which represent the value to follow hierarchical policy and the expected cumulative discounted reward to taking action a to finish the subtask m_i.

Fig. 1 An example of the proposed improved MAXQ learning with 3 robots.

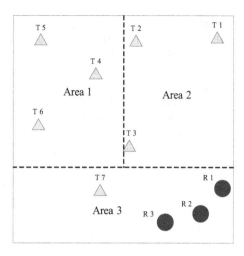

In the cooperative level, cooperative subtasks need to be configured to gain the joint-action values. So for robot R j to complete cooperative subtask m_i , the function C is extend to be denoted as $C_j(i,s,a_1,\cdots a_{j-1},a_{j+1},\cdots,a_n)$, representing the expected discounted cumulative reward to take action a_j in state s. It is the joint completion function for R j. At the same time, other robots are tasking actions a_k, $\forall k \in \{1,\cdots n\}, k \neq j$. Similarly, the action value function can be extended as [4]

$$Q_j(i,s,a_1,\cdots,a_{j-1},a_{j+1},\cdots,a_n) = V_j(a_j,s) + C_j(i,s,a_1,\cdots,a_{j-1},a_{j+1},\cdots,a_n).$$
(1)

Before building the proposed improve MAXQ hierarchy, some matrices are introduced. The parameter S is the set of all essential states, denoted as nodes in the graph. The second parameter E represents the set of edges, while W is the weights matrix. The improved MAXQ algorithm is constructed similarly as an artificial immune network:

1) **Set** S to be the network nodes matrix and W to be the node weight matrix;

2) **For** each node s_i, build the dissimilarity matrix F, choose the nodes with the designed value of the relations to create a memory node matrix G, and delete the nodes with related weights under certain threshold;

3) **Calculate** W, and delete the nodes with related weights under certain threshold; and put $G \leftarrow S$.

After the above steps, each memory node G_j corresponds to a group of states, which have their related weights under G_j. An Option is composed of such a group of states. After integrating the Option into traditional MAXQ, the hierarchy is successfully constructed. Similar as the cooperative multi-agent algorithm in [4], the proposed improved MAXQ algorithm, which is for robot R j with state s to complete subtask m_i, runs as

Initialise the *Sequence* to be {}

   If subtask m_i just composes of basic actions **Then**

      Take action a_i in state s, transit to a new state s' and get a reward r;

      Update the project value V_j, and fill the front item of *Sequence* with (s,a);

   Else

      While subtask m_i compose of primitive subtasks and basic actions **Do**

         If Subtask m_i is a cooperative **Then**

            Take action a_j basing on the policy π_{ij}, and transit to a new state s'

            Pick up an action a^*, and update it;

            Build a new *Subsequence*, and $N \leftarrow 0$;

            For pick up every element in *Subsequence* **Do**

               $N \leftarrow N+1$, and update the discounted cumulative reward $C_j(t+1)$;

            End For

         Else / M_i is not a cooperative subtask/

            Choose action a_j according to $\pi_{ij}(s)$, and build a new *Subsequence*;

            Take action a_j basing on the policy $\pi_{ij}(s)$, and transit to a new state s';

            Update a^*, and $N \leftarrow 0$;

            For each s in *Subsequence* **Do**

            $N \leftarrow N+1$;

            Update $C_j(t+1)$

            End For

         End If

         Fill the front item of *Sequence* with element in *Subsequence*, and $s = s'$;

      End While

   End If

   Feedback *Sequence* to robot R j;

End proposed improved MAXQ.

3 Simulation Studies

In the simulations, the environment space is 20×20 in most cases, except the simulation scenario for 4 robots is 30×30. The detecting distance of the robot sensor is 3 units.

3.1 *Learning Processes*

During learning procedure, three-robot cooperation is chosen. The simulation will be repeated for several times until the trajectories of the robots come to nearly optimal and the simulation time converges under a limited bound ($\pm 5\%$ implemented in the simulations). Two different trials of learning process are shown in Fig. 2. The data from the simulation indicates that the improved MAXQ can provide with a faster leaning speed.

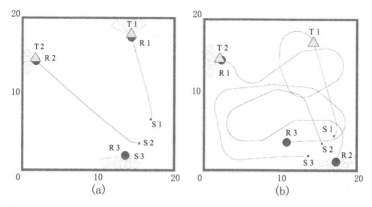

Fig. 2 The 9th trial of learning process when 3 robots reach and cover 2 targets. S i is the start point of the robot R i, $i = 1,2,3$; T j, $j = 1,2,3$, is the target point.(a) improved MAXQ; (b) Q-learning.

In order to explore the unknown environment in an efficient way, two parameters are adopted to lead the robots to work effectively in new tasks. They are dispersion degree and homodromous degree [6]. Dispersion degree is implemented to evaluate how the robot members get closed to each other. It is denoted as L_D. If there are n robots in a $W \times H$ area, the parameter L_D is calculated by a Gaussian function but with a slight revision as $L_D = \exp(-(\delta - \mu)^2 / 2\sigma^2)$, where $\delta = \overline{D}/\sqrt{W^2 + H^2}$; $\mu = \frac{1}{t} \sum_{k=1}^{t} \delta_k$; $\sigma = (\max(\delta_k) - \min(\delta_k))/2$; and $\overline{D}$ is the real-time average distance between the robots. The parameter homodromous degree, denoted by L_H, is implemented to evaluate how the directions robot members are closed to others. Similarly, if there are n robots and the robot directions are $\{\theta_1, \theta_2, \cdots, \theta_n\}$, $0° \leq \theta_i < 360°$, L_H is calculated by $L_H = \frac{1}{m} \frac{2}{n(n-1)} \sum_{i=1}^{n} \sum_{j=i+1}^{n} |\theta_i - \theta_j|$, where m is the number of all possible directions of the robot movements, and $|.|$ is absolute value function. In this research, each possible direction area is regarded as a bound area of $45°$ angle, so there are 8 possible direction areas in the simulations.

3.2 New Tasks in Unknown Environments

During the learning process, the multi-robot system gets the parameters L_D and L_H, and build limited areas of $(L_{D\min}, L_{D\max})$ and $(L_{H\min}, L_{H\max})$. In new tasks shown in Fig. 3, the robots will cooperatively explore the environment based on the previous policies. Different robots with variable targets are tested. The results show that the proposed scheme can lead the robots to reach and cover the targets effectively, when the surroundings are unknown to the robot members.

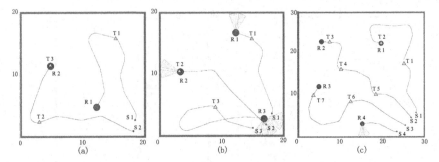

Fig. 3 Cooperations for new tasks. (a) 2 robots for 3 targets; (b) 3 robots for 3 targets; (c) 4 robots for 7 targets.

4   Conclusion

An improved MAXQ algorithm is implemented for multi-robot cooperation to tune the parameters of two designed cooperation rules. The proposed approach is applied to multi-robot target searching. The simulations results demonstrate that the proposed approach can provide the multi-robot system with faster learning and the ability to handle new tasks in unknown environments.

References

1. Ming, G.F., Hua, S.: Course-scheduling algorithm of option-based hierarchical reinforcement learning. In: The 2nd International Workshop on Education Technology and Computer Science, ETCS 2010, Wuhan, Hubei, China, pp. 288–291 (March 2010)
2. Du, X., Li, Q., Han, J.: An analysis and hierarchical decomposition for hams. In: 2009 International Conference on Computational Science and Engineering, Vancouver, BC, Canada, pp. 1050–1054 (August 2009)
3. Mirzazadeh, F., Behsaz, B., Beigy, H.: A new learning algorithm for the maxq hierarchical reinforcement learning method. In: The 2007 International Conference on Information and Communication Technology, Dhaka, Bangladesh, pp. 105–108 (March 2007)
4. Cheng, X., Shen, J., Liu, H., Gu, O.: Multi-robot cooperation based on hierarchical reinforcement learning. In: The 7th International Conference on Computational Science, Beijing, China, pp. 90–97 (May 2007)
5. Juang, C.F., Hsu, C.H.: Reinforcement ant optimized fuzzy controller for mobile-robot wall-following control. IEEE Transactions on Industrial Electronics 56(10), 3931–3940 (2009)
6. Jang, J.R., Sun, C., Mizutani, E.: Neuro-Fuzzy and Soft Computing: a computational approach to learning and machine intelligence. Prentice Hall, New York (1997)

Modeling Specific Energy for Shield Machine by Non-linear Multiple Regression Method and Mechanical Analysis

Qian Zhang, Chuanyong Qu, Zongxi Cai, Tian Huang, Yilan Kang[*], Ming Hu, Bin Dai, and Jianzhong Leng

Abstract. The specific energy, defined as the energy consumption to complete the excavation of unit volume of the soil, can well describe the working efficiency of a shield machine. An identification model of the specific energy is established in this paper by introducing the mechanical analysis of the shield excavating process into the nonlinear multiple regression of the on-site data. The mechanical analysis of the shield-soil system helps to decouple the nonlinear multi-parameter problem and the regression process is conducted based on a group of on-site data of subway project in China. Fairly good consistency between the model results and the on-site recorded datum can be achieved. This work provides a useful tool for the analysis of the energy consumption of shield machines.

Keywords: Shield Machine, Specific energy, Mechanical Analysis, Nonlinear Multiple Regression.

1 Introduction

The shield machine, a kind of giant equipment for tunneling, has been widely applied in constructions with complex geological conditions. The high efficiency during construction process is a greatly important purpose to achieve. It is significant to establish an effective predicted model for the energy consumption of shield machines.

The specific energy, defined as the energy consumption to complete the excavation of unit volume of the soil, can be considered as a reasonable index to describe shield efficiency [R. Teale 1965]. Some studies have analyzed factors of the specific energy [O. Acaroglu et al. 2008, R. Gertsch et al. 2007, D. J. Reddish et al. 1996], but there are still some difficulties in its modeling process. In fact, the

Qian Zhang · Chuanyong Qu · Zongxi Cai · Tian Huang · Yilan Kang · Ming Hu
Key Laboratory of Modern Engineering Mechanics, Tianjin University, Tianjin, China

Bin Dai · Jianzhong Leng
Shanghai Branch Office, China Coal Construction Group, Shanghai, China

[*] Corresponding author.

F.L. Gaol et al. (Eds.): Proc. of the 2011 2nd International Congress on CACS, AISC 144, pp. 75–80.
springerlink.com

shield excavating process is nonlinear, multi-parameter coupling, and can be influenced by many random and uncertain factors. Consequently it is very difficult to obtain a comprehensive and accurate model only depending on a singular method like mechanics method. In recent years, the gradual accumulation of on-site data in shield machine projects and the rapid development of data mining technology provide a new way to improve the modeling process. Therefore much attention has been devoted to the research using on-site data mining in engineering projects [S. Suwansawat et al. 2006, A.G. Benardos et al. 2004, R. Mikaeil et al. 2009, R. L. Ott et al. 2003].

In this paper, a model is found to identify the specific energy of the shield machine during excavation by introducing the mechanical analysis into the nonlinear multiple regression process. Firstly the detailed analysis on the mechanical characteristics of the shield-soil system will be provided. And it shows which factors will be able to affect the energy consumption and how they work. Then, a nonlinear model will be proposed to estimate the specific energy through the combination of the mechanical method and the regression analysis for on-site data of a subway project in China.

2   Nonlinear Multiple Regression Method Combined with Mechanical Analysis

The nonlinear multiple regression, different with traditional linear regression, is a kind of method to find a nonlinear model describing the relationship between the dependent variable and a set of independent variables [R. L. Ott et al. 2003]. Therefore, it has been more and more widely used in complex engineering problems. The model established by nonlinear multiple regression can extract valuable common law from great amount of engineering on-site data, and the results can rapidly provide suggestions back to engineering practice.

However, it is generally not easy to find a reasonable nonlinear model. Especially when there are kinds of parameters coupling with each other, the modeling process is greatly difficult, and the results are usually not satisfactory. So some researchers have to go on with linear regression analysis or reduce the number of variables which results in difficulties to establish an effective model including the influence of complete factors.

The problem above can be well solved though introducing the mechanical analysis focusing on the intrinsic mechanism into the regression process. The detailed mechanical analysis is helpful in finding the pattern of nonlinear multiple regression expression. In this paper, a model for shield specific energy will be established to describe the complex relationship among various kinds of parameters depending on the combination of the mechanical analysis and the nonlinear multiple regression method.

3 Modeling Specific Energy of Shield Machine during Excavation

During tunneling process, the shield machine and its surrounding soil form a coupling mechanical system due to the complex interaction between them. Therefore the detailed analysis on the mechanical characteristics of shield excavation helps to reveal the essential mechanism and the influence of various parameters on the energy consumption.

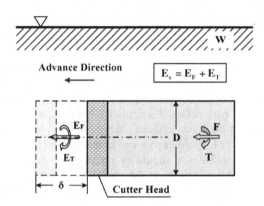

Fig. 1 Schematic representation for the working process of a shield machine

Fig. 1 shows a shield machine working in the earth. The driving load, the geological condition and the operating status are the three main elements of the shield-soil system. The machine is driving by both the thrust F and the torque T during tunneling. So the energy is mainly consumed on the thrust to constantly push the machine forward and the torque to maintain continuously rotating the cutter head to cut the soil. Moreover the geological condition and the operating state are the important determinants of the load. According to the definition of specific energy [R. Teale 1965], the definition expression can be written as

$$E_s = \frac{F\delta + 2\pi T}{0.25\pi D^2 \delta},\qquad(1)$$

where E_s (kJ/m³) stands for the specific energy of shield excavation, F (kN) and T (kN•m) are the thrust and the torque acting on the machine, δ (m/rev) denotes the cutting depth during the cutter head rotating per revolution, and D (m) is the diameter of cutterhead.

According to equilibrium, the thrust and the torque equal to the sum of all kinds of the resistance forces and torques respectively. Therefore, the driving loads are composed of a number of parts. The thrust consists of the frictional force between the shield skin plate and the soil, the earth pressure at rest on excavation face, the squeezing force between the cutter head and the soil to be excavated during advance process, etc. Furthermore the torque includes the frictional rotating torque between the side face of cutter head and the soil, the mixing torque in chamber, the torque caused by the friction as well as cutting between the cutter head and the soil, etc. The load components resulting from different origins, as mentioned above, vary with different characteristics. Some components are relatively stable during excavation and can be approximately considered as having no relationship with operating parameters, such as the shield-soil friction, yet others are determined by the complex interaction between the cutter head and the soil on excavation face, therefore they directly change with the operating adjustment, such as the torque caused by the cutter-soil interaction. Based on the mechanical analysis above, the shield loads can be divided into the external loads and the operating loads in this paper. The external loads are relatively stable during excavation, related to the geological parameters, the shield diameter and the depth of burial, yet can be considered as independent with operating parameters. On the other hand, the operating loads are dynamic and directly related to the interaction between the cutter head and the soil on excavation face, therefore determined by both the operating and geological parameters. Particularly there are approximate exponential relationships between the loads and the shield diameter, which have been widely acceptable in the field of shield design and construction [B. Maidl et al. 1996]. In addition, the dimension balance should be taken into account in modeling to ensure the undetermined coefficients in the model are dimensionless. Consequently, the specific energy model of the shield machine is established as follows:

$$E_s = \frac{(F_e + F_o)\delta + 2\pi(T_e + T_o)}{0.25\pi D^2 \delta}, \tag{2}$$

$$F = F_o + F_e = A_F W \delta^2 + (B_F W + C_F P) D^2, \tag{3}$$

$$T = T_o + T_e = A_T W \delta^3 + (B_T W + C_T P) D^3, \tag{4}$$

$$E_s = (A_E W D^{-2}) \times \delta^2 + (B_E P D + C_E W D) \times \delta^{-1} + (G_E W + H_E P), \tag{5}$$

where F_e and T_e are the external load components, F_o and T_o denote the operating load components, W (kPa) is the ground bearing capacity, P (kPa) is the pressure in chamber characterizing the effect of the depth of burial. In addition, all of A, B, C, G, H are the model dimensionless coefficients to be determined through on-site data regression analysis.

Then further discussion is conducted combined with on-site data out of the project database of 9th line subway of Tianjin in China to determine the dimensionless coefficients in the model. The earth pressure balance shield was used in this project and excavated in the geological condition which mainly consists of silty clay, silt, silty sand and muddy clay with the burial depth of 10 to 15 meters. The real time monitoring of a large number of key parameters during excavation and the professional geological survey reports provide the precise material for the research. The data of total 340 rings are used in the regression analysis.By using the statistical analysis software SPSS (Statistical Product and Service Solutions) to complete the on-site data regression, nonlinear model of shield energy can be determined as

$$E_s = \left(2.2 \times 10^5 WD^{-2}\right) \times \delta^2 + \left(0.15PD + 6.4 \times 10^{-3}WD\right) \times \delta^{-1} + \left(1.5W + P\right). \quad (6)$$

The expression describes the influence of different kinds of parameters on the specific energy. Fig. 2 gives both the calculated and the on-site data of the specific energy of 340 rings, which shows well agreement. The results above indicate the model proposed in this paper based on the mechanical on-site data analysis can be applied to predict and analyze the performance of shield machine.

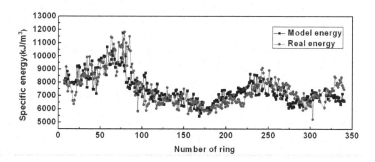

Fig. 2 Comparison between the model energy and the real energy

4 Conclusions

The energy consumption problem of the shield machine is considered in this paper based on the mechanics and regression method. The mechanical analysis of the shield excavating process is very helpful for describing the ways that the parameters work on the specific energy and decouple the nonlinear multi-parameter problem. And it also helps to find the pattern of the regression model. The model of the specific energy can well describe the influence of the geological condition, the operating status and the driving load on the energy consumption. The consistency between the model data and the on-site data shows that the method proposed in this paper can effectively identify the energy consumption during tunneling process. And the work also provides a new way for the analysis of nonlinear multi-factor engineering problems.

Acknowledgments. This work was supported by the National Natural Science Foundation of China (NSFC, Grant no. 11127202), National Basic Research Program of China (973 Program, Grant no.2007CB714001), National High-tech R&D Program of China (863 Program, Grant no. 2009AA04Z423).

References

Benardos, A.G., Kaliampakos, D.C.: Modelling TBM performance with artificial neural networks. Tunnelling and Underground Space Technology 19, 597–605 (2004)

Maidl, B., Herrenknecht, M., Anheuser, L.: Mechanised shield tunnelling, pp. 42–59. Ernst & Sohn, Berlin (1996)

Reddish, D.J., Yasar, E.: A new portable rock strength energy of drilling. International Journal of Rock Mechanics and Mining Sciences & Geomechanics Abstracts 33(5), 543–548 (1996)

Acaroglu, O., Ozdemir, L., Asbury, B.: A fuzzy logic model to predict specific energy requirement for TBM performance prediction. Tunnelling and Underground Space Technology 23, 600–608 (2008)

Gertsch, R., Gertsch, L., Rostami, J.: Disc cutting tests in Colorado Red Granite: Implications for TBM performance prediction. International Journal of Rock Mechanics & Mining Sciences 44, 238–246 (2007)

Ott, R.L., Longnecker, M., Zhang, Z.Z. (trans.): An introduction to statistical methods and data analysis, pp. 583–589. Science Press, Beijing (2003) (in Chinese)

Mikaeil, R., Naghadehi, M.Z., Sereshki, F.: Multifactorial fuzzy approach to the penetrability classification of TBM in hard rock conditions. Tunnelling and Underground Space Technology 24, 500–505 (2009)

Teale, R.: The concept of specific energy in rock drilling. International Journal of Rock Mechanics and Mining Sciences & Geomechanics Abstracts 2(1), 57–73 (1965)

Suwansawat, S., Einstein, H.H.: Artificial neural networks for predicting the maximum surface settlement caused by EPB shield tunneling. Tunnelling and Underground Space Technology 21, 133–150 (2006)

Design and Realization of College Finance OLAP Analyzer Based on MDX

Peng Cheng and Tong Qiuli

Abstract. University financial data can be used to analyze university operational status and guide decision-making. In order to browse financial data from multiple dimensions, we proposed a financial analyzer based on the MDX (multidimensional expression) technology, realized querying financial data from multiple dimensions and the function of data analysis, then test the query performance of the analyzer and compared with SQL language in operation efficiency and complexity.

1 Introduction

University financial data contains various dimension information such as project, subjects, department, finance employee in charge, etc. When dealing with financial data, university financial department needs several of these dimensions at the same time to help analysis financial status better. However, traditional financial data report only contains two dimension, thus cannot satisfy the need for multiple analysis.

Most of the current college financial data processing are based on the two-dimensional data table query mode. At present, college mostly use ERP or other traditional software which cannot satisfy the needs of multiple financial data queries or help decision-making. Multidimensional analysis of financial data will gradually replace traditional two-dimensional statement analysis form.

In order to query the financial data from multiple dimensions, a financial analyzer based on MDX using MOLAP is proposed which realizes establishing cube from source database, querying financial data from multiple dimensions, and realizes the function of data analysis. Additionally, this paper realized deriving result to Excel, user and group management, database access control and table name setting to help management, use and decision making.

Peng Cheng
Department of Computer Science and Technology, Tsinghua University, Beijing, China
email: pc@cic.tsinghua.edu.cn

Tong Qiuli
Department of Computer Science and Technology, Tsinghua University, Beijing, China
email: tql@cic.tsinghua.edu.cn

F.L. Gaol et al. (Eds.): Proc. of the 2011 2nd International Congress on CACS, AISC 144, pp. 81–86.
springerlink.com   © Springer-Verlag Berlin Heidelberg 2012

On this basis, test the query performance of OLAP analyzer, further compare with SQL language in operation efficiency and complexity to analyze the advantage of multidimensional inquiry system to traditional two-dimensional data table query.

2 OLAP System Design

A. Overall System Structure

OLAP (On-Line Analysis Processing), is proposed in 1993 by E.F.Codd, the father of the relational database.[1] This technique can analysis data from multiple angles with a rapid speed and accuracy, and realize the multidimensional analysis of multidimensional database.

In this paper, we choose MOLAP from the consideration of structure and query performance, it is suitable for the scenario of frequently use and require fast response. For ROLAP, it is suitable for large database query with a low query rate.

Fig. 1 The system is mainly divided into four parts: the database store part, multi-dimension data generated part, OLAP parser part and user interaction part. Database store part is used to store database instance, Multi-dimension data generated part is mainly used to generate multidimension data which the help of Analysis Services, OLAP parser part converts user input into MDX query language, then operates multidimension data which MDX statements, and show query results to front interface.

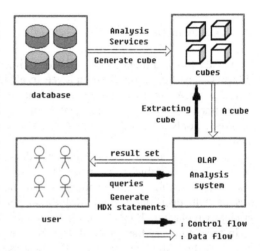

B. OLAP Analysis System Structure

The whole system structure is shown in Figure2. The core part of the parser is MDX query and SQL query. According to the MDX statements and SQL statements received the parser will send query command to multi-dimension data or resource database and get needed results. In the query time record modules, we can get the running time of every inquiry.

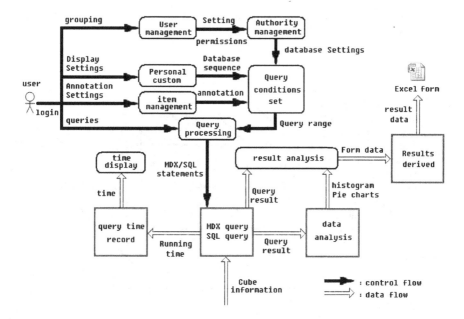

Fig. 2 OLAP parser structure.

The front-end interface mainly includes query results, the analysis of results, query condition settings and time display. Comparing to other OLAP systems which only show the MDX query results, the OLAP system designed in this paper contains data analysis part which help users analyzes the query results. In this part, user can choose two or more metrics in the result table and design calculations on them.

Besides, the OLAP system designed in this paper contains a few auxiliary functions, including authority administration, user and group management, table item management, personal customization, etc.

In addition, the OLAP system designed in this paper also includes results export modules, which can output result data to Excel, for users to store and share search results easier.

C. MDX and SQL Language Comparison

MDX (Multidimensional Expressions) is a multi-dimensional scripting language based on description, it is the grammar used to define, manage and inquire multi-dimensional object and data . MDX and SQL are alike in many ways, but MDX is not a simple extension of SQL language. MDX provides various and powerful grammar used to acquire and operate multidimensional data.

The object MDX inquires is multi-dimension data, namely multidimensional data organization generated from multidimensional database, and we can describe it as follows:

Fig. 3 The data has 3 dimensions, each fact data is stored as a cube. Of course, the dimensions of data it can also be other numbers, they all use the way of multi-dimensional array to store. The biggest difference of MDX and SQL lies in that MDX is multi-dimensional, each dimension represents a axis but SQL only has two dimensions.

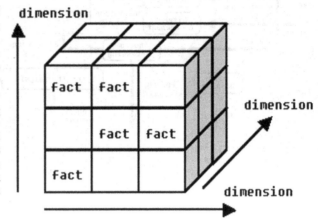

MDX inquires is more simple, and the equivalent groups SQL statement collection is more complicated. If use SQL language to realize MDX language functions, its complexity equals designing a ROLAP system. Thus, the multidimensional inquiry function MDX provides can satisfy user needs for data analysis better.

3 OLAP System Implementation

In order to realize the MDX inquires, we need to establish cube from the original database, then query the multi-dimension data. Multidimensional data can be established by Analysis Services. During the process of generating multi-dimension data, Analysis Services will first scan the structure of data sources. The IntelliCube technology will analyze primary key and foreign key relations between tables of data source view, thus distinguish fact table and dimension table.

The OLAP parser designed in this paper mainly realizes the function of the MDX inquires to multidimensional data sets. During the inquiry, users can choose database and multi-dimension data. By defining needed dimensions, users can set conditions for queries.

In the data analysis part, users can analyze the data in the result table, by selecting the data items need to analyze and conditions for classification, users can generate a pie chart or bar chart to reflect the analysis result based on the specified conditions they select, thus get the needed information.

4   Comparing with SQL

In order to compare the querying speed of two languages, we realized executing queries in both MDX language and SQL language, and record the execution time of them. Table 1 shows the average execution time(ms) of different types of queries of 10 times, we can compare MDX with SQL language on runtime comparison.

We test a database of 60MB, including 22 dimension table, 8 fact sheets on CPU of Genuine Intel(R) T2050. In table 1 below we can see, under same conditions, MDX language is faster than the SQL language. MDX language is more efficient than SQL language, and the grammar is more suitable for multidimensional operation, the only insufficient of MDX is it has to spend some time and space in generating cube. As the database becomes larger and more complex, multi-dimensional inquires enables users to analyze data better. In the trend of financial analysis transferring from 2d to multi-dimensional, MDX language will get more extensive application.

Table 1 Efficiency Comparison of MDX and SQL

	1 metrics		4 metrics		16 metrics	
	MDX	SQL	MDX	SQL	MDX	SQL
1 dim,1 table	10.938	46.875	29.688	83.125	87.5	92.188
2 dim,1 table	15.625	46.875	31.25	104.688	89.063	165.625
3 dim,2 tables	20.313	50.0	37.5	103.125	101.5	281.25
4 dim,2 tables	37.5	46.875	48.438	130.125	134.375	287.5

5   Closing

In this paper, we analyze the existing financial processing software, point out the insufficiency, and the demand of multidimensional analysis at present. The OLAP parser designed in this paper can realize multidimensional query and other multi-dimensional processing operations, and can form a table of results to interface, which users can check and sort. Besides it realized the function of data analysis.

The OLAP analyzer is a multidimensional query tool based on MDX technology, and has essential difference to two-dimensional tool, with the passage of time, the multi-dimensional query tool will gradually replace two-dimensional query tool, the research in this paper will gradually become an important part of actual financial analysis system.

References

[1] Shumate, J.: Practical Guide to the Microsoft Olap Server. Addison-Wesley Longman Publishing (2000)

[2] Bellatreche, L., Giacometti, A., Marcel, P., Mouloudi, H., Laurent, D.: A personalization framework for olap queries. In: DOLAP, pp. 19–28 (2005)

[3] Gunderloy, M., Sneath, T.: SQL Server Developer's Guide to Olap with Analysis Services (Developer's Handbook Series). In: SYBEX, pp. 240–250 (2001)

[4] Harinath, S. Quinn, S.R.: Professional SQL Server Analysis Services 2005 with MDX. Hungry Minds Inc., pp. 68–73 (2005)

[5] Whitehorn, M., Zare, R., Pazusmansky, M.: Problem solving for MDX for SQL Server 2005, pp. 105–131. Springer, New York (2007)

[6] Morfonios, K., Konakas, S., Ioannidis, P., Kotsis, N.: ROLAP implementations of the data cube. ACM Computing Surveys (CSUR) 39, 12–15 (2007), doi:10.1145/1287620.1287623

[7] Kalnis, P., Papadias, D.: Multi-query optimization for on-line analytical processing. Information Systems 28, 230–234 (2003), doi:10.1016/S0306-4379(02)00026-1

[8] Giacometti, A., Marcel, P., Negre, E., Soulet, A.: Query recommendations for OLAP discovery driven analysis. In: Proceeding of the ACM, pp. 81–88. ACM Press (November 2009), doi:10.1145/1651291.1651306

[9] Niemi, T., Nummenmaa, J., Thanisch, P.: Constructing OLAP cubes based on queries. In: Proceedings of the ACM, pp. 9–15. ACM Press (November 2001), doi:10.1145/512236.512238

[10] Dehne, F., Eavis, T., Rau-Chaplin, A.: Parallel querying of ROLAP cubes in the presence of hierarchies. In: Proceedings of the 8th ACM International Workshop, pp. 89–96. ACM Press (November 2005), doi:10.1145/1097002.1097019

[11] Chen, J.V.: The Perspectives of Improving Web Search Engine Quality. In: Calero, C. (ed.) Handbook of Research on Web Information Systems Quality, pp. 481–490 (2008)

[12] Li, C.-H.: Application of OLAP Technology in Production Evaluation. Computing Technology and Automation 28, 133–137 (2009); 李才华. OLAP技术在生产评价中的应用[J]. 计算技术与自动化 28(4): 133–137 (2009)

[13] Lan, J., Jin, H.-M.: The Design and Realization of OLAP Based Data Warehouse Analysis Model in Enterprises. Process Automation Instrumentation 27, 9–12 (2006); 蓝 箭, 金红梅. 基于OLAP的企业数据仓库分析模型设计与实现. 自动化仪表 27(5):9-12 (2006)

[14] Jiang, N., Gao, W., Zhang, L.-Q.: Research and Application of Data Mining Model based on Analysis Services. Machinery Design & Manufacture 4, 83–85 (2007); 姜楠, 高 巍, 张丽秋. 基于Analysis Services的数据挖掘模型的研究与应用. 机械设计与制造, 年4期:83–85 (2007)

[15] Chen, Q.-M., He, C.-B., Liu, H.: OLAP-based collaborative educational decision. Computer Applications 29, 304–305 (2009); 陈启买, 贺超波, 刘 海. 基于OLAP的高校教学协同决策. 计算机应用 29(1): 304–305 (2009)

Data Mining: Study on Intelligence-Led Counterterrorism

Wang Shacheng

Abstract. This paper establishes a relatively complete theoretical framework, with multi-disciplinary perspective. Intelligence-led counterterrorism framework based on data mining can be divided into four function modules: data storage, data pretreatment, data analysis and data visualization processor. The data mining process of counterterrorism's information analysis includes problem identification of counterterrorism, data preparation, data mining, model assessment and knowledge representation. Bayesian Belief Network is a very useful tool for data mining in intelligence-led counterterrorism.

Terrorists more or less will leave the corresponding clues in premeditated, planning and implementing crime. However, it is harder than looking for a needle in a bottle of hay to use the existing technology and method to extract the anti-terrorist information from the clues! Because this kind of terrorism information which is covered in the magnitude of the general information is indefinite and unknown in time series and space. The traditional database retrieval system cannot respond to this inquiry. Artificial identification, mutatis mutandis, monitoring technology and the implementation of known goal detection all need lots of manpower, and it is powerless to search unknown target in voluminous information. But the Data Mining can do!

1   Data Mining and Terrorist Information Processing

In brief, data mining is to dig or find knowledge hidden in the large amounts of data. It is an analysis process of exploring and discovering the useful information hidden in mass data. Data mining is not only a very active interdisciplinary branch of learning, absorbing lots of the novel ideas, which synthesized database technology, artificial intelligence, neural network, pattern recognition, statistics, decision tree, Bayesian Analysis, genetic algorithm, Fuzzy Logic, rough theory and other base areas, but also a key link in the process of KDD (Knowledge Discovery in Database), sometimes regarded as synonyms of Knowledge Discovery. From the case analysis of current international data information

Wang Shacheng
Institute of Defense Economics and Management, Central University of Finance and Economics, Beijing, P.R. China 10008
e-mail: 007@pku.edu.cn

F.L. Gaol et al. (Eds.): Proc. of the 2011 2nd International Congress on CACS, AISC 144, pp. 87–93.
springerlink.com   © Springer-Verlag Berlin Heidelberg 2012

system, data mining is a high-efficiency practical technology of terrorism information detection.

Intelligence-led counterterrorism framework based on data mining can be divided into four function modules: data storage, data pretreatment, data analysis and data visualization processor. Data storage function module use data warehouse technology to store all kinds of original data and intermediate results. The first function of data pretreatment function module is data cleaning, smoothing noise data, identifying and deleting isolated points, solving inconsistent and so on. Second is integrating data and merging of multiple data storage. Third is data transformation——transform the data of different sources, different channels and different formats into the data format which this system is unified for analyzing and disposing. If there are conditions, we also screen data as well. Data analysis is the core function module of this framework——mainly uses various data mining methods mentioned above to accomplish data analysis operation. Data visualization processing function module transforms the models or knowledge which this framework has analyzed into the display formats suitable for user to understand. Another way of saying that is the framework can choose different forms of knowledge expression according to content, such as rules, tables, charts, pictures, decision trees and data cubes, etc. Concrete structure can be seen in figure 1. [2]

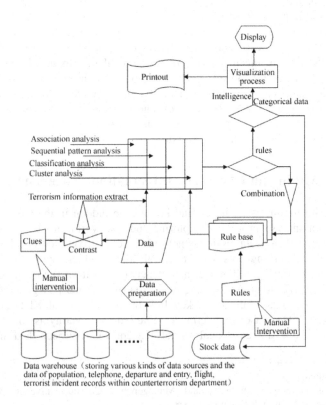

Fig. 1 Work Frame of Intelligence-led Counterterrorism

2 Data Mining Process of Terrorism Information Analysis

Data mining process of terrorism information analysis generally consist of four main stages: the identification of terrorism problems, data preparation, data mining, model assessment and knowledge representation, as shown in figure 2 [3].

Specific terrorism problems must first be identified before data mining, which is to determine to do what counterterrorism operations and counterterrorism decision-making, to achieve what kind of counterterrorism effect and target, etc. Data mining can not only individually but also synthetically utilize various kinds of data mining methods to analyze the data. Firstly determine the aim of data mining, then select algorithm of data mining——select a particular data mining algorithm (such as collection, classification, regression, clustering, etc) for searching data patterns. Model assessment evaluates the discovered rules, trend, classification and model, thus ensure screening out meaningful results which accord with counterterrorism demand. Knowledge representation is making use of the technology of visualization and knowledge expression to show counterterrorism users the mining results. In the following, we will focus on a data mining tool suitable for work of intelligence-led counterterrorism——Bayesian belief network.

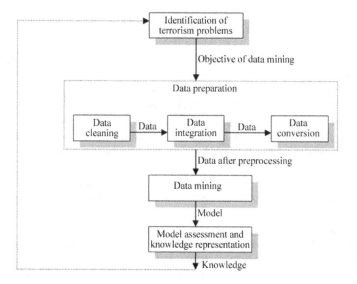

Fig. 2 Data Mining Process of Counterterrorism Activities

3 Application of Bayesian Belief Network in Terrorism Information Analysis

Bayesian belief network is a probability reasoning method, which is able to make reasoning from incomplete, inaccurate and uncertain knowledge and information, and deal with incomplete data sets with noise, so as to solve the problem of inconsistency and even independence between data. The solid theoretical foundation, natural expression of knowledge structure, flexible reasoning ability and convenient decision-making mechanism of Bayesian belief network make itself more and more widely applied.

Using graphical methods to describe the relationship between data, Bayesian belief network has clear semantics and strong intelligibility, which contribute to use the causal relationship between data to do forecast analysis. Bayesian belief network is a directed acyclic graph, which is the graphic representation of complex joint probability distribution. A Bayesian belief network $S= \langle G, \theta \rangle$ consists of two parts: topology structure and partial probability distribution set. In part one, a directed acyclic graph G shows a conditionally independent assertion network structure of a group of variable X = {clamps its X1, X2... Xn}, which is conditionally independent, set in modeling field, and these variables are expressed with nodes, while the correlations between variables are expressed with node connections. In part two, θ is the set of p $(X_i|\pi_{xi})$——local conditions probability distribution associated with each variable. π_{xi} is the father node of Xi in G, and p $(X_i|\pi_{xi})$ is the conditional probability distribution of node Xi under the condition of father node π_{xi}. The two parts together have defined the unique joint distribution probability p (X), and there:

$$p(X) = \prod_{i=1}^{n} p(Xi \mid \pi xi)$$

Bayesian belief network use the graph mode to express joint probability between variables, use nodes to express variables, use directed edge to express dependent relationships between variables, and use the weight measured by probability to describe the relationship between data, thus it provide a natural method of showing probabilistic information.

Information analysis of personnel activities is an important field of counterterrorism intelligence judgment, the essence of which is a kind of classification. In the following, we will discuss personnel information analysis with Bayesian methods. Let us assume that, in actual counterterrorism activities, we found that there is a certain relationship between personnel activities information of suspected terrorists who performed abnormally in short periods of time (such as suddenly going to a city frequently, suddenly contacting frequently with another terrorist suspects, etc.) and the persons mentioned above who have suspected terrorist activities, and there are statistical information as follows:

The proportion of persons on suspicion of terrorists in transients' information (or communication postal information, etc.) is:

P (f=Y) =a;

The proportion of persons not on suspicion of terrorists in transients' information (or communication postal information, etc.) is:

P (f=N) =b;

The proportion of persons on suspicion of terrorists, who become suspected terrorists once again, can also be figured out with statistical methods:

$$P\ (h|f=Y) =c;$$

And the proportion of persons not on suspicion of terrorists, who become suspected terrorists this time, is:

$$P\ (h|f=N) =d.$$

Likewise, we can also count the proportion of temporarily living information in a short period (or the information of frequently contact with a suspected terrorists, etc.).

The formula below can be used to express various activities information of the actor:

$$D\ (P_i) = (A_1,\ W1;\ A2,\ W2;...;\ An;\ Wn)$$

In the formula, P_i represents some people, A_i represents the activities information of some people, W_i represents the weight of this behavior in judging actor information, and $D\ (P_i)$ is the vector representation corresponding to some people P_i.

Now we use simple Bayesian method to calculate $P\ (I|D\ (P_i))$——the probability of some people on suspicion of terrorists, and $P\ (L|D\ (P_i))$——the probability of some people not on suspicion of terrorists. Obviously there is a formula below:

$$P\ (I|D\ (P_i)) +P\ (L|D\ (P_i)) =1$$

Therefore, according to any kind of calculation results of the formula 1, we can judge whether a person is on suspicion of terrorists. But in fact, we need to adjust the judgment of the serious situation between judging non-terrorists for terrorists and misjudging terrorists. Consequently, an adjustment quantity ε should be added in formula 1 (formula 2). ε can be negative, and as long as we adjust the size of ε, can realize different seriousness proportion.

$$P\ (I|D\ (P_i)) <50\%$$
$$P\ (L|D\ (P_i))>50\% \qquad \text{(Formula 1)}$$
$$P\ (I|D\ (P_i)) < P\ (L|D\ (P_i))$$

$$P\ (I|D\ (P_i)) <50\% +\varepsilon$$
$$P\ (L|D\ (P_i))>50\%+\varepsilon \qquad \text{(Formula 2)}$$
$$P\ (I|D\ (P_i)) + \varepsilon< P\ (L|D\ (P_i))$$

Based on Bayesian theorem, the category of unknown sample can be predicted by the training sample, as (formula 3) shown. Through prior probability and conditional probability obtained in sample set, we can calculate posteriori probability which is the probability of unknown personnel on suspicion of terrorists or not.

$$P(I \mid D(Pi)) = \frac{P(D(Pi) \mid I) * P(I)}{P(D(Pi))}$$

$$P(L \mid D(Pi)) = \frac{P(D(Pi) \mid L) * P(L)}{P(D(Pi))} \qquad \text{(Formula 3)}$$

In the (formula 3), P (I) represents the prior probability of someone on suspicion of terrorists, P (L) represents the prior probability of someone not on suspicion of terrorists. These probabilities can be figured out with statistical methods through correlation matching with temporary living information, communication postal information and other information.

The personnel information in vector D (Pi) can be seen as a personnel attribute. Hence the essence of personnel identify technology based on simple Bayesian method is to do Bayesian classification on people (legal and illegal). The amount of computing P (D (P$_i$)|L) and P (D (P$_i$)|I) is very large, but if we assume that each attribute is mutually conditional independent——there is no dependence relationship between personnel attributes, conditional probability P (D (P$_i$)|L) and P (D (P$_i$)|I)can be calculated according to the formula 4:

$$P(D(Pi) \mid I) = \prod_{i=1}^{n} P(A_i \mid I)$$

$$P(D(Pi) \mid L) = \prod_{i=1}^{n} P(A_i \mid L) \qquad \text{(Formula 4)}$$

Probability components in formula 4 the probability can be approximately calculated based on training sample. Specific calculation method can be seen in formula 5:

$$P(A_i \mid I) = \frac{|A_i(I)|}{|I|}$$

$$P(A_i \mid L) = \frac{|A_i(L)|}{|L|} \qquad \text{(Formula 5)}$$

After having figured out some people on suspicion of terrorists through calculation, we will compare and cluster the information of related personnel, find out people corresponding to these persons, and implement key monitoring and reviewing [4].

4 Conclusions

With the information environment changing quickly, the concept of traditional national security information management and information service has been challenged. This paper try to establish a relatively complete theoretical framework, with multi-disciplinary perspective of computer science, information science, and national defense economics etc, and discusses the problems in this research field at a higher theoretical level.

Acknowledgment. The research in this paper is supported by the National Social Science Fund of China: Terrorism and Intelligence-led Counterterrorism in China (Program No: 10CGJ003).

References

1. Wu, S., Gao, X., Buster, M.: Data Warehouse and Data Mining. The Metallurgical Industry Press, Beijing (2003) (in Chinese)
2. Jin, G., Qian, J., Qian, J., et al.: The Crime Risk Prediction Model Based on Data Mining Decision Tree. Computer Engineering 29(9), 183–185 (2003) (in Chinese)
3. Wang, R., Ma, D., Chen, C.: The Data Mining Technology and Analysis of its Application Status. Computer Application Technology (2), 20–23 (2007) (in Chinese)
4. Li, Y.: The Data Classification Mining Algorithm Based on Bayesian Belief Network. Computer Science 33(10), 157–158 (2006) (in Chinese)

References

Keywords Weights Improvement
and Application of Information Extraction

Yang Junhui and Huang Chan

Abstract. In keywords extraction approach, TF-IDF algorithm was commonly used as a formula for calculating the weighting of keywords, the algorithm was relatively simple and had higher precision and recall rate, but it exits many defects. This article based on the traditional TF-IDF formula to calculation weighting, put forward improvement TF-IDF formula based on the weighting of the location and the keyword length, through the experimental result inspects show that the proposed method outperforms TF-IDF in precision and recall.

1 Introduction

Information extraction was an important development direction for natural language processing technology. And the keywords became the most important information extraction approach.

The essence of the keywords extraction was to extracted the feature words from the text content, then calculate its weight by calculating the text frequency (TF) and inverted document frequency (IDF)of the feature words, And sort the weight of each feature words to select the threshold value greater than a given thresholds as the keyword of the text. Therefore, the size of keyword weight will directly affect the efficiency of information extraction. The current method of calculating the weight was primarily TFIDF algorithm, the algorithm was relatively simple and has higher precision and recall rate, but it was an empirical formula, could not adapt to a variety of environments, so various improvements for its study has been conducted[1-6]. Based on the traditional TF-IDF formula adopted for weight calculation, improved TF-IDF algorithm is proposed in this article according to the location and length of the keyword and the feasibility of improved TF-IDF algorithm had been verified by experiments.

Yang Junhui
Jiangxi University of Science and Technology Ganzhou, Jiangxi, China
email: jwcjhy@126.com

Huang Chan
Department of Computer GanNan Teach College Ganzhou, Jiangxi, China
email: huan@163.com

F.L. Gaol et al. (Eds.): Proc. of the 2011 2nd International Congress on CACS, AISC 144, pp. 95–100.
springerlink.com © Springer-Verlag Berlin Heidelberg 2012

2   Shared Design Based on Ontology

TF-IDF, Key words weight quantitative methods, was the most widely used weighting methods in text processing. It was proposed by Salton and the effectiveness of information retrieval was proved [7]. In fact, three factors were taken into account when using TF-IDF to calculated term weights. There are term Frequency, IDF(Inverse Document Frequency)and Normalization Factor. So based on the three factors, the weight can be calculated as follows:

$$W_{ik} = \frac{TF_{ik} \times IDF}{\sqrt{\sum_{k=1}^{n} (TF_{ik})^2 \times (IDF)^2}} \tag{1}$$

$$IDF = \log(\frac{N}{n_k} + 0.01) \tag{2}$$

3   TF-IDF Algorithm Analyses

A.   Location of Feature Words

The position information of the words was not reflected in the traditional TF-IDF method, whereas the structural features of HTML should be reflected in the weight calculating methods in view of web texts. If the feature words in a different tag reflect the different degrees in the content of the article, the calculation of the weights should be different.

The survey results of the United States P.E.Baxendale showed that the first sentence of a paragraph was a topic sentence make up 85% probability while the last sentence of a paragraph barely amounts to 7% [8]. And the features of article forms in news reports decided that the first paragraph of the article was generally revealing the main content of the article. The most structure of internet text information was HTML, for the different positions feature words in the structure of the Web text, they have different capabilities to express text content or distinctions the category of text, so the position information of the term should be reflected at the weight of feature words. As a result, the feature words of the different positions of in pages should be given different coefficients, then multiplied words frequencies of the feature words, so as to enhance the of effectiveness of the text representation.

B.   Influence of the Feature Words on the Document or the Length of the Words

The traditional TF-IDF formula was primarily used to calculate the weight of a single words in a single document, in facts, information extraction was not confined to calculation the weight of single words (ti) in the N-documents (d1, d2, ... dN), instead, for frequently for certain phrase in the text, the shorter the was, the more generic the semantic was.

In general, the longer word length of a term in Chinese tends to reflect more specific and lower concept while the shorter terms were more likely to express the relatively abstract and upper concept. so it could be believed that short words has higher frequency and more meaning which was function-oriented. Relatively, long words have lower frequency which was content-oriented, so the increased weights of the long words will favor word segmentation and accordingly could more accurately reflect the importance of the feature words in the article. Discovery could be made at practical application that keywords are usually some combination of specialized academic vocabulary, and it was longer than the general vocabulary. Considering the length of the candidate word, the role of length words could be highlighted. Length of the term can also use the logarithmic function to smooth out the dramatic difference in length between words. Generally speaking, long terms contain more clearly meaning and they could better reflect the text topic, suitable for Keywords. Hence, the number of words in phrases could play a role at weight, that is to say, the longer the phrase was, the more likely it was the concept of the field, and the effect of artificial assessment would increase accuracy and recall of the keyword extraction.

4   TF-IDF Improvement Ideas

For the feature words in a different tag reflect on the different levels for content of the article, the calculation of the weights should also be different. As a result, feature words of the different locations in the web page should be given different coefficient, and then multiplied words frequencies of the feature, so as to enhance the effectiveness of text representation. The position information of the words was not presents at traditional TF-IDF formula, for the web text, structural features of HTML should be illustrates in the calculation method of the weight, so $Block_i$ was use to express feature words structure location coefficient in the Web text.

TF-IDF formula was used to calculate the weight wik of a single word in a single document, but this paper want to calculate weight of a single word (ti) within N document (d1, d2, ...dN), and the concept of the paper was not a single word, it was a phrase, the number of the words in phrase should play a role at weight, that was to say, the long the phrase was the more likely it is the domain concept. In general, L was use to express length of the document phrase ti at the document definition, when length of the candidate concept increase, the weight should increase accordingly. So the weight of the concept length without consideration multiplied by the function log2(L), this function was monotonically increasing, and when L=2 the function value was 1, which indicate weight need not be added when length of candidate phrase was 2; but when L>2,the candidate phrase length increase, and the weight increase too. The w_totali reflect the keywords weight at the whole field of document, the higher the value was, the more capability it reflects the characteristics of the document at domain, on the other hand the capability was weaker.

Based on the above two points, improved TF-IDF formula could be obtained.

$$W\_total_i = \log_2(L)\sum_{k=1}^{n}\frac{Block_i \times tf_k \times IDF}{\sqrt{\sum_{k=1}^{n}(Block_i)^2 \times (tf_k)^2 \times (IDF)^2}}$$

(3)

To verify influence of the feature words on document or on the length of the words, sixteen interrelated feature words of different lengths or different locations in certain field to calculated its weights before and after the TF-IDF formula has been improved, contrasted as a curve on Figure 1.As can be seen from Figure 1, the weight of feature words was superior to the before improved.

Fig. 1 Compared weight of feature words extracted with improved before and after the TF-IDF

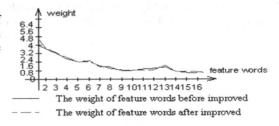

5   Performance Analysis of Extraction Results

A.   *Extraction System Performance Evaluation*

Precision and recall were two indicators to assess the performance about information extraction system [9].

Recall rate equal to the proportion of all possible correct results about the system correctly extracted.

$$R = \frac{\text{Correct extract structure of the system}}{\text{All the correct results should be extract}} \times 100\%$$

(4)

Accuracy rate equal to the proportion of all extraction results about the system correctly extracted.

$$P = \frac{\text{Correct extract results of system}}{\text{Results of the system extract}} \times 100\%$$

(5)

In order to comprehensively assess system performance, usually calculated weighted geometric mean about recall rate (REC) and accuracy (PRE) [10], that is F index, the calculated formula as follows:

$$F-MEASURE = \frac{((beta)^2 + 1.0) \times PRE \times REC}{(beta)^2 \times PRE + REC}$$

(6)

Among the formula, beta was the recall rate and accuracy of the relative weight, as beta equal to 1, recall and accuracy both equally important; as beat greater than 1, the accuracy was more important; as beta less than 1, recall was more important, in generally Beta value was 1,1 / 2,2.

B.   Test Result

Training and testing corpus arrange by Natural Language Processing Group of International Database Center of department of Computer Information and Technology of Fudan University, in which five categories was select including 5 000 training documents , 5 000 test document, each class had 1 000 and 1 000 training documents and test documents. In test, 40K words was singled out from each training text, similarity threshold evaluate 0.8. Compared taking effect with TFIDF improved and Tradition TF-IDF, compared the results as Table1.

Table 1 Improved results taken before and after the TF-IDF

	Improve TF-IDF			Tradition TF-IDF		
	R(%)	P(%)	F(%)	R(%)	P(%)	F(%)
sports	81.37	87.35	84.25	78.89	86.57	82.41
politics	76.35	88.52	81.42	68.38	76.50	72.18
agriculture	78.95	82.75	80.77	74.52	79.67	74.88
agriculture	78.95	82.75	80.77	74.52	79.67	74.88
environment	68.95	83.86	75.69	65.87	79.14	71.25

New TFIDF refers to improved TFIDF algorithm, tradition TFIDF refers to the traditional TFIDF algorithm, R, P, F respectively indicates the recall rate, precision, F1 assessed value. As can be seen from Table 1, the extracted effects of improved TFIDF algorithm are superior to traditional TFIDF algorithm concerning recall, precision and F1 assessed value.

6   Conclusion

The size of keyword weight would directly affect the efficiency of information extraction; this paper analyzed and improved the TFI-DF algorithm from factors of length and location of key words which would be extracted to obtain the improved weight formula. Compare the improved weight formula with the traditional TF-IDF formula through experiments, the results show that the improved formula of the weight enhanced the extract accuracy of the algorithm.

References

[1] How, B.C., Narayanan, K.: An empirical study of feature selection for text categorization based on term weight age. In: Proceedings of the 2004 IEEE/WIC/ACM International Conference on Web Intelligence, pp. 599–602. IEEE Computer Society, Washington, DC (2004)

[2] Mladenic, D., Grobelnik, M.: Feature Selection for Unbalanced Class Distribution and NaYve Bayees. In: Proceedings of the 6th International Conference on Machine Learning, pp. 258–267. Morgan Kaufmann, Blrf (1999)

[3] Luo, X., Sun, M., Tsou, B.K.: Covering ambiguity resolution in Chinese word segmentation based on contextual information. In: Pleadings of the 19th International Conference on Computational Linguistics, pp. 1–7. Association for Computational Linguistics, Morristown (2002)

[4] Hulth, A.: Improved automatic keyword extractiongiven more linguistic knowledge. In: Proceedings of the Conference on EmpiricalMethods in Natural Language Processing, EMNLP, Sapporo, pp. 216–223 (2003)

[5] Yu, H.: SVMC: Single-class classification with support vector machines. In: Proc. of IJCAI, pp. 415–422 (2003)

[6] Qu, S., Wang, S., Zou, Y.: Improvement of Text Feature Selection Method based on TFIDF. In: International Seminar on Future Information Technology and Management Engineering, pp. 79–81 (2008)

[7] Salton, G., Fox, E.A., Wu, H.: Extended boolean information retrieval. Communications of the ACM 26(11), 1022–1036 (1983)

[8] Borko, H., Bernier, C.L.: Abstracts of the concepts and methods. Academic Press, America New York (1991)

[9] Taniar, D.: Web Information Systems, pp. 25–58. Idea Group Publishing, London (2004)

[10] Douthat, A.: The Message Understanding Conference Scoring Software User's Manual. In: Proceedings of the Seventh Message Understanding Conference (1998)

Scientometric Study of the IEEE Transactions on Software Engineering 1980-2010

Brahim Hamadicharef

Abstract. In this paper a scientometric study on the IEEE Transactions on Software Engineering 1980-2010 is presented. Using the full records from the Thomson Reuters (ISI) Web of Science (WoS), the journal's bibliometric measures are examined in terms of growth of literature, authorship characteristics, country of origin, distribution of articles' citations and references, and finally graph network of the research collaborations. A keyword occurrence analysis was carried out on the articles' title and the results used to create TagClouds showing perceptually their research importance. Furthermore, these TagClouds were created for the full 3 decades, shorter 5-year periods and most recent years, providing insights into potential research trends and help to relate them to major historic contributions in software engineering.

1 Introduction

The journal IEEE Transactions on Software Engineering (TSOFT) has been publishing article for more than 30 years and it was timely to carry out a scientometric study on its body of literature. The usefulness of such studies has been demonstrated as it allows not only to identify key bibliometric aspects of a particular research topic [1][2], a journal [3][4], or a series of conferences [5][6]. More than just providing bibliometric measures, it can also help young novice in their field to learn about eminent researchers in their field, identify key papers to read (typically reviews and surveys), journal to publish in. Furthermore, keyword analysis also allow to create network of research based on co-authorship (i.e. collaboration), research topics, etc.

The remainder of the paper is organized as follows. In Section 2 we detail the analysis methodology and results from the bibliometric study. Finally in Section 3, we conclude the paper.

Brahim Hamadicharef
Tiara, 1 Kim Seng Walk, Singapore 239403
e-mail: bhamadicharef@hotmail.com

F.L. Gaol et al. (Eds.): Proc. of the 2011 2nd International Congress on CACS, AISC 144, pp. 101–106.
springerlink.com   © Springer-Verlag Berlin Heidelberg 2012

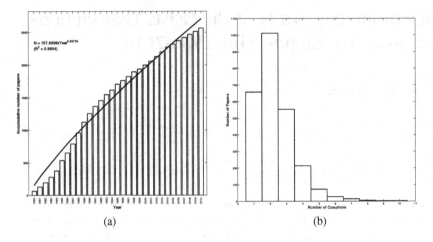

(a)   (b)

Fig. 1 Growth of the TSOFT literature (a) and (b) Overall distribution of co-authorship

2   Methodology and Results

The Thomson Reuters ISI Web of Science (WoS) online interface was used to re-
trieve the records (with abstract) of the journal. Each year has one file with all the
entries, saved as text files, and subsequently processed using MATLAB scripts. We
present the growth of literature, authorship, citations, countries of origin, number of
references, and international collaborations.

2.1   Growth of Literature

The growth of the TSOFT literature over the last 30 years (1980 to 2010) is shown
in Figure 1(a). A quasi-linear growth can be observe, which is different from other
studies [7], which usually showed power law characteristics.

   Although it is interesting to observe such growth, one useful bibliometric measure to calculate is its *obsolescence*. In the 70s, researchers established mathematical
law for describing the temporal loss of utility [8], an indicator of obsolescence [9]
called half-life h (associated to an annual loss), that can be calculated for a set of
documents. Using the 30 years of records, the half-life (h) for the TSOFT literature
is equal to $h = 15.62$ with an annual loss of 4.34%.

2.2   Authorship

In Figure 1(b) the distribution of co-authors is shown with a peak at two co-authors.
Looking at the (color-coded) yearly basis distribution revealed more details, as
shown in Figure 2(a). For e.g., the contributions from recent years show a slightly
larger number of co-authors compared to early years. In the 80s, in particular, a
large number of articles had only one or two co-authors, whereas in recent years

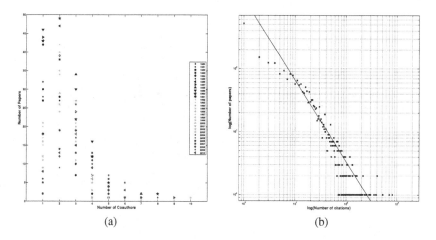

(a)   (b)

Fig. 2 Yearly distribution of co-authorship (a) and (b) Citations of the TSOFT literature

this seems to increase more papers having 5, 6 or 7 co-authors. One paper on *Detection of Mutual Inconsistency in Distributed Systems* has the highest number (10) of authors [10].

2.3 Citations

In Figure 2(b), the citation distribution (i.e. the number of citations versus the number of papers) is shown as a loglog plot. Such typical distribution can be fitted with a regression model of the form $y = a \times x + b$, with a=-1.32 and b=7.20 (goodness-of-fit R^2= 0.90). Notice that the model is a power law but shown as linear due to the loglog scales. The most cited paper is cited 811 times, TSOFT papers are on average cited 22.22 times (median 9.00 times). Some 13.38% of these publications have never been cited.

2.4 Countries of Origin

There are in total 46 countries contributing to the TSOFT literature (WoS has 1632 entries with the field country of origin) and include USA (932, 57.11%), Canada (108, 6.62%), Italy (103, 6.31%), U.K. (82, 5.02%), Germany (56, 3.43%), France (42, 2.57%), Norway (35, 2.14%), Japan (30, 1.84%), Taiwan (21, 1.29%), Australia (19, 1.16%), India (18, 1.10%), Israel (17, 1.04%), China (14, 0.86%), Belgium (13, 0.80%), Korea (13), Spain (11, 0.67%), Sweden (11), Greece (10, 0.61%), etc. In Figure 4(a), the graph of international collaborations is shown, with link between countries. Of all countries, the main strength is between the USA and Canada, and with U.K., Italy, Germany and France. In Figure 3(b), the international collaboration is shown at researchers' level (100 top researchers publishing in TSOFT).

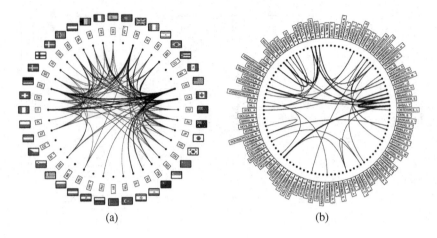

Fig. 3 TSOFT (a) countries collaboration and (b) researchers collaboration network

The evolution of accumulated contributions per countries is shown in Figure 4(a). From the very early years, the USA has the most significantly contributed to the TSOFT literature. It is worth noticing the increasing contributions from Canada, U.K, Italy, Germany and France, as well as Norway and Japan.

2.5   Number of References

From 1980 to 2010 the yearly average number of references per article grew from 12.89 to 42.85 (median 9.96 to 41.07). The article with the greatest number of reference is by M. Jørgensen and M. Shepperd on *a systematic review of software development cost estimation studies* [11] with 311 references. On the same year, J. Hannay and colleagues also published a TSOFT paper on *a systematic review of theory use in software engineering experiments* [12] with 182 references.

2.6   Keywords Analysis

The tool WORDLE [13] was used to to create TagClouds from the keywords distribution of the TSOFT articles' title. As shown in Figure 5, there are many keywords which are all equally sized, showing their importance in the research published in TSOFT. The 30 years of contribution was split into smaller 5-year periods, to be able to compare different advancements in software engineering from the 80s, the 90s and the first decade after the year 2000 (e.g. Figure 5(a) and Figure 5(b)). Focus was given to the recent years as well as an overall picture for the 30 years of publications as shown in Figure 5(c).

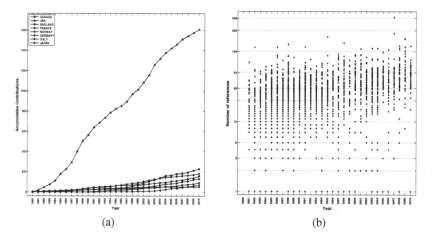

(a)   (b)

Fig. 4 Accumulated contribution to the TSOFT literature per countries (a) and (b) Distribution of the number of references

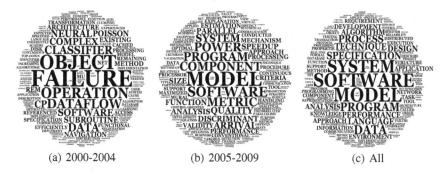

(a) 2000-2004   (b) 2005-2009   (c) All

Fig. 5 TagClouds of TSOFT articles' title

3 Conclusions

After 30 years of publications, it was timely to present a scientometric study of the IEEE Transactions on Software Engineering. In this paper, based on bibliometric measures, characteristics such as growth of literature, authorship, country of origin, citations, references, were detailed. A graph was created to present the journal's network of collaborations between authors. Keywords from on the articles' title, and in particular their occurrence, were analyzed and the results used to create TagClouds. These are showing the key research topics within the software engineering research field. Furthermore, using the full 3 decades, shorter 5-year periods and most recent years, insights into potential research trends could be investigate and thus help to relate these topics to major historic contributions in software engineering.

References

1. Patra, S.K., Mishra, S.: Bibliometric study of bioinformatics literature. Scientometrics 67(3), 477–489 (2006)
2. Chiu, W.-T., Ho, Y.-S.: Bibliometric analysis of tsunami research. Scientometrics 73(1), 3–17 (2007)
3. Hamadicharef, B.: Scientometric Study of the Journal NeuroImage 1992–2009. In: Proceedings of the 2010 International Conference on Web Information Systems and Mining (WISM 2010), Nanjing, China, October 23-24, pp. 201–204 (2010)
4. Thanuskod, S.: Journal of Social Sciences: A Bibliometric Study. Journal of Social Sciences 24(2), 77–80 (2010)
5. Hamadicharef, B.: Bibliometric Study of the DAFx Proceedings 1998-2009. In: Proceedings of the 13th International Conference on Digital Audio Effects (DAFx 2010), Graz, Austria, September 4–6, pp. 427–430 (2010)
6. Bartneck, C., Hu, J.: Scientometric Analysis of the CHI Proceedings. In: Proceedings of the Conference on Human Factors in Computing Systems (CHI 2009), Boston, USA, April 4–9, pp. 699–708 (2009)
7. Hamadicharef, B.: Brain–Computer Interface (BCI) Literature – A Bibliometric Study. In: Proceedings of the 10th International Conference on Information Science, Signal Processing and their applications (ISSPA 2010), Kuala Lumpur, Malaysia, May 10–13, pp. 626–629 (2010)
8. Brookes, B.C.: The Growth, Utility, and Obsolescence of Scientific Periodical Literature. Journal of Documentation 26(4), 283–294 (1970)
9. Line, M.B.: The 'half-life' of periodical literature, apparent and real obsolescence. Journal of Documentation 26(1), 46–54 (1970)
10. Parker, D.S., Popek, G.J., Rudisin, G., Stoughton, A., Walker, B.J., Walton, E., Chow, J.M., Edwards, D., Kiser, S., Kline, C.: Detection of Mutual Inconsistency in Distributed Systems. IEEE Transactions on Software Engineering 9(3), 240–247 (1983)
11. Jørgensen, M., Shepperd, M.: A Systematic Review of Software Development Cost Estimation Studies. IEEE Transactions on Software Engineering 33(1), 33–53 (2007)
12. Hannay, J.E., Sjoberg, D.I.K., Dyba, T.: A Systematic Review of Theory Use in Software Engineering Experiments. IEEE Transactions on Software Engineering 33(2), 87–107 (2007)
13. Viegas, F.B., Wattenberg, M., Feinberg, J.: Participatory Visualization with Wordle. IEEE Transactions on Visualization and Computer Graphics 15(6), 1137–1144 (2009)

Query-Based Automatic Multi-document Summarization Extraction Method for Web Pages

Qi He, Hong-Wei Hao, and Xu-Cheng Yin[*]

Abstract. In order to overcome the shortcomings of the incomprehensive of traditional automatic summarization, this paper proposes the automatic multi-document summarization extraction method based on user's query for web pages. The key technology in our method is the sentence importance weight calculation, which takes varieties of impact factors into account to score the candidate sentence importance weight in the retrieval results. These impact factors include the segmentation results weight, characteristics of sentence structure, length of sentence and the mutual information of search terms. On the basis of our method, this paper gives a description of the automatic summarization process. Then, the comparative experimental results show that our method is more effective on the Precision and Recall than others in abstract extraction.

1 Introduction

Today, along with booming development in network technology, the Internet has become an important source of information. In these years, how to extract useful and needed information more quickly from vast Internet information has attracted more and more researchers at home and abroad. The technology of automatic abstract extraction is the foundation of data mining [1], which obtains abstract from the web pages automatically and generates a brief and indicative summary of the page content to the users intuitively, and it saves time for accessing the information [2].

Automatic abstract has been researched for a long process, and also puts forward all kinds of methods for automatic abstracting [3, 4]. Generally speaking, automatic

Qi He
University of Science and Technology Beijing, Xueyuan Road No.30, Haidian District, Beijing 100083, China
e-mail: heqi8247@163.com

Hong-Wei Hao · Xu-Cheng Yin
University of Science and Technology Beijing, Xueyuan Road No.30, Haidian District, Beijing 100083, China
e-mail: {hhw,xuchengyin}@ustb.edu.cn

[*] Corresponding author.

F.L. Gaol et al. (Eds.): Proc. of the 2011 2nd International Congress on CACS, AISC 144, pp. 107–112.
springerlink.com   © Springer-Verlag Berlin Heidelberg 2012

abstract research is including two parts, one is single document automatically abstract and the other is multi-document summarization [5, 6]. The multi-document summarization extraction algorithm is mainly researched in this paper.

One of the methods based on single document abstract technology is the automatic extraction algorithm based on the original text. Its advantages include unlimited fields, fast speed of extracting the abstract and the length of abstract is adjustable. On the basis of this method, this paper presents the query-based automatic multi-document summarization method for web pages, which takes varieties of impact factors into account.

2   Summarizition Extraction System

The query-based automatic multi-document summarization extraction flow chart is shown in Fig. 1.

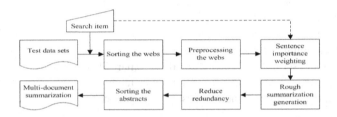

Fig. 1 Flow chart of query-based automatic multi-document summarization extraction.

As shown in Fig. 1, query-based automatic multi-document summarization extraction method can be divided into the following steps:

- Using the open source web crawler Heritrix to download webs randomly, then these webs are saved in the local hard drive and regarded as the test data sets. Their contents involve education, military, political, economic and other fields.
- Input the search term, then retrieval from the web test data.
- The title of the relevant web page is calculated by the Lucene's document scoring algorithm, and the search results ordered by scores from high to low are output. Then adopting the previous 50 retrieval results as the experimental web data for the automatic multi-document summarization extraction.
- These 50 pages are preprocessed. According to the web tags, this method obtains the title, text and date information through analyzing the web by the web block algorithm.
- The web title and text are segmented and filtered by the ICTCLAS(Institute of Computer Technology, Chinese Lexical Analysis System), which is also used to part-of-speech tagging, and then the sentence importance weight is calculated by the formula (5).

- The candidate sentences are ordered by the sentences importance weight from high to low, after that the rough summarization is formed through extracting the candidate sentences according to the specified compression ratio.
- The similarity of the sentences among the rough summarization is calculated. By setting a threshold that confined the correction value, removing the rough abstracts redundancy [7].
- The abstract is sorted by the web's creation time after reducing abstract redundancy, afterwards output them according to the original order, and then the final multi-document summarization is attained.

3 The Key Technology of Multi-document Summarization Extraction

Document Scoring Algorithm

In order to make users to see the retrieval results more humane, when user inputs search terms, the relevance between the retrieval results and search terms is calculated by the Lucene's document scoring algorithm, if the score is higher, the document is more relevant, and then output the retrieval results ordered by scores from high to low.

In this paper, calculating the related document by the Lucene's document scoring algorithm after comparing the segmentation results of search terms and the web title, the formula is

$$T = \sum \sqrt{tf(q_i,T)} \times idf \times lengthNorm \qquad (1)$$

Where q_i is the ith segment word item of search terms; $\sqrt{tf(q_i,T)}$ is the frequency of q_i appears in the web title T ; idf is the inversed document frequency, because of the scoring algorithm aims at the web title, this coefficient in practical application can be set 1.0; $lengthNorm$ is the length factor, which is decided by the reciprocal of the length of T when it is calculated.

Sentence Importance Weight

After automatic segmentation and filtering the single character, we regard the segmentation results as feature items and use vector $S(w_1,w_2,\cdots,w_m)$ to represents the candidate sentence, where w_i denotes the ith segment word item of S , $1 \le i \le m$.

- Segmentation Results Weight: Except from the word frequency, part-of-speech is also considered when calculating the candidate sentence weight. Generally speaking, the main content of webs is reflected by the nouns and the verbs, but the adverbs and others can't be given important significance. Moreover, the analysis shows that a sentence contains more low frequency words is more possible to be regarded as the abstract than the others includes a high one. So the calculation formula of segmentation results is shown as follows:

$$W(S_i) = \left[\sum_{j=1}^{m} tf(w_{ij}) / m \right] \times pos(w_{ij}) \times \left[\sum w_{ij} / \sqrt{|w_{ij}|} \right].$$ (2)

Where the vector space model of the candidate sentence S_i is $S_i(w_{i1}, w_{i2}, \cdots, w_{im})$; $tf(w_{ij})$ is the word frequency of w_{ij}; m is the number of S_i segmentation results; $pos(w_{ij})$ means the part-of-speech coefficient of the w_{ij}, $0 < pos(w_{ij}) < 1$; $\sum w_{ij}$ is the sum of all different segmentation results weight in S_i; $|w_{ij}|$ is the count of all different segmentation results appear in S_i.

- Characteristics of Sentence Structure: The characteristics of sentence structure consist of the sentence's position in the original web pages and the syntax structure. According to the statistics, the sentence can be able to represent the theme of the whole article is about 85% appear in the first paragraph and about 7% in the end of the article. So, in this paper, the weight is increased when calculation of the sentence importance weight if the sentence is in the first paragraph. $P(S_i)$ is the coefficient of location, which is defined as follows: if S_i is the first paragraph, $P(S_i) = 1$; the others $0 < P(S_i) < 1$. Moreover, in general article, the declarative sentence can basically describe the important information of the document, while the interrogative sentence and the exclamatory sentence are rarely regarded as abstracts. $S(S_i)$ represents the coefficient, when S_i is the declarative sentence, $S(S_i) = 1$; the others $0 < S(S_i) < 1$.

- The Length of Sentence: Usually, the short sentences are not fully reflect the main content of the article, and the long sentences contains redundant information, so they are not suitable for the abstract. For this reason, the normal distribution model is used in calculating the sentence weight, as in

$$L(S_i) = e^{\frac{(x_i - \mu)^2}{2\delta^2}} / \sqrt{2\pi}\delta$$
$$\delta^2 = \sum_{j=1}^{n} (x_j - \mu)^2 \times \frac{1}{n}$$ (3)

In formula (3), $L(S_i)$ is the length weight coefficient of candidate sentence S_i; x_i is the length of S_i; μ denotes the average length of all sentence; δ is the mean square deviation between the length of every sentence and the average length of all sentences; x_j is the length of the jth sentence; n represents the number of the sentence set.

- The mutual information of search terms: The mutual information is used to measure the correlation between two objects. The similarity between search terms and candidate sentence is calculated in extracting the multi-document summarization, because of our method is based on user's query. The mutual information weight of search terms is determined by formula (4)

$$MI(S_i) = \sum match(q, S_i) / \sum \sum match(q, S_i) / m .$$ (4)

Where $\sum match(q, S_i)$ is the number of search terms q appears in the candidate sentence S_i; $\sum \sum match(q, S_i)$ represents the total number of q in all candidate sentences; m denotes the count of S_i segmentation results.

- Calculating Sentence Importance Weight: As the above-mentioned technology, during calculating the sentence importance weight, the following factors are considered: segmentation results weight, characteristics of sentence structure, length of sentence and the mutual information of search term. This paper proposes the calculation formula is

$$Total(S_i) = \lambda_1 W(S_i) + \lambda_2 [P(S_i) + S(S_i)] + \lambda_3 L(S_i) + \lambda_4 MI(S_i)$$ (5)

Where $\lambda_i (1 \le i \le 4)$ is an adjusted parameter, $\lambda_1 + \lambda_2 + \lambda_3 + \lambda_4 = 1$, and they all not equals to 0.

4 Experimental Results and Conclusions

In this paper, the internal evaluation methods are used to estimate the experimental results, which judge the coherence and integrity of automatic summarization by comparing the manual-marked abstract with the automatic-marked abstract. The evaluation indexes are the Precision and Recall.

In this paper, the comparative experiments use the following three methods. The first method: extract the abstract by our method, which calculates sentence importance weight by formula (5). The second method: cancel the factor of sentence length, that is to say, $\lambda_3 = 0$. The third method: call off the mutual information of search term, namely, $\lambda_4 = 0$.

Setting the compression ratios are 10%, 15%, 20%, 25%, 30%, the experimental results are shown as follows.

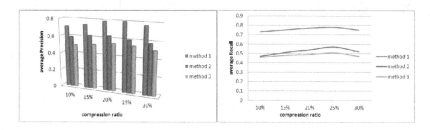

Fig. 2 Comparative results of average Precision and Recall.

As shown in Fig. 2, when adopting the technology of the query-based automatic multi-document summarization extraction, the experimental results are more significantly improved in Precision and Recall than the after two methods.

The query-based automatic multi-document summarization extraction algorithm is proposed in this paper, which takes into account of segmentation results weight, characteristics of sentence structure, length of sentence and the mutual information of search term. The internal assessment results demonstrate that the proposed algorithm has good performance on multi-document abstract extraction than the other comparative experiments.

However, the algorithm is based on original text without considering deep information. Therefore, the future work is necessary for the article syntax, semantic and contextual analysis. So as to improve the automatic summarization, then can make users easily to judge the web pages relevance through browsing the abstract rather than the original web document, in order to enhance the efficiency.

Acknowledgments. This work is supported by the R&D Special Fund for Public Welfare Industry(Meteorology) of China under Grant No. GYHY201106039 and the Fundamental Research Funds for the Central Universities under Grant No. FRF-BR-10-034B.

References

1. Han, J., Kamber, M.: Data Mining Concepts and Techniques. Morgan Kaufmann Publishers, San Francisco (2001) ISBN 1558604898
2. Luhn, H.P.: The automatic creation of literature abstract. IBM Journal of Research and Development 2(2), 159–165 (1958)
3. Wang, J., Wu, G., Zhou, Y., Zhang, F.: Research on Automatic Summarization of Web Document Guided by Discourse. Journal of Computer Research and Development, 398–405 (2003)
4. Tadashi, N., Yuji, M.: A New Approach to Unsupervised Text Summarization. In: Proceedings of ACM SIGIR 2001, pp. 26–34 (2001)
5. Conroy, J.M., Schlesinger, J.D.: CLASSY 2007 at DUC 2007. In: Proceedings of the 2007 Document Understanding Conference (DUC 2007), New York (2007)
6. Gong, Y.H., Liu, X.: Generic Text Summarization Using Relevance Measure and Latent Semantic Analysis. In: Processing of ACM SIGIR 2001, pp. 19–25 (2001)
7. Gong, Y., Liu, X.: Generic Text Summarization Using Relevance Measure and Latent Semantic Analysis. In: Proceedings of ACM SIGIR 2001, pp. 19–25 (2001)

Top-k Algorithm Based on Extraction

Lingjuan Li, Xue Zeng, and Guoyu Lu

Abstract. Algorithms for top-k query are widely used in massive data query, which return k most important objects based on aggregate functions. The classical Threshold Algorithm (TA) is one of the most famous algorithms for top-k query. It requires sequential and random accesses to the lists. The time cost of TA will be very high when data is massive. This paper proposes a new algorithm TABE (Top-k Algorithm Based on Extraction) to minimize the query time. TABE first extracts the objects which have higher ranking on each attribute, and then execute the Threshold Algorithm on these objects. Test results show that TABE has high accuracy to meet the general query requirements, and the experimental results of comparing TABE with NRA (No Random Accesses) show that our proposed algorithm TABE can largely reduce the query time.

1   Introduction

Top-k query returns k objects with the highest combined score of each attribute. It is quite efficient in querying massive data. For this reason, Top-k is widely used in many areas. Currently, there are some Top-k algorithms such as TA (Threshold Algorithm) [1], NRA (No Random Accesses) [2-4], LARA [5]. TA is the classical algorithm for solving Top-k problem. It is divided into two steps: the first is finding the threshold and the second is looking for objects that fit the threshold. When dealing with massive data, every random access needs to scan every list, and the cost of random access is quite expensive. NRA does not need random access. It is also divided into growing and shrinking phases, and it is better than TA [4].

Lingjuan Li
College of Computer Science, Nanjing University of Posts and Telecommunications, Nanjing, China
e-mail: fqlilj@163.com

Xue Zeng
College of Computer Science, Nanjing University of Posts and Telecommunications, Nanjing, China
e-mail: zengxue666@hotmail.com

Guoyu Lu
Department of Information Engineering and Computer Sceince, University of Trento, Trento, Italy
e-mail: guoyu.lu@studenti.unitn.it

F.L. Gaol et al. (Eds.): Proc. of the 2011 2nd International Congress on CACS, AISC 144, pp. 113–118.
springerlink.com   © Springer-Verlag Berlin Heidelberg 2012

In this paper, by improving TA, we propose a Top-k Algorithm Based on Extraction (TABE) to get higher query efficiency on massive data.

2   Design of Top-k Algorithm Based on Extraction

Definitions related to the algorithm are as follows:

A table T contains N tuples and each tuple has M+1 attributes ID, A_1, A_2,..., A_m. Every attribute is stored as a sorted list $L_i(ID, A_i)$, where ID is the tuples' identifier and A_i is the value of a tuple's attribute. For a tuple t satisfying $\forall t \in T$, if $t.ID=id_i$ and $t.A_i=a_i$, then (id_i, a_i) belongs to L_i. Each list is sorted by the decreasing order of the attribute value. Given W_i as the weight of attribute A_i, the definition of the score function F will be

$$F(t) = \sum_{i=1}^{m} (w_i \times t.A_i) \qquad (1)$$

Top-k query is to find a sub-set K which satisfies |K|=k, and the restrictions are given in (2):

$$\left\{ \begin{array}{l} \forall\ t_i \in\ K \\ \forall\ t_j \in\ T\ -\ K \\ F\ (t_i)\ \geq\ F\ (t_j) \end{array} \right\} \qquad (2)$$

The main idea of the classic Top-k algorithm TA is to sequentially access objects of m lists. According to the query result, every access gets an object as well as the corresponding attribute value, followed by random access of the other m-1 attribute values. Through this way, we can get this object's score. Assuming the objects finally accessed in all the lists are $(id_1,a_1),...,(id_m,a_m)$, the threshold will be calculated as $\tau=F(a_1,...,a_m)$. This threshold will be updated when new object comes up. For a new object t_{new}, which has never arisen before, if t_{new}'s score $F(t_{new})$ is smaller than τ, the algorithm will end up. When dealing with massive data sets, if m lists cannot be put into memory, the cost of accessing disc for TA becomes extremely high. At the same time, if every attribute score of the object is not high, the possibility that the object is among the top-k is also small.

From this point, if the top R objects can be chosen in each sorted list first, then at most m*R objects will be selected, which is less than the objects number N. In that case, the algorithm efficiency will be largely improved. From the application aspect, the query result will be k' after using the algorithm, which brings mistake compared with k. However, as long as the mistake is controlled under a certain degree, most applications can still accept it.

According to this idea, we improve TA, and design a new algorithm for top-k query to reduce the time cost. The algorithm is based on extraction, called TABE (Top-k Algorithm Based on Extraction).

TABE extracts top R attribute values in each list, which reduces the searching scope in certain objects. Then these objects can be accessed in memory at the same time, so that the defect of TA algorithm is avoided. We first build an empty table Temp containing m+1 columns, using ID as the object identifier. Each object contains an ID and m attributes, represented as A1, A2,..., Am. After this step, the top R attribute values in A1 and the corresponding object ID in the first list L1 will be read into the A1 and ID position of the Temp table. Then the top R attribute values in L2 which are not repeated in Temp. ID are read into Temp's ID and A2. This process continues until the access operation on Am is finished. Since Temp is updated from front to back, A1, A2,..., Am in Temp all have NULL value. In order to make table Temp complete, we update every Ai value in each column. Finally, the original TA algorithm is applied to table Temp to get k objects. Now, our operation and query scope is restricted in table Temp. The algorithm TABE can be described as follows:

Algorithm. Top-k Algorithm Based on Extraction (TABE)

Inputs: $L_j(j=1,2,...,m)$ /*m list files*/
Outputs: Temp/* the table storing extracted objects including the first r objects
                        extracted from each list*/
                  k' /* the final query result */

Steps:		9:	end for
1:	$I_{id}=0$;/* Store the identifier that represents whether the object id has been accessed or not*/	10:	end for
		11:	for j=1 to m do
		12:	Scan I_{id},
2:	n=0;/* number of objects accessed*/	13:	If $I_{id}==1$, then
3:	for j=1 to m do	14:	Read the object's value of this attribute and write it into Temp
4:	for i=1 to r do		
5:	Read object's ID and Aj	15:	end if
6:	if $I_{id}==0$, then	16:	end for/* all the top r objects having attribute value will be kept in Temp*/
7:	n++, write ID and Aj into Temp.ID and Temp.Aj		
		17:	Run TA algorithm on Temp to achieve k'
8:	end if		

---

3   Experiments and Result Analysis

According to above description, we know TABE is better than TA. Because NRA is also better than TA, we design experiments to test the query accuracy of algorithm TABE and to compare its performance with NRA algorithm in order to validate the feasibility and the validity of TABE. NRA and TABE are implemented in Java language with Eclipse SDK (version 3.5.0). PC configuration is: 2.10GHz CPU, 2GB memory, Windows XP operating system. For facilitating the evaluation process, we assign the weight of each attribute to 1 and every attribute is distributed independently. The attribute value is chosen randomly among [1,100].

3.1   Accuracy Test

First of all, we test the accuracy to make sure that our algorithm TABE is feasible.

We choose 1 million data, and adopt two schemes to divide them: 5 lists and 4 lists. Selecting the top 5%, 4%, 3% of the data and changing the value k of Top-k, we test the algorithm accuracy respectively. The results are shown in Fig. 1.

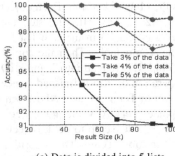
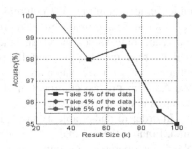

(a) Data is divided into 5 lists   (b) Data is divided into 4 lists

Fig. 1 Algorithm accuracy test

In Fig. 1(b), when selecting 5% and 4% data, the accuracy is always 100%, the line representing 4% data is overlapped by the line representing 5% data.

The experiment results prove that TABE can achieve at least 90% accuracy. The accuracy will change along with the data selection amount, the more data, the higher the accuracy. And when k becomes smaller, the accuracy increases. Meanwhile, the accuracy will also go up when the number of lists m gets smaller.

3.2 The Algorithm's Time Performance Analysis

Firstly, we observe the affection of lists number m. We choose 1 million data, i.e. N=1,000,000, and choose the number of lists m as 3, 4. The comparison results of execution time between NRA algorithm and TABE are shown in Fig. 2 and Fig. 3.

From Fig. 2, we can get to know that even if we take 10% data, the time-consuming of TABE is just 1/10 of NRA algorithm. When k is 25, this number becomes 3%. From this point, it can be seen that, compared with NRA algorithm, TABE's time performance does not change a lot when k increases.

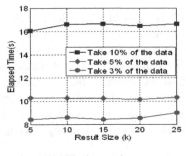
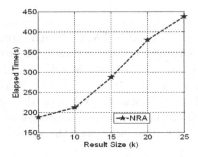

(a) TABE's time performance   (b) NRA's time performance

Fig. 2 Time performance comparison when m is 3

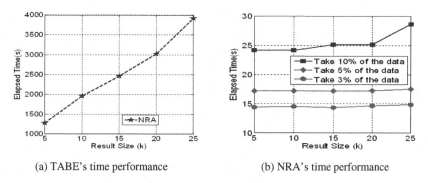

(a) TABE's time performance   (b) NRA's time performance

Fig. 3 Time performance comparison when m is 4.

From Fig. 3, when the lists number m gets larger, the difference of time performance between NRA and TABE will also increase significantly.

In a word, from the experiment results, we can get the conclusion that the query efficiency of TABE is much higher than that of NRA.

Now we consider the affection of objects number N to the experiment result. When m is 4 and N is set to 500,000 and 1,000,000 respectively, the time performance of TABE and NRA are plotted in Fig. 4.

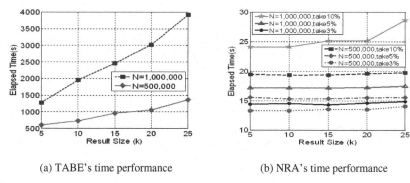

(a) TABE's time performance   (b) NRA's time performance

Fig. 4 Time performance comparison when m is 4 and N is set to 500,000 and 1,000,000

In Fig. 4(a), the experiment result proves that when N is 1,000,000, the time consumed by TABE is just 20% to 30% higher than the consumed time when N is 5,000,000. And the time difference for unequal N does not change along of the increase of k.

In Fig. 4(b), when objects number N is 1,000,000, the running time of NRA is over twice higher than the running time when N is 500,000. Especially, when k increases, the consumed time also increases substantially.

The experiment results show that TABE algorithm has higher query efficiency and is stable to the change of returned objects number k. When objects number and lists number increase, TABE's superiority is more obvious.

As we have shown in accuracy analysis, when m is 3, the accuracy can still achieve 100%. In that case, we can choose to take 3% data, which has greatly reduced the running time.

We also compare TABE with TA. When N is 5,000,000, m is 3, and k equals to 5, the execution time of TA algorithm is 4099,531s. Since the time difference between TABE and TA is significant, we do not do further compare.

4   Conclusion

Since TA is expensive when dealing with massive data sets, this paper proposes a new Top-k algorithm TABE. It first extracts objects that are most likely to be the final result and then obtains the final k objects among all extracted ones. The experimental results have shown TABE has reliable accuracy and high query efficiency. And it is more efficient to perform under large data set, compared with NRA and TA. For facilitating result analysis, we assume the attributes are selected randomly in a certain range and all the attributes are distributed independently.

Acknowledgments. This paper is supported by the National Basic Research Program of China (973 Program: No.2011CB302903) and National Natural Science Foundation of China (No.60863001, No.61073189).

References

1. Fagin, R., Lotem, A., Naor, M.: Optimal aggregation algorithms for middleware. In: Proceedings of the 20th ACM SIGACT-SIGMOD-SIGART Symposium on Principles of Database Systems (PODS 2001), California, USA, pp. 102–113 (2001)
2. Fagin, R., Lotem, A., Naor, M.: Optimal aggregation algorithms for middleware. Journal of Computer and System Sciences 66(4), 614–656 (2003)
3. Pang, H., Ding, X., Zheng, B.: Efficient processing of exact Top-k queries over disk-resident sorted lists. VLDB Journal 19(3), 437–456 (2010)
4. Han, X., Yang, D., Li, J.: TKEP: an efficient Top-k query processing algorithm on massive data. Chinese Journal of Computers 33(8), 1405–1417 (2010)
5. Mamoulis, N., Cheng, K.H., Yiu, M.L., Cheung, D.W.: Efficient aggregation of ranked inputs. In: Proceedings of the 22nd International Conference on Data Engineering (ICDE 2006), Atlanta, GA, USA, pp. 72–83 (2006)

Keyword Extraction Based on Multi-feature Fusion for Chinese Web Pages

Qi He, Hong-Wei Hao, and Xu-Cheng Yin[*]

Abstract. In order to overcome the shortcomings of the incomprehensive of traditional keyword extraction, this paper proposes a keyword extraction based on multi-feature fusion for Chinese web pages. First, the part-of-speech and the position information of candidate words are combined in the improved TF-IDF algorithm. Second, the mutual information of the web title is taken into account to calculate the weight of candidate words. Third, the multi-feature fusion technology is formed by the linear combination of the improved TF-IDF method and mutual information. Thus, our method is proposed based on this multi-feature fusion technology for keyword extraction. Comparative experiments show that extracting keywords generated by our method has higher precision and recall compared with the classical TF-IDF algorithm.

1   Introduction

Keywords are applied to reflect the theme or central content of the article. Extracting keyword from Chinese web pages is helpful to have a general idea of the webs. Keyword extraction plays a key role in Chinese language processing, and it is the foundation of data mining [1]. Over the years, this method has been widely used in text retrieval, categorization, clustering, abstract generation and other fields.

This approach extracts keyword according to the frequency of candidate word in the article [2], which is simple, practical and used diffusely, but its precision is low. There are many common methods based on statistics, such as TF-IDF [3], Pat-tree, word co-occurrence [4, 5] and so on.

This paper extracts keywords from the web pages, due to so many realms the content of webs concerned, building the training data set in the process of machine

Qi He
University of Science and Technology Beijing, Xueyuan Road No.30, Haidian District, Beijing 100083, China
e-mail: heqi8247@163.com

Hong-Wei Hao · Xu-Cheng Yin
University of Science and Technology Beijing, Xueyuan Road No.30, Haidian District, Beijing 100083, China
e-mail: {hhw,xuchengyin}@ustb.edu.cn

[*] Corresponding author.

F.L. Gaol et al. (Eds.): Proc. of the 2011 2nd International Congress on CACS, AISC 144, pp. 119–124.
springerlink.com   © Springer-Verlag Berlin Heidelberg 2012

learning is difficult and cannot be completed within a short period of time. Based on the analysis of the advantages and disadvantages of the existing algorithms, the classical TF-IDF method is improved, and then a new keyword extraction method based on multi-feature fusion for Chinese web pages is put forward in this paper. The word frequency, part-of-speech of candidate word [6, 7], position information and mutual information of title are combined into this approach.

2   Keyword Extraction

Multi-feature Fusion Keyword Extraction

In this paper, the keyword extraction method is fused with multiple features, which includes the improved TF-IDF algorithm and mutual information together. Fig. 1 gives a description of calculating the candidate words weight as follows.

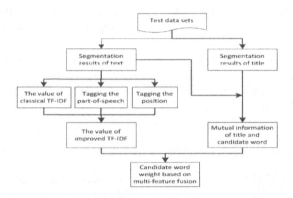

Fig. 1 Process of calculating the candidate words.

Fig. 1 shows that based on classical TF-IDF method, the improved TF-IDF algorithm considers the two other factors of the part-of-speech and the position information of candidate word. The mutual information is used to measure the correlation between the web title and candidate word. The weight of candidate word is defined by formula (1) as in

$$W(w_i) = k \times TFIDF_{imp}(w_i) + (1-k) \times MI(w_i) . \tag{1}$$

Where k is the representation of a parameter which can be adjusted and determined by experiments, $TFIDF_{imp}(w_i)$ is the TF-IDF value of w_i, $MI(w_i)$ is the value of mutual information between web title and the candidate word w_i.

Improved TF-IDF Algorithm

TF-IDF is a classic keyword extraction algorithm based on statistics. The main idea of this algorithm is that if a word in a document appears with high frequency and in other document rarely appears, then this word is regarded as this document's keyword. In this paper, the TF-IDF is determined as follows [8]:

$$TFIDF(w_i) = \left[f(w_i) \times \log_2(N / n_i + 1) \right] / \sum_{j=1}^{m} \left[tf(w_j) \times \log_2(N / n_j + 1) \right]. \qquad (2)$$

Where $tf(w_i)$ is frequency of the candidate word w_i in the document d, $\log_2(N / n_i + 1)$ is the inverse document frequency, N is the whole number of the documents and n_i is the number of the documents which contains w_i.

In formula (2), only the word frequency is focused, but the other factors which can affect the keyword weight are not considered. The lexical category feature of candidate word and its location in web pages should be more attention when calculating the candidate word weight.

Generally speaking, the main content of Chinese webs is reflected by the nouns and the verbs, but the adverbs, the pronouns and so on can't be considered as the keywords. In addition, some single character continually appear in the webs, nevertheless it has no definite meaning, so that they can't give expression to the theme of the web pages. For this reason, the single words can be filtered by the stop word list when calculating the weight of candidate words, and endow the different $pos(w_i)$ which is the weight coefficient of part-of-speech according to the lexical category feature of Chinese words, as a matter of course, the nouns and the verbs have the higher coefficients than the others. When w_i is noun or verb, $pow(w_i) = 1$; the others $0 < pow(w_i) < 1$.

According to the position of the candidate word in the web pages, the $loc(w_i)$ which denotes the location weight coefficient is given. The candidate word in the title is the most important because of that the web title summarizes the whole text and it has the largest amount of information. In the web text, the first paragraph and the last have the higher weight coefficients than the other paragraphs, and the same to the first sentence and the last in the paragraph on account of they are normally the summary sentences. So the $loc(w_i)$ is determined by formula (3) as in

$$loc(w_i) = \begin{cases} 1, & if\ S_i\ is\ the\ title \\ e^{\left|\left|\frac{S_{total}}{2}\right| - S_i\right|} / e^{\left|\frac{S_{total}}{2}\right|}, & if\ S_i\ is\ in\ the\ first\ paragraph\,. \\ \beta \times e^{\left|\left|\frac{S_{total}}{2}\right| - S_i\right|} / e^{\left|\frac{S_{total}}{2}\right|}, & others\ and\ 0 < \beta < 1 \end{cases} \qquad (3)$$

Where S_i is the location of the sentence where is the first position the candidate word w_i appears. S_{total} is the count of all sentences in the paragraph which contains S_i.

As a result, on the basis of the original TF-IDF algorithm, this paper proposes the improved TF-IDF method, which considers the two factors of the part-of-speech of candidate words and their position in Chinese web pages. Using this approach to calculate the weight of candidate word as in

$$TFIDF_{imp}(w_i) = TFIDF(w_i) \times \left[1 + pos(w_i) + loc(w_i) \right]. \qquad (4)$$

Where $pos(w_i)$ is the weight coefficient of part-of-speech of the candidate word w_i, $loc(w_i)$ represents the location weight coefficient.

Mutual Information

It is known that the web content is almost reflected by its title which is the succinct generalization of the web page, through analyzing the relationship between the web content and its title. Consequently, the mutual information of web title is drawn into calculating the candidate word weight.

The mutual information is used to measure the correlation between two objects, it specially indicates the mutual information of the web title in this paper, so it particularly measures the relativity of the candidate word and the segmentation results of title. After automatic segmentation of the web title and filtering by the stop word list, we use vector $T(t_1, t_2, \cdots, t_m)$ to represents the web title, where t_i denotes the ith segment word items of title. The formula of calculating the mutual information of title and candidate word as in

$$MI(w_i) = \log_2(1 + \sum_{j=1}^{m} \frac{p(w_i, t_j)}{p(w_i) \times p(t_j)}). \tag{5}$$

Where $p(w_i, t_j)$ is the frequency of the candidate word w_i and t_j appear in the same sentence, $p(w_i)$ is the frequency of w_i turns up in the web page, $p(t_j)$ is the frequency of t_j appears in the web page.

Realizing the Keyword Extraction Algorithm

The keyword extraction flow chart is shown in Fig. 2.

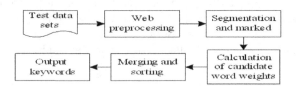

Fig. 2 Flow chat of keyword extraction.

- Preprocessing: Normally, the structures of HTML tags in different web sites and different web pages have different characteristics. The symbol information of blocked pages is confirmed by the features of Chinese web pages in specific site. According to the blocked information of the web source files, the web title and text are extracted by the web block algorithm.
- Segmentation and marked: The web page W can be regarded as the muster which contains a number of candidate words, which is used as vector, after automatic segmentation of the web title and text through the ICTCLAS and filtered by stop word list, at the same time, the candidate words list is produced through tagging the part-of-speech and position.

- Calculation of candidate word weights: Using the formula (1) to calculate the weights of candidate words.
- Merging and sorting: The same candidate words are merged and their weights are repeated addition, then they are sorted by the final weights.
- Output keywords: The keywords are obtained from the top 3% words.

3   Experimental Results and Conclusions

In this paper, the effect of extracting the keywords is evaluated by calculating the Precision(P) and Recall(R)[9]. Precision is gained by the matched number divides the number of the manually marked keywords. Recall is obtained by the matched number divides the number of the automatic keywords.

Using the Heritrix to download 200 webs randomly, then these webs are saved in the local hard drive and regarded as the test data sets. The web pages content involve education, military, political, economic and other fields. The web titles and texts of these 200 test webs are obtained through be preprocessed.

In order to make the experimental results easy to analysis, setting the number of keywords is 5. We use the same test data sets in the following 4 experiments.

First, extracting the manually marked keywords from the test data sets, and the number takes 3 to 7. In experiment 1, keywords are extracted by using the classical TF-IDF algorithm, the value of candidate word is calculated by formula (2). In experiment 2, the keyword extraction method takes into account of the lexical category feature of Chinese words, which uses formula (5) and sets $loc(w_i) = 0$. In the third experiment, the improved TF-IDF algorithm uses formula (5), which considers the two factors of the par-of-speech of candidate word and their position in web pages.

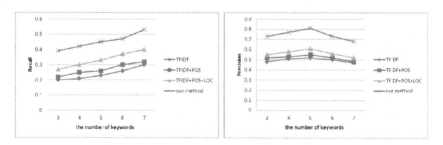

Fig. 3 Precision and Recall of the all comparative results.

In Fig.3 the change trend of precision and recall, which manifest that the results of the last experiment using our method obviously outperform the other three methods. And they show that, when the number of manually marked keywords takes 3 to 7, the proposed keyword extraction method based on multi-feature fusion significantly improves the precision and recall compares to the front three experiments.

In this paper, on the basis of the classical TF-IDF algorithm, a method for extracting keywords is proposed by using the keyword extraction based on multi-feature fusion for Chinese web pages. In this method, the part-of-speech, the position in web pages and the mutual information of web title are combined adequately. The experimental results show that the proposed method is outperformance of the comparative experiments.

However, in the domain of data mining, the effect of Chinese word segmentation system is not well, and only the external performance of candidate word is considered in this paper. Therefore, the future work is improving Chinese word segmentation algorithm and introducing semantic information into keyword extraction, so as to advance the effectiveness of keyword extraction.

Acknowledgments. This work is supported by the R&D Special Fund for Public Welfare Industry(Meteorology) of China under Grant No. GYHY201106039 and the Fundamental Research Funds for the Central Universities under Grant No. FRF-BR-10-034B.

References

1. Han, J., Kamber, M.: Data Mining Concepts and Techniques. Morgan Kaufmann Publishers, SanFrancisco (2001) ISBN 1558604898
2. Edmundson, H.P., Oswald, V.A.: Automatic Indexing and Abstracting of the Contents of Documents. Planning Research Corp., Document PRC R-126, ASTIA AD No. 231606, Los Angeles, pp. 1–124 (1959)
3. Li, J., Fan, Q., Zhang, K.: Keyword Extraction Based on tf/idf for Chinese News Document. Wuhan University Journal of Ntural Sciences 12(5) (2007)
4. Matsuo, Y., Ishizuka, M.: Keyword extraction from a single document using word co-occurrence statistical information. Int. J. Aritif. Intell. 13(1), 157–169 (2004)
5. Csomai, A., Mihalcea, R.: Linguistically motivated features for enhanced back-of-the-book indexing. In: Proc. ACL, pp. 932–940 (2008)
6. Hulth, A.: Improved automatic keyword extraction given more linguistic knowledge. In: Proc. EMNLP, pp. 216–223 (2003)
7. D'Avanzo, E., Magnini, B., Vallin, A.: Keyphere extraction for summarization purposes: The LAKE system at DUC-2004. In: Proc. Document Understanding Conf. (2004)
8. Kolcz, A., Prabakarmurthi, V., Kalita, J.: Summarization as Feature Selection for Text Categorization. In: Processings of the 11th International Conference on Information and Knowledge Management, pp. 365–370. ACM Press, USA (2001)
9. Gong, Y.H., Liu, X.: Generic Text Summarization Using Relevance Measure and Latent Semantic Analysis. In: Processing of ACM SIGIR 2001, pp. 19–25 (2001)

Word Semantic Orientation Calculation Algorithm Based on Dynamic Standard Word Set[*]

Yun Sha, Ming Xia, Huina Jiang, and Xiaohua Wang

Abstract. Orientation calculation of words is the basis of orientation calculation on the large-scale text. Except some words, most of these word orientations are complex, and should be analysis in specific domain. The traditional algorithm of word orientation calculation relies mainly on the standard word set. The word orientation is calculated according to its semantic distance between words in the standard set. In this paper, a dynamic standard word set is build based on domain and part of speech information. The word orientation is calculated based on the corresponding standard set according to its domain and part of speech. The experiment results show that the proposed algorithm is reasonable. The domain and part of speech is useful for the orientation calculation.

1 Introduction

"The 27th China Internet Development Statistics Report [1]" shows that: by December 2010, the number of Internet customers adds up to 457 million and the Internet penetration rates climb to 34.3%. Internet becomes the main stream of information flow. Due to exchange and speaking freely on the Internet, diverse views and attitudes to commodities or matter exist in Internet information. Faced with such a mass of evaluated information, it is very difficult to artificially arrange them. Consequently, that, how to use computer to acquire orientation of these evaluated texts, has been an urgent problem to be solved; Research on text orientation is born. The word orientation is the base of text orientation calculation.

Yun Sha
Information Engineering, Beijing Institute of Petrochemical Technology, Beijing, China
e-mail: shayun@bipt.edu.cn

Ming Xia
Information Engineering, Beijing Institute of Petrochemical Technology, Beijing, China
e-mail: shayun@bipt.edu.cn

[*] This project is supported by the Outstanding Young Foundation of Beijing (No. 2009D005005000004).

Usually, word orientation includes two aspects: polarity (positive or negative) and degree. There are two problems in word orientation calculation: (1) Words appear different polarity in different domain. So lots of word should be put in a domain before we judge its orientation. (2) It is likely that a word has more than one different meaning. The words have different meaning and semantic orientation in different domain.

Usually there are two kinds of method for orientation calculation: statistical method and method based on knowledge library. Statistical method is that: the semantic orientation of words is calculated according to concurrent probability of them. Early in 1997, Hatzivassiloglou[2] made use of conjunctions to gain word orientation on adjectives. Turney and Litman[3] computed the word's similarity between labeled words to get the emotional polarity of words. Wang[4] focus on the role of orientation words in sentence context, instead of individual word only.

Method based on knowledge library is as follows: word semantic distance is calculated based on knowledge library (such as HowNet or WordNet). In 2002, Kamps[5] used structure diagram of synonymous words to get semantic distance between the word to be estimated and selected standard word set to calculate the word's emotional polarity. In 2005, Zhu [6] presented two methods of word's semantic orientation based on meaning similarity grounded on HowNet. Du[8] build an undirected graph by the use of word similarity computing technology.

In traditional algorithms the word orientation is calculated by the similarity of current word with each word in a standard word set. Usually, it is selected by artificial or semi-artificial. And the semantics of word in standard set must be clear. Number of positive standard word and negative word should be balance.

But in these standard word sets, the domain and part of speech divergence is not considered. Moreover, a fixed standard word list can't adapt lingual diversity.

In this paper, a word orientation algorithm based on a dynamic standard word set is proposed.

2 Construction of Dynamic Standard Word Set

2.1 Construction of Domain Dynamic Standard Word Set

The dynamic standard word set which are built based on domain and part of speech must follow a number of principles as follows:

- Semantics of standard wordlist must express the current domain information and part of speech, which is ready for orientation judgment.
- Dynamic standard word set must obtain from large-scale corpus and maintain two opposite balance sides.
- Semantics of word in standard set must be enough clear (not having any ambiguity), and be marked by extreme intensity of emotions.

The domain dynamic standard word set is:

$$A= \{A_1, ..., A_i, ... A_n | n \geq 1\}$$

In which, the set A_i is the i'th domain; n is the number of domains in data set. Words in A_i should be the often occurred ones in large scale database in the i'th domain.

Each domain key word set is:

$$I = \{I_1, ..., I_i, ... I_n | n \geq 1\}$$

The subset I_i is the i'th key word set, n is the domain number. The relationship of I_i and A_i is one to one. $|I_i|$ is the number of key word in the i'th domain.

Web search engines provide an efficient interface to this vast information. Page count is a useful information sources provided by most web search engines. Page count of a query is the number of pages that contain the query words. Page count for the query P AND Q can be considered as a global measure of co-occurrence of words P and Q. For example, the page count of the query "apple" AND "iPad" in Google is 559,000,000, whereas the same for "grape" AND "iPad" is only 5,160,000. The more than 108 times more numerous page counts for "apple" AND "computer" indicate that apple is more semantically similar to ipad than is grape.

The constructing of standard word set process is: given A^+ is the positive word set, in which are positive orientation words. $a \in A^+, \exists w_i \in I_i$, x_{a,w_i} is the number of search engine return when the query is "a" AND "w_i". Sort each word from A^+ by x_{a,w_i} in descending order, and then extract the first M word to compose the i'th domain positive standard word set A_i^+, and same as A^-.

2.2   Construction of Part of Speech Dynamic Standard Word Set

Word's part of speech has much to do with word's meaning orientation and same word with different part of speech contextually may lead to different semantic orientation. Usually, the text orientation is expressed by adjectives and verbs. The dynamic standard word set is built according to adjective and verb words based on domain dynamic standard word set.

$$A_i = A_i^+ \cup A_i^- \; , \; A_i^+ = A_{i,adj}^+ \cup A_{i,v}^+ \; , \; A_i^- = A_{i,adj}^- \cup A_{i,v}^-$$

The dynamic standard word set A_i for the i'th domain includes four child word sets: positive adjective set: $A_{i,adj}^+$, negative adjective set $A_{i,adj}^-$, positive verb set: $A_{i,v}^+$, negative verb set $A_{i,v}^-$.

3   Words Semantic Orientation under Dynamic Standard Word Set

The method of semantic orientation calculation of words based on dynamic standard word set using similarity degree between current word and corresponding standard word set according to word's semantic orientation in its domain and part of speech. So the formula of semantic orientation calculation of words based on dynamic wordlist is following.

- Given adjective word in the i'th domain $w_{i,adj}$, formula of semantic orientation calculation of it is as following:

$$Orient(w_{i,adj}) = \sum_{j=1}^{M}(D_{i,adj}^{j,+}) - \sum_{j=1}^{M}(D_{i,adj}^{j,-}) \qquad (1)$$

In which, $D_{i,adj}^{j,+}$ is the semantic distance of word $w_{i,adj}$ with the j'th standard word: $w_{i,adj}^{j,+}$ in the standard word set $A_{i,adj}^{+}$, so as the $D_{i,adj}^{j,-}$ is the distance of word $w_{i,adj}$ with the $w_{i,adj}^{j,-}$.

- Given verb word in the i'th domain $w_{i,v}$, formula of the semantic orientation calculation of word $w_{i,v}$ is following:

$$Orient(w_{i,v}) = \sum_{j=1}^{M}(D_{i,v}^{j,+}) - \sum_{j=1}^{M}(D_{i,v}^{j,-}) \qquad (2)$$

In which, $D_{i,v}^{j,+}$ is the semantic distance of word $w_{i,v}$ with the j'th standard word: $w_{i,v}^{j,+}$ in the standard word set $A_{i,v}^{+}$, so as to the $D_{i,v}^{j,-}$.

The $Orient(w_{i,adj})$ and $Orient(w_{i,v})$ are semantic orientation calculation value of adjective word $w_{i,adj}$ and the verb word $w_{i,v}$ in the i'th domain respectively.

Semantic similarity of words means that the degree that in a context both two words can be used interchangeably and syntactic semantic structure isn't changed. Greater the semantic similarity degree of two words, closer their semantic orientation. Consequently word orientation is calculated by dynamic standard word set and its semantic similarity degree with words in standard set.

Semantic orientation of words under dynamic standard word set build on semantic similarity degree of words, so it is necessary for using form of structural to describe complicated semantic information. HowNet[8] is a common knowledge base, which takes concept which be expressed by Chinese word and English word as descriptive object and relationship between among concepts and concept's possessing attributes as major content. Therefore we compute word's semantic similarity degree founded on HowNet.

In this paper, the semantic similarity between words is calculated by the formula in Xia's method[9].

4   Experiment and Result Analysis

4.1   Selection of Dynamic Standard Word Set

As mentioned before, the dynamic standard word set is composed according to the word current in large scale database, and the page count of its query with domain keywords by search engine.

We select two domain comments from the: KTV and restaurant domain. And use "KTV" as KTV domain key word, and the "MEISHI" as the restaurant domain word. These comments come from the http://www.dianping.com/.

We select only one keyword in each domain. Finally, we got four standard word set for each domain, they are positive adjective standard word set $A_{i,adj}^{+}$, negative adjective set $A_{i,adj}^{-}$, positive verb set:$A_{i,v}^{+}$, negative verb set $A_{i,v}^{-}$.

So 4 standard word sets are established in this experiment. They are: (1) adjective standard word list in KTV domain; (2)verb standard word list in KTV domain; (3) adjective standard word list in restaurant domain; (4) verb standard word list in restaurant domain, each set has two child sets: positive word set and negative word set.

4.2 The Experiment Setting

In the paper, the test set for experiment is derived from positive (negative) emotion word list and positive (negative) appraisal word list provided by HowNet. They are divided into adjective test group and verb test group according to their part of speech and words without any semantic orientation on a part of speech are excluded. By filtrating and sorting, the test set for experiment contains 2619 positive adjectives, 2085 negative adjectives, 776 positive verbs and negative verbs, in total, 4704 adjectives, 1635 verbs, 6339 words in test set.

We construct two domain data sets from KTV comments and restaurant comments from Internet. Then we classified these comments into positive or negative comments artificially. There are 144 positive comments and 50 negative comments in KTV domain. In restaurant, 101 positive comments and 96 negative.

4.3 Word Orientation Based on Its Part of Speech

This experiment only calculation the word orientation without any sentence or domain information, so the standard word set used by here is only on the adjective and verb standard word set.

Dynamic part of speech standard set is used to analysis semantic orientation of words. The experimental results show the difference between the accurate rates of words semantic orientation on different standard word set, and the dynamic standard set outperform others in table1. The last row is average result.

Table 1 Accurate rate of word orientation based on different standard word set

	Standard word set 1	Standard word set 2	Dynamic Standard Word set
w_{adj}^{+}	99.92%	98.82%	92.78%
w_{adj}^{-}	80.34%	82.69%	90.36%
w_{v}^{+}	47.90%	64.73%	82.78%
w_{v}^{-}	78.40%	97.33%	90.22%
w_{adj}	55.78%	91.67%	91.71%
w_{v}	63.63%	81.55%	86.67%
$\overline{w}$	57.80%	89.06%	91.20%

In this experiment, zero is viewed as threshold, that is to say, the word, whose semantic orientation measurement is greater than zero, is considered as positive, otherwise negative.

We choose two fixed standard word sets to compare with dynamic standard word set. Standard words set 1(fixed) comes from Liu's[10] paper, which is obtained by sorting returning Hit number of word through logging in Google Search Engine; standard word set 2 by Yahoo Search Engine, and the dynamic standard word set proposed in this paper.

According to the experimental results, it is evident that, regardless of the method of semantic orientation calculation of words based on dynamic standard word set or fixed set, adjective words perform better than verb test set. That is because the number of "Sememe" in adjective semantic expression ("Meaning") is more than verb's in HowNet. Therefore, computing result of adjective semantic similarity is more accurate than verb's.

4.4   Dynamic Standard Word Set Used in Different Domain

Figure1 shows that the orientation accurate of the KTV and Restaurant comments based on different standard word set. In the figure 1, KTV is means KTV data set, restaurant means restaurant data set. The traditional standard word set is called Fix, and the dynamic standard word set is Dyn.

Clearly, the dynamic standard set performs better than the fixed standard set. This is for the dynamic standard word set can include more "detailed" information for word orientation.

Fig. 1 The accurate rate of KTV comments orientation analysis based on the dynamic word set is 71.36%, while 57.84% is the fixed standard set. And 67.08% vs. 39.54% is for the restaurant comments.

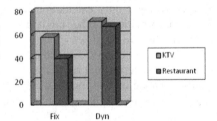

5   Conclusions

Word orientation is the bases technology on text orientation. In this paper, an algorithm of word orientation calculation is proposed based on the part of speech and domain information. In the experiment, a comparative study of the method of semantic orientation calculation of words based on different standard reveals that: the method of semantic orientation calculation of words based on dynamic standard word set may have more stable for word in different domain. The method based on dynamic word set performs better than that based on fixed word set. By contrasting the work on adjective test set and verb test set, it is found that: regardless of the method of semantic orientation calculation of words based on dynamic benchmark set or based on benchmark word set, they show more effectively on adjectives than verbs.

References

1. The 27th China Internet Development Statistics Report, http://www.chinaewindow.net/blog/
2. Hatzivassiloglou, V., McKeown, K.R.: Predicting the semantic orientation of adjectives. In: ACL 1998, Proceedings of the 35th Annual Meeting of the Association for Computational Linguistics and EACL 1997, pp. 174–181 (1998)
3. Turney, P.D.: Thumbs up or thumbs down? Semantic Orientation Applied to Unsupervised Classification of Reviews. In: Proceedings of the 40th Annual Meeting of the Association for Computational Linguistics, pp. 417–424 (2002)
4. Wang, G., Zhao, J.: Orientation analysis of Chinese word (in Chinese), http://nlpr-web.ia.ac.cn/2006papers/gnhy/nh4.pdf
5. Kamps, J., Marx, M., Mokken, R.J., Rijke, M.D.: Using WordNet to Measure Semantic Orientations of Adjectives. In: LREC 2004, Proceedings of the 4th International Conference on Language Resources and Evaluation, vol. IV, pp. 1115–1118 (2004)
6. Zhu, Y.-L., Min, J., Zhou, Y.-Q., Huang, X.-J., Wu, L.-D.: Semantic orientation computing based on HowNet. Journal of Chinese Information Processing 20(1), 14–20
7. Du, W., Tan, S., Yun, X., Cheng, X.: A new method to compute semantic orientation. Journal of Computer Research and Development 46(10), 1713–1720 (2009)
8. Dong, Z., Dong, Q.: HowNet, http://www.keenage.com
9. Xia, M., Sha, Y., Wang, X., Jiang, H.: Semantic orientation calculation of words with different part of speech. In: ICNC 2011, The 7th International Conference on Natural Computation, pp. 939–943 (2011)
10. Liu, Q., Li, S.: Word Similarity Computing Based on How-net. In: 3th Chinese Lexical Semantics Workshop (2002) (in Chinese)

A KNN Query Processing Algorithm over High-Dimensional Data Objects in P2P Systems

Baiyou Qiao, Linlin Ding, Yong Wei, and Xiaoyang Wang

Abstract. In this paper, we propose a quad-tree based distributed multi-dimensional data index (QDBI) for efficient data management and query processing in P2P system. In which, each peer use a MX-CIF quad-tree to generate index items of their high-dimensional data, and index items are organized into one dimensional ring structure, super-peers dynamically join the ring, the routing connections among super-peers are created, which are similar to Chord ring, thus forms a semantics-based structured super-peer network, and an efficient KNN query processing algorithm is proposed for high-dimensional data objects based on the index structure. Experiments show that the index structure and the KNN query algorithm have well search performance and scalability.

Keywords: KNN query, MX-CIF quad tree, High-dimensional data objects.

1 Introduction

Recently, P2P systems are used to manage and process high-dimensional data objects, such as text, images, videos, and how to process complex queries over these high-dimension data objects becomes an important research issue in P2P system. Many works have been done to process complex queries over high-dimensional data objects, such as range queries [1, 2, 7], indexing techniques [4, 8, 9, 10], K-Nearest Neighbor (KNN) query [3, 5, 6]. From the research status of high-dimensional data query processing, how to enhance query ability is the key problem. [3] proposed a framework which based on super-peer structure, and a NR-tree index was proposed to process multi-dimensional data objects, due to based on R*-tree, the maintaining cost is high. [4] Based on the proposed indexing structure SSW, implements the related KNN query algorithm, and proposes the concepts of Local Partition Tree (LPT) and Global Partition Tree (GPT). Each peer maintains a LPT about the partition course itself, so it may lead to root node bottleneck when the data volume increasing. [10] proposes a distributed quad-tree index (DQI) for P2P network, which adapts the MX-CIF quad-tree to index multi-dimensional data objects, but the quad-tree blocks are uniformly hashed into the

Baiyou Qiao · Linlin Ding · Yong Wei · Xiaoyang Wang
Information Science and Engineering School, Northeastern University, Shenyang, China
e-mail: Qiaobaiyou@ise.neu.edu.cn, linlin.neu@gmail.com

F.L. Gaol et al. (Eds.): Proc. of the 2011 2nd International Congress on CACS, AISC 144, pp. 133–139.
springerlink.com   © Springer-Verlag Berlin Heidelberg 2012

ring, this loosens the semantics of data objects, and makes routing cost increase. In this paper, we propose a quad-tree based distributed multi-dimensional data index (QDBI) for efficient complex query processing which is similar to DQI [10]. In QDBI, each peer use a MX-CIF quad-tree to generate index items of their multi-dimensional data, each index item acquires a code according to the MX-CIF quad-tree, all index items are organized into one dimensional ring according to their code, and super-peers dynamically join the ring according to the requirement, the routing connections among super-peers are created, which is similar to Chord ring, and thus forms a semantics-based structured super-peer network. Based on QDBI structure, an efficient KNN query processing algorithm is proposed for high-dimensional data objects. Experiments on synthetic data sets demonstrate that the proposed index structure and the KNN query processing algorithms have well search performance and scalability.

2   Quad-Tree Based Distributed Index

Now we first introduce the system structure, which consists of two layers, the bottom layer is composed of peers and the top layer is composed of super-peers, super-peers are connected into a ring overlay. System uses MX-CIF Quad-tree to describe data objects and generates index data. The whole quad tree consti-tutes the data semantic space, and its sub quad tree represents the corresponding semantic subspace. Each Super-peer and the connected peer together form a se-mantic cluster, each semantic cluster only maintains data and index items in its semantic subspace, and is responsible for answering queries fell into its seman-tic subspace. In a cluster, data is stored in the peers; index items and routing in-formation are maintained by the Super-peer. Peers within a cluster communicate directly; communications between clusters are forwarded by Super-peers. Peers may join the different semantic clusters according to their containing data ob-jects. Queries are only sent to the related semantic clusters according to their semantics, which can significantly reduce the number of peers and messages the query involved. The system combines efficient routing mechanism of structured P2P networks and the manageability of centralized network, so it has good search performance.

2.1   MX-CIF Quad-Tree Based Index

MX-CIF quad-tree structure has been widely used in spatial database and game programming, which is capable of facilitating spatial data query processing. It starts with a big rectangle. The rectangle is the root of MX-CIF quad-tree and is divided into four congruent sub rectangles. The four congruent rectangles are the children of the original rectangle. In turn, each of these rectangles is divided into four until it reaches the terminal condition of the dividing course. Each rectangle represents the certain data range, which is uniquely identified by its centroid named a control point. In our system, for each data object o, the dividing course stops encountering a block b such that o overlaps at least two child blocks of b or upon reaching a maximum level of dividing course. Fig. 1 is an example of partitioning data space using MX-CIF quad-tree approach. In the figure, the two-dimensional square-shaped space is recursively divided into thirteen quad-tree blocks, and the multidimensional data A to I is distributed to the blocks. Each block is identified by a control point; the control point is associated with a super-peer and presents the block, the super-peer is responsible for storing the index data. A quad tree node is a control point which is labeled with a code. The root code is "0". In each dividing course, control point codes of the top right, the top left, lower left and lower right blocks are parent control point code respectively attach to the "0", "1", "2" and "3". Fig. 2 is the corresponding MX-CIF quad-tree of figure 1, we can see that the data C and I are stored at node 01, A and B are stored at nodes 00 and 003. Multi-dimensional data objects are stored at pees and the index data are logically stored at the quad tree nodes which are actually store at the super-peers associated with the tree nodes.

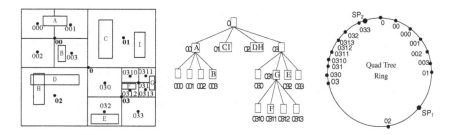

Fig. 1 An example of data space partition using MX-CIF quad-tree

Fig. 2 The corresponding MX-CIF quad-tree of Fig. 1

Fig. 3 The distribution of quad tree nodes showed in Fig. 2 in the ring

2.2   Distributing of Quad Tree Based Index

In a super-peer system, how to distribute index items to different super-peers is very important to improve system performance and implement efficient query processing. In the system, when a peer joins the network, it firstly builds a local MX-CIF quad-tree index to manage the data objects, and then generates the corresponding global index items. A global index item is a three tuple (C_O, C_C, P_{ID}), where C_O is the central point of the data object, C_C is the code of the control point which minimal containing the data object, P_{ID} is the peer's ID. All peers' global index items constitute the global quad tree index, and are distributed to the different super-peers; this forms the distributed quad tree index. In the system, the codes of quad tree node i.e. codes of quad tree blocks are use to implement the mapping of index items to the ring. Multiple quad tree nodes and hence quad tree blocks are mapped to the ring according to their value of code, and index items can be stored with each quad tree node. Super-peers join the corresponding position in the ring to store quad tree nodes and corresponding index items according to the load balance requirement, and are responsible for all index queries associated with these quad tree nodes. Fig. 3 shows the distributing of quad tree nodes shown in figure 2 in the ring. From the figure 3, it can be seen that quad tree nodes are placed to the ring according to value of their codes, that is, index items are organized according to the quad tree, and stored in the ring in accordance with the depth-first traversal. So the order of index data in the ring is preserved. According to the quad tree nodes encoding, index belonging to the quad tree none-leaf node are always stored with its offspring nodes adjacent in the ring. This keeps better semantic relationships between index data. When processing a query, as long as find the quad tree node with respect to the query, it can easily obtain the index items stored in its descendant nodes along the ring, which can greatly reduce the transformation cost to search, thereby enhance the overall network performance.

3   KNN Query Processing Algorithm

In the system, each super-peer stores connections to other super-peers, through these connections, network connectivity and search routing services are provided. Each super-peer maintains two connections to the predecessor super-peer and the successor super-peer; it also maintains a routing table which stores the connections to other neighboring super-peers which is similar to Chord ring. The routing connections of a super-peer is established when the super-peer joins the system and the routing mechanism is also resemble to the Chord ring. For paper size constrains, the detail description of routing connections and routing mechanism is omitted.

For point query and range query processing is relative simple, here we only give the KNN query processing algorithm. A KNN query is denoted as $Q(o, k)$, which returns the k nearest data to the point o, here o is the query point. When a super-peer receives a KNN query, it firstly routing the query to the super-peer which storing its associative control point (quad tree block), and then the query is continuously processed by the super-peers. For the index items is stored in the ring in clockwise according to their code value of its control points, we can get the whole answers from the super-peer to other several super-peers along the ring in Counterclockwise. The idea is that the query starts at a small range and enlarges the search space accordingly. In each search space, we find the k nearest data to the query point and replace the old data with a nearer data at all search space. The whole course stops when nearest data number is greater than K and not gets more nearer data longer. The KNN query processing algorithm is shown in Algorithm 1. In each dividing course, a query region is generated with key as the center and distance between key and the dividing center as radius to obtain the nearest k results of the query.

Algorithm 1. KNN query Processing algorithm.			
Input: Q(O,K): KNN query, O is query point and K:is the number of KNN query results;			
SP: current super-peer;			
Output: Rset: KNN results set			

1:	If O ⊄ Sp.index-range then	10:	D ← the distance from O to the control
2:	Forwards Q(O,K) to a neighbor super-		point Ci, which contains query point O;
peer		11:	C← the circle region with O as the cen-
	whose distance to O is shortest;	ter	
3:	End if		and D as the radius;
4:	If received the query Q(O,K) before then	12:	Replace data objects in Rset with data
5:	Drop Q(O,K);		objects which are nearest to query point O in
6:	End if	C;	
7:	Rset←Null	13:	If |Rset| =k then
8:	L ←the quad tree level of point O lies in;	14:	exit
9:	For (i = L, i>0, i--) do	15:	End if
		16:	End for

4 Performance Evaluation

In order to verify validity of the proposed quad-tree based distributed index (QBDI) and the KNN query processing algorithm, we evaluate the performance by simulating experiment which is based on the two type synthetic data set. In the experiments, all operations are run on Bamboo DHT, which is DHT based efficient open source software. The two synthetic data set are based on uniform distribution and Zipf-like distribution respectively; each data set consists of 50K to 100K data objects which randomly generated in [0, 1] and has 2 to 5 attributes. For simplicity, we assume each peer only generates one index item and each super-peer only stores not great than 100 index items, so the number of index items on super-peers is the number of peers. We compare our work with the representation distributed quad-tree index (DQI)[10] from the index scalability and KNN query performance. The detail comparison results are as following.

1) Scalability of QBDI. To evaluate the scalability of QBDI, the number of peers is varying from 256 to 16384, peers satisfy Zip-file like distribution, and 100 2NN queries are sent randomly from each super-peer. The query path length and maintaining cost is used to evaluation

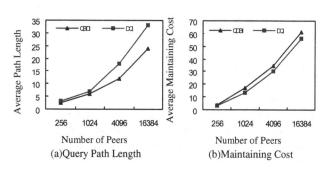

(a)Query Path Length   (b)Maintaining Cost

Fig. 4 Scalability Comparison of index structure

scalability of QBDI; query path length is the logical hops from the issuing query super-peer to the destination super-peers. The query path length of the two distributed index is showed in Fig. 4(a). It can be seen the query path length of QBDI is

smaller than that of DQI, the reason is that in QBDI quad-tree nodes are adjacently stored in the ring according to their semantics, this realizes the semantic cluster
for the index items and thus reduces the search hops. Fig. 4(b) gives the maintaining cost of both QBDI and DQI. The maintaining cost proportional to the number
of indexes and neighbors maintained at each super-peer. In QBDI, the neighbors
of each super-peer are increasing with the peers joining, so the cost should bigger
than DQI, but the test result shows that the maintaining cost of QBDI is approximate to that of DQI. Above experimental results show the QBDI has good
scalability.

2) KNN Query
Performance. KNN
query is widely
used as a complex
query. For a KNN
query, the value of
K affect the query
processing per-
formance directly,
a large K implies
more results to re-
trieve and hence
more nodes to vis-
it. So we use

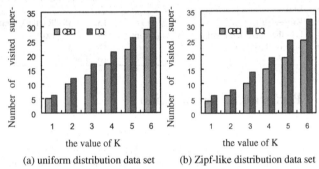

(a) uniform distribution data set   (b) Zipf-like distribution data set

Fig. 5 Search Cost vs. KNN query size

search cost to evaluate the proposed KNN algorithm performance, the search cost
is measured by the average number of visited super-peers. Fig. 5 shows the search
cost varying with K value varying, which is based on two type data set consists of
80K data objects. From the Fig. 5(a), we can see the number of visited super-peers
of the two algorithms increase slowly with K value increasing, and the number of
visited super-peers of the KNN query processing based on DQI is bigger than that
based on QBDI. This is because In QBDI, quad-tree nodes having similar semantics are adjacently stored in the ring, and each super-peer has more neighbors, this
can reduce the query search hops, and thus reduce the number of visited
super-peers. In DQI, the neighbors of each super-peer is relative less, and quad-
tree nodes are uniformly hashed into the ring, this loosens the semantics of data
objects, and thus makes routing cost and the number of visited super-peers
increase. Fig. 5(b) is the result based on Zipf-like (α=0.8) distribution data set,
it can be seen that the number of visited super-peers per query slowly
increase with k value increasing almost in consistent with that of Fig. 5(a), this
shows the algorithm is not sensitive to data sets. From the above experimental
results, it can be seen that the KNN algorithm based on QBDI has good query
performance.

5   Conclusions

In P2P systems, to efficiently process complex query over high-dimensional data
objects becomes very important. But current P2P systems cannot support complex

queries over high-dimensional data objects efficiently, which seriously limits their practical value. In this paper, we propose a quad-tree based distributed multi-dimensional data index (QBDI), in which each peer uses MX-CIF quad-tree to generate index items of their multi-dimensional data, all index items are organized into one dimensional ring, and forms a semantics-based structured super-peer network. Based on QDBI structure, an efficient KNN query algorithm is proposed. Experiments on synthetic data sets show that the QDBI structure and the proposed KNN query processing algorithms have good scalability and performance.

Acknowledgment. The research was supported by the National Natural Science Foundation of China (No. 61073063), the fundamental research funds for the central universities (No. N090404012).

References

1. Bharambe, A.R., Agrawal, M., Seshan, S.: Mercury: supporting scalable multi-attribute range queries. In: Proc. of SIGCOMM, pp. 353–366 (2004)
2. Jagadish, H.V., Ooi, B.C., Vu, Q.H.: Baton: A balanced tree structure for peer-to-peer networks. In: Proc. of VLDB, pp. 661–672 (2005)
3. Liu, B., Lee, W.-C., Lee, D.L.: Suppporting complex multi-dimensional queries in p2p systems. In: Proc. of ICDCS, pp. 155–164 (2005)
4. Li, M., Lee, W.-C., Sivasubramaniam, A.: Semantic Small World: An Overlay Network for Peer-to-Peer Search. In: Proc. of ICNP, pp. 238–248 (2004)
5. Li, M., Lee, W.-C., Sivasubramaniam, A., Zhao, J.: Supporting K nearest neighbors query on high-dimensional data in P2P systems. In: Proc. of Front. Comput. Sci., China, pp. 234–247 (2008)
6. Jagadish, H.V., Ooi, B.C., Vu, Q.H., Zhang, R., Zhou, A.: Vbi-tree: A peer-to-peer framework for supporting multi-dimensional indexing. In: Proc. of ICDE, pp. 34–43 (2006)
7. Chen, H., Jin, H., Wang, J., Chen, L., Liu, Y., Ni, L.: Efficient Multi-keyword Search over P2P Web. In: Proc. of WWW, pp. 989–997 (2008)
8. Crainiceanu, A., Linga, P., Machanavajjhala, A., Gehrke, J., Shanmugasundaram, J.: P-Ring: An Efficient and Robust P2P Range Index Structure. In: Proc. of SIGMOD, pp. 223–234 (2007)
9. Tang, Y., Zhou, S., Xu, J.: LigHT: A Query-Efficient yet Low-Maintenance Indexing Scheme over DHTs. In: Proc. of TKDE, pp. 59–75 (2010)
10. Tanin, E., Harwood, A., Samet, H.: Using a distributed quadtree index in peer-to-peer networks. The VLDB Journal 16(2), 165–178 (2007)

Encouragement of Defining Moderate Semantics for Artifacts of Enterprise Architecture

Shoichi Morimoto

Abstract. Enterprise Architecture (EA) is a method for the holistic optimization of organizational resources, including enterprise strategic directions, infrastructures, and facilities of information technology. It is effective to manage enterprise resources and decision-making with documentation of the whole of enterprises. However, EA provides no notation and methodology for the documentation; it leaves the notation to discretion of each enterprise. Moreover, EA requires various artifacts for the management. These trigger off difficulty of creation and maintenance of the documentation. Therefore, we propose formal semantics of the artifacts which is not swayed by each notation and does not obstruct unfettered selection of the modeling methodology. The semantics can provide rigorous traceability of the artifacts for any notation, thus efficiency of the maintenance will improve at a tremendous rate.

1 Introduction

Enterprise Architecture (EA) is a key method of optimizing interrelationships among business environments and information technologies. In the government areas, it has been common to implement EA for optimal effects through business processes well within an information technology environment. EA has two sides; both a management program and a documentation method that together provides an actionable, coordinated view of an enterprise's strategic direction, business services, information flows, and resource utilization [1].

An enterprise which implements EA needs to model enterprise components, e.g., strategic goals and initiatives; business products and services; information flows, knowledge warehouses, and data objects; information systems, software applications, enterprise resource programs, and web sites; voice, data, and video networks; and supporting infrastructure including buildings, server rooms, wiring runs/closets, and capital equipment. The modeled component is called an EA artifact, which

Shoichi Morimoto
Senshu University, Kanagawa, 214-8580, Japan
e-mail: morimo@isc.senshu-u.ac.jp

F.L. Gaol et al. (Eds.): Proc. of the 2011 2nd International Congress on CACS, AISC 144, pp. 141–149.
springerlink.com

is a documentation product, such as a text document, diagram, spreadsheet, brief slides, or video clip [1]. The artifact is both a plan/program and a bird's-eye view which are needed by an enterprise to close an existing performance gap or support a new strategic initiative, operational requirement, or technology solution. When the components are modeled, then it can visually manage the direction of the enterprise.

However, EA is a concept of the implementation and a guideline of the artifacts; it does not provide concrete modeling methods and notations for the artifacts. An enterprise must select the method and notation for each artifact. An enterprise can select their skillful method and notation, e.g., BSC (Balanced Score Card), SWOT (Strength, Weakness, Opportunity, Threat) analysis, text scenarios, UML (Unified Modeling Language), BPMN (Business Process Modeling Notation), DFD (Data Flow Diagram), ERD (Entity Relationship Diagram), their original notation, and so on. Thus, there are no unified modeling procedures and rules.

This liberal selection is a demerit as well as a merit. The notation is not fixed, so that semantics for the artifacts and the relationships among them cannot be defined. When an artifact of EA has to be modified, it is difficult to know how the correction will have an effect on the other artifacts. That is, the maintenance of the artifacts consumes much time and effort.

The primary purpose of EA is not modeling enterprise components rigorously; it is that such constant activities take root in the enterprise. However, the semantics is important, because traceability which is derived from semantics of modeling notation is actually effective for the correction, such as the case of UML. Therefore, this paper proposes formal, but *moderate*, common semantics for artifacts of EA, which can be utilized by any notation using in EA. The definition helps to mitigate the labor of the maintenance. Moreover, it allows artifacts to be formally verified and validated. Consequently, it vitalizes the implementation of EA.

2   Related Works

There are some definitions of semantics about EA or modeling notation.

2.1   EA Framework

The EA documentation framework identifies the scope of the architecture to be documented and establishes relationships between the architecture's areas [1]. The best-known EA framework is the Zachman framework, which is said to be the origin of EA. The framework is a schema for classifying and organizing the topics related to managing the enterprise, as well as to the design, development, and manifestation of the enterprise [6]. The Zachman framework is organized as 36 cells arranged in a six-by-six matrix—six rows and six columns (Fig. 1). It is a two-dimensional schema, used to organize the detailed representations of the enterprise. Each cell defines a fundamental piece of the enterprise resources. If such elements are distinguished, the functioning enterprise will be working appropriately. Each

Aspects / Perspectives	What (Data)	How (Process)	Where (Network)	Who (Role)	When (Timing)	Why (Motivation)
Scope (Planner)	List of things important to the business	List of processes the business performs	List of locations in which the business operates	List of business responsibilities	List of events significant to the business	List of business goals/strategy
Business model (Owner)	Semantic model	Business process model	Logistics network	Work flow model	Master schedule	Business plan
System model (Designer)	Logical data model	Application architecture	Distributed system architecture	Human interface architecture	Processing structure	Business rule model
Technology model (Builder)	Physical data model	System design	Configuration design	Presentation architecture	Control structure	Rule design
Detailed representations (Subcontractor)	Data definition	Program	Network architecture	Security architecture	Timing definition	Rule specification
Functioning enterprise	Data	Process	Network	Organization	Schedule	Strategy

Fig. 1 The Zachman framework

element is modeled in some notation and the modeled element corresponds to an artifact.

The Zachman framework describes definitions of each cell. The original cell definitions found in Appendix C of [8]. The definitions describe the following contents of each cell: Definition Description, Sample Diagram, Measurements, Use of the Model, Techniques for Modeling, Tools, Content Owners (Subject Matter Experts, Data Stewards), Definition Owner (Metamodel Owner), Metamodel, Custodian, Primary Users, Development Consultant (Modeling Expert), and Reference Materials. However, the definitions are written in natural language and they do not give any fixed notation and semantics to each artifact. Therefore, it is difficult to establish formal and rigorous traceability among the artifacts.

2.2 Formalizing EA Framework

Some efforts have tried to introduce formalism into EA. Reference [4] presents a formalism for the structure and semantics of the Zachman framework. Their ultimate goal is to formalize the framework development process; formalizing the artifacts produced by such development is an essential precursor for the process. In the paper, the authors have formalized only the Zachman framework's structure. The paper has said that the framework structure is always defined in terms of two fixed sets, R and I, and a domain of discourse D. For $i \in I$ or $r \in R$, they have written of "column i" or "row r", respectively. R is a fixed set of roles and I is a fixed set of interrogatives. Typical vocabularies used for R are {"Planner", "Owner", "Designer", "Builder", "Subconstractor"} and for I are {"what","how","where","who","when","why"}. D is a finite set, the domain. When they decomposed a frame, they first formed a grid of cells indexed by $R \times I$ and then designate components of a cell according to values in D. Thus each frame is connected to its subframes by triples of the form $\langle r,i,d \rangle$, which they have called edge labels. A succession of edges forms a path and thus a

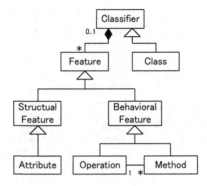

Fig. 2 UML metamodel of the class diagram

sequence of edge labels is a path label $L\{\langle r,i,d\rangle : r \in R, i \in I$, and $d \in D\}$. They have also defined a finite set of frames, paths for a framework, frame details, their abbreviations, annotations, indirections, and constraints. For further details see the paper.

2.3 UML Semantics

The UML is a modeling language, "and the way the world models not only application structure, behavior, and architecture, but also business process and data structure."[1] The objective of UML is to provide system architects, software engineers, and software developers with tools for analysis, design, and implementation of software-based systems as well as for modeling business and similar processes [5]. One of the primary goals of UML is to advance the state of the industry by enabling object visual modeling tool interoperability. UML is often used for modeling EA artifacts.

UML is the unified notation which includes many diagrams representing various aspects and structures. Each diagram is intricately intertwined with the others. Thus, in order to distinguish and regulate them and their relationships, the metamodel has been defined. It describes the constituents of all well-formed models that may be represented in the UML, using the UML itself. Fig. 2 shows the semantics of parts in the class diagram. It is possible to define the relationships (i.e., traceability) formally by the metamodel (i.e., UML semantics). The traceability can help the maintenance of masses of UML diagrams.

3 The Semantics

The original of the Zachman framework does not regard formalism as important, because it is not suitable for describing and understanding the cells [7]. It has also said

---

[1] http://www.uml.org/

that few formalisms are available for some cells. On the other side, formal semantics and traceability of models are essential in the fields of the model-based development or model-driven architecture. Therefore, we propose a formal semantics which is not too rigorous and never attenuates the merits of EA.

EA artifacts which must be implemented are specified, nevertheless there is no fixed notation for them. Therefore, before defining the semantics, first we have surveyed various EA frameworks and their artifacts: the Zachman framework, RM-ODP, EABOK, FEAF, TEAF, DoDAF, TOGAF, and so on. Next we have found and abstracted common essential elements of EA artifacts based on the survey. Finally, we have defined the semantics from the elements by following the UML semantics example.

3.1 Common Elements of EA

The fact in common among all frameworks is to have layers classifying artifacts. If we categorize the layers roughly, it will be four types; business entities, data/information, applications/systems, and technology/infrastructure. Business entities include enterprise missions/visions, strategies/plans, conceptual business resources, business products/services, and business activities/processes. Data/information includes logical/physical enterprise data entities/objects, data relationships, processed or reorganized meaningful data (information), and established information (knowledge). Applications/systems include information systems, software functions/applications, databases, and web services. Technology/infrastructure includes IT solutions, networks, and hardware which implement the category of applications/systems. For the formalization, we have adopted these four categories of artifacts, named Business, Data, System, and Technology.

3.2 Abstract Syntax

We have given formal semantics to EA artifacts by following the examples [2] and [3] of UML semantics. Let *Artifact* denote the set of EA artifacts. An EA artifact $artf \in Artifact$ is a pair

$$artf = (name, Parts)$$

where

1. *name* $\subseteq$ *String* is the name/identifier of the artifact being drawn.
2. *Parts* $\subseteq$ *ArtifactParts* is the set of all parts in the artifact.

Strings are sequences of characters. They are enclosed in double quotation marks.

An artifact is composed by parts of a diagram in some notation. A part, an indivisible component of an artifact, $prt \in Parts$ is a pair

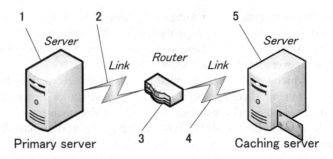

Fig. 3 Example 1: network diagram

$$prt = (text, rel)$$

where

1. *text* $\in$ *String* is the description on/in the part.
2. *rel* $\subseteq$ *Relationship* is the set of the other parts which relate to the part.

Relationship is a subset of *Parts*. Each part does not need to be named; however, for dealing with the syntax of artifacts, we find it simpler to assign it a number, e.g., 1, 2, etc.

Moreover, an artifact must belong to one of the four categories mentioned above. The categories are disjoint, respectively. Thus,

$$ArtifactParts = Business \cup Data \cup System \cup Technology$$

where

1. $Business \cap Data = \{\}$ 4. $Data \cap System = \{\}$
2. $Business \cap System = \{\}$ 5. $Data \cap Technology = \{\}$
3. $Business \cap Technology = \{\}$ 6. $System \cap Technology = \{\}$

The four sets (*Business*, *Data*, *System*, and *Technology*) bring an advantage to our proposal. The semantics is very flexible, because we have not given the details of definition to the sets. If the sets are instantiated, the semantics can formalize EA artifacts in any notation. We will demonstrate how to utilize the semantics in the next section.

3.3 Examples

To show the advantage of the semantics, we present two examples.

EA requires that networks, backbone routers/switches/hubs, equipment rooms, wiring closets, and cable plants should be described in detail. Fig. 3 is an example of this type of documentation. The diagram is composed by the symbols of three types (*Server*, *Link*, and *Router*). Moreover, the diagram is included in the Technology category. Therefore, the abstract syntax for this network diagram can be seen as follows:

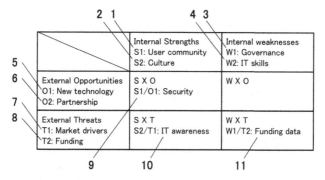

Fig. 4 Example 2: SWOT analysis

$Artifact = (name, Parts)$ $Link = \{2,4\}$
$name =$ "Network Diagram" $Router = \{3\}$
$Parts = \{1,2,3,4,5\}$ $1 = ($"Primary server", $\{2\})$
$Business = \{\}$ $2 = (none, \{1,3\})$
$Data = \{\}$ $3 = (none, \{2,4\})$
$System = \{\}$ $4 = (none, \{3,5\})$
$Technology = \{Server, Link, Router\}$ $5 = ($"Caching server", $\{4\})$
$Server = \{1,5\}$

Moreover, we can define the following constrains.

1. $Server \cap Link = \{\}$
2. $Server \cap Router = \{\}$
3. $Link \cap Router = \{\}$
4. $|\forall srv \in Server.Relationship| \geq 1$
5. $|\forall rtr \in Router.Relationship| \geq 1$

6. $|\forall ln \in Link.Relationship| = 2$
7. $\forall srv \in Server.Relationship \subseteq Link$
8. $\forall rtr \in Router.Relationship \subseteq Link$
9. $\forall ln \in Link.Relationship \subseteq Server \cup Router$

Server, *Link*, and *Router* are disjoint. The symbols of *Server* and *Router* must relate to another symbol at least. The symbol of *Link* has always two symbols which relate to it. The symbols of *Server* and *Router* can relate only to the symbol of *Link*. The symbol of *Link* can relate to the symbols of *Server* or *Router*.

The second example is SWOT analysis. The SWOT analysis is one of the activities the enterprise performs in developing a strategic plan. The SWOT analysis assesses both internal and external aspects of doing the enterprise business, which is often presented in a matrix form. Fig. 4 is an example of the matrix visualizing the analysis. The matrix is composed by eight cells and they are included in the Business category. Therefore, the abstract syntax for this matrix can be seen as follows:

$Artifact = (name, Parts)$ $\quad\quad\quad\quad S \times O = \{9\}$

$name = $ "SWOT Analysis" $\quad\quad\quad W \times O = \{\}$

$Parts = \{1,2,3,4,5,6,7,8,9,10,11\}$ $\quad S \times T = \{10\}$

$Business = \{InternalStrengths,$ $\quad\quad W \times T = \{11\}$

$\quad\quad\quad\quad InternalWeaknesses,$ $\quad\quad 1 = $ ("User community", $\{9\}$)

$\quad\quad\quad\quad ExternalOpportunities,$ $\quad 2 = $ ("Culture", $\{\}$)

$\quad\quad\quad\quad ExternalThreats,$ $\quad\quad 3 = $ ("Governance", $\{11\}$)

$\quad\quad\quad\quad S \times O, W \times O,$ $\quad\quad\quad 4 = $ ("IT skills", $\{3,5\}$)

$\quad\quad\quad\quad S \times T, W \times T\}$ $\quad\quad\quad 5 = $ ("New technology", $\{9\}$)

$Data = \{\}$ $\quad\quad\quad\quad\quad\quad\quad 6 = $ ("Partnership", $\{\}$)

$System = \{\}$ $\quad\quad\quad\quad\quad\quad 7 = $ ("Market drivers", $\{10\}$)

$Technology = \{\}$ $\quad\quad\quad\quad\quad 8 = $ ("Funding", $\{11\}$)

$InternalStrengths = \{1,2\}$ $\quad\quad 9 = $ ("Security", $\{1,5\}$)

$InternalWeaknesses = \{3,4\}$ $\quad 10 = $ ("IT awareness", $\{2,8\}$)

$ExternalOpportunities = \{5,6\}$ $\quad 11 = $ ("Funding data", $\{3,8\}$)

$ExternalThreats = \{7,8\}$

Moreover, we can define the following constrains.

1. $InternalStrengths, InternalWeaknesses, ExternalOpportunities, ExternalThre$-$ats, S \times O, W \times O, S \times T$, and $W \times T$ are disjoint.
2. $|\forall so \in S \times O.Relationship| \geq 2$
3. $|\forall wo \in W \times O.Relationship| \geq 2$
4. $|\forall st \in S \times T.Relationship| \geq 2$
5. $|\forall wt \in W \times T.Relationship| \geq 2$
6. $\forall so \in S \times O.Relationship \subseteq InternalStrengths \cup ExternalOpportunities$
7. $\forall wo \in W \times O.Relationship \subseteq InternalWeaknesses \cup ExternalOpportunities$
8. $\forall st \in S \times T.Relationship \subseteq InternalStrengths \cup ExternalThreats$
9. $\forall wt \in W \times T.Relationship \subseteq InternalWeaknesses \cup ExternalThreats$

The descriptions of $S \times O$ are primitive factors for a strategic plan which seizes one or more descriptions of $ExternalOpportunities$ by one or more descriptions of $InternalStrengths$. Similarly, the descriptions of $W \times O$ are the factors for a strategic plan which prevents a loss of one or more descriptions of $ExternalOpportunities$ caused by one or more descriptions of $InternalWeaknesses$. The descriptions of $S \times T$ are the factors for a strategic plan which overcomes one or more descriptions of $ExternalThreats$ by one or more descriptions of $InternalStrengths$. The descriptions of $W \times T$ are the factors for a strategic plan which prevents a serious situation caused by a combination of one or more descriptions of $InternalWeaknesses$ and one or more descriptions of $ExternalThreats$.

Such syntax of the examples can help to formally verify consistency of the artifact itself. The syntax is not dependent on the types of artifacts and notations, thus it can describe any artifact in any notation. Moreover, the correspondence of the parts in an EA artifact between the parts in the other artifacts can be formally defined by the syntax, i.e., traceability. For instance, a strategy for the security ($S \times O$) in Fig. 4

may be achieved by components of the network in Fig. 3. This constraint can be easily derived by the syntax.

4 Conclusion

In this paper, we have proposed a flexible semantics for EA artifacts, based on EA frameworks and the UML semantics. First we have surveyed EA frameworks and elicited common essential elements of EA artifacts. Next we have imitated the UML semantics and defined the moderately formal semantics for EA artifacts.

The semantics can formalize any artifact in any notation and create traceability of the artifacts. We have also shown the examples using our proposal. The semantics never loses the original advantages of EA and can cope with both feasibility and formalism. It excludes ambiguities or omissions in the artifacts, thus it can formally verify whether the artifacts satisfy their constraints and define flexibly new constraints. It leads to mitigation of the load of the activities of EA. Consequently, implementation of EA becomes easy.

We are validating the syntax by applying it to further practical descriptions and verifications of various artifacts now.

References

1. Bernard, S.A.: An Introduction to Enterprise Architecture, 2nd edn. AuthorHouse, Bloomington (2010)
2. Crane, M.L., Dingel, J., Diskin, Z.: Semantics of UML Class Diagrams: Abstract Syntax and Mapping to System Model. STL: UML 2 Semantics Project, Queen's University (2006)
3. Dingel, J., Crane, M.L., Diskin, Z.: Semantics of UML Activity Diagrams: Abstract Syntax and Mapping to System Model. STL: UML 2 Semantics Project, Queen's University (2006)
4. Martin, R., Robertson, E.L.: Formalization of multi-level Zachman frameworks. Department of Computer Science, Indiana University Technical Report, No. 522 (1999)
5. Object Management Group, Inc., OMG Unified Modeling Language (OMG UML), Infrastructure Version 2.4 (2010)
6. O'Rourke, C., Fishman, N., Selkow, W.: Enterprise Architecture Using the Zachman Framework. Course Technology, Florence (2003)
7. Sowa, J.F., Zachman, J.A.: Extending and formalizing the framework for information systems architecture. IBM Systems Journal 31(3), 590–616 (1992)
8. Zachman, J.A.: The Zachman Framework for Enterprise Architecture: Primer for Enterprise Engineering and Manufacturing. Zachman Framework Associates, Toronto (2005)

Development of PC-Based Radar Signal Simulation Test System

Li Xiaochun, Ding Qingxin, and Tian Fei

Abstract. In allusion to the composition of radar signal simulation test system (RSSTS) for the control of multiple instruments based on PC, this paper, based on the construction of RSSTS, proposed the principle for the achievement of instrument driver library(IDL) and instrument softpanel(ISP), and presented a PC-based test system(PCBTS) that adopted a distributed approach for integrating a variety of instrument software; in addition, developed IDL, ISP, test report modules and various instrument integration software. Finally, an experiment based on aforesaid principles were selected to conduct a comprehensive evaluation for the PCBTS, and the results showed that this test system could fully control the physical test system(PTS) in a PC; could control the PTS to achieve a variety of radar signal simulations and outputs; and could support the report feature of test results.

Keywords: radar signal simulation, PC-based test system, instrument driver library, softpanel, IQ modulation.

1 Introduction

A test system composed of multiple electronic measuring instruments is generally installed in a fixed cabinet, and thus the process for the tester to measure is quite intricate. A variety of electronic measuring instruments are generally support GPIB programmed control or LAN remote control, therefore, it is possible to control the test system composed of multiple instruments by virtue of a PC-based test system [1]. As a consequence, the whole testing process is available in a single configured PC, and the analysis of test results and the preparation of test reports can be integrated in the corresponding test system.

   In general, radar signal is derived from the modulation of radio frequency (RF) carrier signal by pulse baseband signal. The frequency range of a RF signal generator developed by some company is 250kHz ~ 6GHz, covering the commonly-used RF band, which could provide high-performance complex analogue simulation signals for radar test [2]. This paper utilizes this signal generator and

Li Xiaochun · Ding Qingxin · Tian Fei
Faculty of Mechanical and Transportation Engineering, China University of Petroleum (Beijing), 102249. Beijing, China
e-mail: ammelichun@yahoo.com.cn, dindqx@263.net, tfsofei@yahoo.com.cn

F.L. Gaol et al. (Eds.): Proc. of the 2011 2nd International Congress on CACS, AISC 144, pp. 151–156.
springerlink.com   © Springer-Verlag Berlin Heidelberg 2012

arbitrary waveform generator and real-time spectrum analyzer to comprise a radar signal simulation test system. In the meanwhile, a PC-based test system was developed for the control of the whole physical test system. All the instrument software mainly include three parts: instrument driver library, softpanel software, and test report modules, as well as the system integration of various instrument software composed of this test system, namely, PC-based test system.

2   Principle of Test System

2.1   Overview of Signal Generator

Signal generator is the signal source of excitation for electronic measurement. I/Q modulation and baseband will be described as follows:

1. I/Q Modulation

I/Q are the abbreviation of In-phase/Quadrature, in which, I component represents the In-phase components, and Q component represents the quadrature components, and these two components are orthogonal and unrelated. When I and Q are represented X axis and Y axis in the Cartesian coordinate of vector signals respectively, it will form the planisphere of vector signal [2].

2. Baseband

Baseband is the abbreviation of inherent frequency bandwidth given by the original source signals, and the base-band signal is the original electrical signal issued by the signal source. Baseband signal is usually featured in lower frequency, which could support the effective transmission over long distances via radio upon being modulated by carrier signal [2].

2.2   Introduction to Arbitrary Waveform Generator and Real-Time Spectrum Analyzer

Arbitrary waveform generator can be used to excite or simulate test signal, and its main functions include: (1) simulate complex actual signal; (2) generate complex waveform set by user; and (3) the waveform can be modified. Real-time spectrum analyzer has the feature of simultaneous measurement of amplitude and phase, and could measure the relation between magnitude, phase versus frequency, and could measure the relation between magnitude, phase versus time.

2.3   Components of Test System

The test system consists of signal generator, arbitrary waveform generator, real-time spectrum analyzer and a variety of connection cables and other components. The main functions of the test system are as follows: (1) signal generator is the self-contained complete excitation source of acquisition instrument, forming a complete measurement solution; (2) arbitrary waveform generator can generate arbitrary baseband signals, used for the baseband modulation signal for generating radar signal of the signal generator, or the signal generator utilizes its internal baseband signal generator to generate baseband modulation signal; and (3) real-time spectrum analyzer is applied   for the analysis of the previous generated radar signals. The structure diagram of test system is shown in Figure 1.

3   Principles of Instrument Driver Library

In general, all instruments meet VXIplug & play (VPP) specifications can be completely controlled by the computer, and individual functions can only be programmed by computer. With respect to the program-controlled of various instruments, it must firstly select the appropriate GP-IB command statement [3], and then add specific instrument (SCPI, analyzer) program-controlled command codes in the selected statement, in order to achieve the specified function.

Instrument driver program is a set of subroutines that can be invoked by the user library. Instrument driver library(IDL) is to encapsulate the programmed commands of the instrument one by one as the interface functions in the dynamic link library(DLL), when the DLL is called by the softpanel software, it will be able to exchange data with instruments, thereby achieving PC program-controlled equipment. IDL is available to be realized by IVI standard DLL developed by LabWindows/CVI 9.0, and the structure diagram of the IDL is shown in Figure 2.

4   Principles of PC-Based Test System

4.1   Components of Softpanel Software

Softpanel software is the integration of instrument softpanel and instrument driver library, used for human-computer interaction of operators, setting the instrument parameters of various menus and viewing the instrument current work status. Thus, softpanel software is the master control software of the physical test system [4]. For example, the softpanel of the signal generator is shown in Figure 3. Instrument softpanel is to control the instrument by virtue of calling the instrument driver library, and to display the current state of the instrument, and this approach is to program-control or remote control the instrument. Instrument softpanel is developed by MFC project created by VS2005.Net. For instance, with regard to the

Arbitrary waveform generator   RF signal generator (I/Q modulation)   Radar signal output   Real-time spectrum analyzer

Fig. 1 Structure Diagram of the Test System

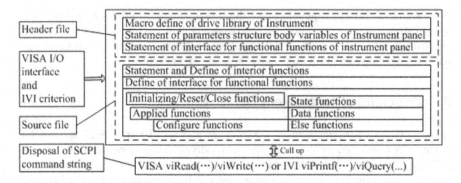

Fig. 2 Structure diagram of Instrument Driver Library

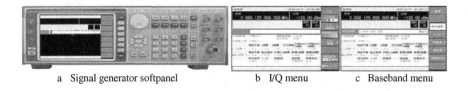

    a   Signal generator softpanel   b   I/Q menu   c   Baseband menu

Fig. 3 I/Q and Baseband Menu Interface Bitmap

signal generator, The drawing area of softpanel in Figure 3a is used to draw the menu bitmap of button operation in the softpanel, including frequency menu, power menu, I/Q menu, baseband menu and so forth, in which, I/Q and digital baseband menu interface bitmaps are shown in Figure 3b and Figure 3c.

4.2 Components of PC-Based Test System

In the PC-based test system(PCBTS), various instrument software are mainly divided into three parts: instrument driver library(IDL), softpanel software(SPS) and test report modules, as well as distributed system integration for the various instrument software that composed of the test system [5]. The testing process is as follows: (1) set various parameters in a variety of instrument soft panel interfaces, and control the instruments by virtue of the corresponding instrument driver library; (2) control the instrument for testing and obtain the test results; and (3) Test report module will automatically analyze the test results, forming the test data report. The structure diagram of PCBTS is shown in Figure 4, and its main functions are as follows:

- The PCBTS is able to programmed-control or remote control a variety of instruments in test system. Therefore, during the testing, it only needs to operate the PCBTS software, which can basically do not operate the instrument.
- The test report module can analyze various test results and present statistics, and generate test data report automatically.

- The SPS could control the instrument through the IDL, in general, the IDL supports the same type of instruments, and hence the PCBTS can be ported to a test system composed of the same type of instruments.
- Various instrument software are independent, when the test system composes of diverse instruments, it just needs to integrate the relevant instrument software and then could achieve the appropriate PCBTS.

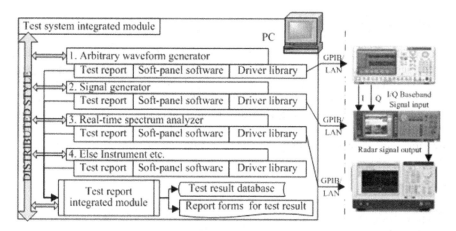

Fig. 4 Structure Diagram of PC-Based Test System

5   Vector Precision Test Experiments of Test System

Set parameters in the softpanel interface(SPI) of the arbitrary waveform generator to generate the baseband signal, and input I/Q signals into the signal generator through the output channel; Set parameters in the SPI of the signal generator to modulate the base-band signal, and then input the modulated signal into the real-time spectrum analyzer; Set parameters in the SPI of the real-time spectrum analyzer to analyze the input signal, and then generate the test report by virtue of the test report module. The test results of real-time spectrum analyzer are shown in Table 1, the analysis chart versus 2100.0 MHz is shown in Figure 5, and the test report module can automatically formulate the data of Table 1 and Figure 5 into a test report. As shown in Table 1 and Figure 5, the vector accuracy of the physical test system is higher, which can meet the system requirements.

Table 1 Test Results of Real-Time Spectrum Analyzer

Set/Measured Parameters	Parameter Measurement Value			
Set Frequency dots f/MHz	1000.0	2100.0	3900.0	5400.0
Error vector magnitude (EVM) v/%(RMS)	0.374	0.365	0.387	0.368
Amplitude error ΔA/%(RMS)	0.198	0.220	0.219	0.213
Phase error ΔP/°(RMS)	0.182	0.167	0.183	0.172

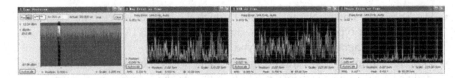

Fig. 5 Analysis diagram versus the frequency of 2100.0 MHz

6 Conclusions

The conclusions are as follows: (1) The PC-based test system(PBTS) is able to fully control a variety of instruments composed of the whole physical test system(PTS) on a PC; (2) The PBTS is able to control the PTS to achieve a variety of radar signal simulation and output; (3) The PBTS is able to support report feature of test results; and (4) The PBTS adopts distributed approach to integrate a variety of instrument software, which could be easily extended to other test systems, and the same type of instrument may share the same type of instrument software.

References

1. Li, X., Ding, Q., Tian, F.: Development of PC-based virtual radio frequency signal generator. In: ICISE 2010 (2010), doi: 10.1109/ICISE, 5688702
2. Zhang, H., Chen, S.: Fundamentals of modern communication systems, 2nd edn. Higher education press, Beijing (2009) (in Chinese)
3. Jiménez, F.J., De Frutos, J.: Virtual instrument for measurement, processing data, and visualization of vibration patterns of piezoelectric devices. Computer Standards and Interfaces 27(6), 653–663 (2005)
4. Cao, Y., Qian, G., Tang, H.: Software design of dense radar signal simulator. Journal of UEST of China 34(5), 622–625 (2005) (in Chinese)
5. Zhi, C., Gao, Y., Li, X.: A distributed system for aircraft hydraulic system simulation test based on LXI. In: IEEE AUTOTESTCON (2008), doi:10.1109/AUTEST.2008.4662609

Decision Support System for Agribusiness Investment as e-Government Service Using Computable General Equilibrium Model

Arif Imam Suroso and Arief Ramadhan

Abstract. Investment is one activity that is quite tricky. Investment activity should be based on accurate calculations. Government as the regulator of investment activity within a country, can build an e-Government services in the form of a Decision Support System (DSS) to assist the investment decision-making. Investment activity in one sector can influence other sectors. One model that can be used to see the influence is Computable General Equilibrium (CGE) model. Some softwares has been available to perform complex CGE calculations. However, these softwares are generally not easy to use and not attractive. This paper presents the results of research in creating a system called Decision Support System for Agribusiness Investment (DSSAI). This system can perform complex CGE calculations, easy to use and has an attractive interface. These systems can be used as a benchmark for the user to determine whether he can invest in one of agribusiness sector or not. Users can perform a simulation by changing one variable, then see the effect on some sectors in agribusiness. The results of the DSSAI simulation are suitable with the opinion and the experience of economic experts, especially with the expert in agribusiness investment.

Keywords: Decision Support System, e-Government, Computable General Equilibrium, Agribusiness, Investment.

1 Introduction

Investing in a country is generally governed by a special government agency. Along with the development of information technology these days, the government agency can build an e-Government system to improve its services.

e-Government is the use of Information Technology (IT) by public sector organizations [1]. e-Government is different from ordinary information system

Arif Imam Suroso
Faculty of Economy and Management, Bogor Agricultural University, Bogor, Indonesia
e-mail: arifimamsuroso@ipb.ac.id

Arief Ramadhan
Faculty of Computer Science, University of Indonesia, Depok, Indonesia
e-mail: arief.ramadhan@ui.ac.id

F.L. Gaol et al. (Eds.): Proc. of the 2011 2nd International Congress on CACS, AISC 144, pp. 157–162.
springerlink.com   © Springer-Verlag Berlin Heidelberg 2012

that is generally targeting the private sector. The main orientation of e-Government is the accessibility of information by the public sector, rather than income [1]. Currently, e-Government specifically appears in various paradigms, such as e-Democracy, e-Participation, e-Service etc. In this paper, we will propose one form of e-Service or e-Government service, that is e-Government service in the areas of agribusiness investment.

Increased growth in agriculture-based business (agribusiness) is one of priority directions of development in agricultural country, for example in Indonesia. The growth of agribusiness activities can be driven by sustainable capital investment. There are so many sectors, in the field of agribusiness or associated with it, that can be used as investment objects, such as rice, soybeans, corn and others. Investment in one sector, will affect other sectors, so it requires the right economic model to observe the general influence. Computable General Equilibrium (CGE) can be considered as a model for doing so.

Since the early 1960s, CGE models have been used to analyze an enormous variety of issues ranging from the effects of agricultural and trade policies to tax policies, environmental policies, regional policies, etc [2]. The applied CGE model has been widely used in both developing and the developed countries [3]. CGE is a macroeconomic model that integrates the microeconomics and macroeconomics [7]. This model are generally non-stochastic and strongly nonlinear [4].

CGE models have many advantages. These models have the advantage of taking into account crowding-out effects on other sectors [5]. CGE models combine some of the advantages of econometric and IO models, strengthening the theoretical basis of the modeling effort and enabling examination of a wider set of policy issues [6]. The structural CGE models are built with the basics of the theory of microeconomics in which the behavior of economic agents and the details specifically set forth in the form of the equation system [7].

There are so many equations that will be used in CGE models. In order to simplify the calculation of CGE models, it would require special software to do that. One of the commonly used software is General Equilibrium Modelling Package (GEMPACK) developed by Monash University and based on RunGEM interface. The data processing in GEMPACK is quite complicated, because the software is in the form of a software suite that consists of several programs. In addition, the output (the simulation results) of GEMPACK are still produced in a format that not easy to be interpreted and can only be read by GEMPACK. Other common software that usually being used by people or economic researcher, such as Microsoft Excel, can't read directly the simulation result of GEMPACK. Therefore, it is necessary to build other software that is easy to use and have an attractive interface. As noted in [8], that the software interface is the most important in computer systems. In this case, the new software should make it easier for users to read data, perform processing, update the data and do the simulation, as well as reading the simulation results.

In this paper, we will propose a new e-Government system that can calculate CGE, easy to use and have an attractive interface. This system is built in the form of Decision Support System (DSS) and it is intended to use in agribusiness investment decision. This system is called Decision Support System for Agriculture

Investment (DSSAI). This system can be used as a benchmark for the user to determine whether he can invest in one of agribusiness sector or not. Users can perform a simulation by changing one variable, such as the tax rate, then look at its effect on some sectors in agribusiness.

2 Research Methodology

This research is conducted in Indonesia, so that the data and the stakeholders involved are the data and stakeholders that are related to agribusiness investment in Indonesia. In addition, we also use a simple methodology to built the DSSAI. This methodology was adopted from the linear sequential model that was proposed in [9]. There are four phase in this methodology, i.e. system analysis, system design, system implementation and system testing.

To conduct the system analysis and system design phases, we engage several experts and academicians in the field of economics. Several stakeholders in the field of agribusiness investment are also involved in this phase. One major stakeholder that is involved is the Indonesian Investment Coordinating Board.

System implementation is performed based on the results of system analysis phase and system design phase. Because of the target of e-Government is the public sector, then DSSAI will be implemented based on web technology in order to reach its users quickly and widely.

Several experts and stakeholders later re-engaged in the system testing phase. The things being tested are functionality, reliability, and usability of DSSAI. In addition, the simulation results from DSSAI are compared with the economic expert opinion and the economic expert experience.

3 System Analysis and Design

First, we define some key principles to be achieved in the development of DSSAI. Some of the key principles are: (1) the system is built to support users in making decisions about investments in the field of agribusiness, (2) the system must be easy to use, (3) the data that is used in the system should be able to be manipulated quickly, (4) users should be able to change some values of existing CGE variables in the system, (5) users should be able to perform simulations using the system, (6) the results of the system, should be presented in a form that can be easily understood by users.

The ability to change the variable value is a very important principle. This conforms to what was raised in [4], that the results obtained by simulating CGE model rely on several assumptions, pertaining both to the behavior of agents and to the choice of exogenous variables (the "closure" of the model). In [4], it is also being stated that the values assigned to the parameters of the behavioral functions, which underlie the "calibration" of the model, are also crucial.

Second, we determine the business architecture in DSSAI using the Business Use-Case Diagram. This diagram shows the relationship between the users (actors) with DSSAI. As implied in [10], there are three actors who can engage in the use of an e-Government system, i.e. Government (G), Business (B) and Citizen (C). The key actor who seems to be the most widely using DSSAI is the business, especially business people in the field of agribusiness. Therefore, G2B relationship will be the most powerful relationship in DSSAI.

Third, we define the data and the model for DSSAI. In [4] it is stated that the nature and quality of the available data also affect the results, whether these are base-year data in static models (e.g., the reference year in the social-accounting matrix) or the stationary equilibrium in dynamic models. There are 5 key data are used in DSSAI, i.e.: Social Accounting Matrix (SAM), Armingthon & CET Elasticity, Premier Input Substitution Elasticity, Labor Substitution Elasticity, and Household Expenditure Elasticity. In accordance with one of principles prescribed in the first stage, that the data should be easily accessible, then the five data is designed to be stored in Microsoft Excel files. Therefore, the data administrator as an internal user, that is an expert in econometrics, can change the five data easily. Besides using the data stored in Microsoft Excel files, DSSAI also using CGE models. The model was built and will be run using the General Algebraic Modeling Software System (GAMS) application. GAMS is selected because the software is very reliable in calculating econometrics, especially the CGE.

Data and models are linked by a centralized system that also the DSSAI interface. External users, such as business or citizen, are associated with this centralized system without being connected directly to the data or the models. When the external user wants to change the data, he can simply enter it through a centralized system. When the external user wants to run the model, then he simply order it through the facilities that available within the centralized system. This kind of configuration makes it easy for external users in using the system.

Fourth, we determined the application architecture of DSSAI. This application architecture is illustrated using the Application Communication Diagram as shown in Fig. 1.

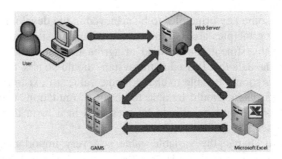

Fig. 1 Application Communication Diagram of DSSAI.

4 System Implementation

Besides using Microsoft Excel to store the data and using GAMS to create and run the model, there are several other softwares associated with DSSAI. First, DSSAI need a Windows 32 bit operating system. This occurs because the version of GAMS that being used is a version that can only be run on the Windows operating system. Second, DSSAI is developed using the triad that commonly being used in the web system development, i.e. PHP, Apache and MySQL. The reasons for this three, is because all of them are open source, thus saving the cost of implementation. We also use AJAX and CSS in the program code. Third, DSSAI require Adobe Acrobat Reader software, so users can open .pdf file format that being provided in the system. And lastly, DSSAI users must have a web browser to run it.

5 System Testing

The whole functions in DSSAI have been tested using a black-box testing method. The testing was done by involving several experts and stakeholders in agribusiness investment. Some of the end-user of DSSAI also being involved in this testing phase.

There are three regions of the Capital Investment Coordinating Board in Indonesia, which became the location of test, i.e. North Sumatra, North Sulawesi and Bali. Within each region, there are 20 end-users involved as a tester. The things being tested are functionality, reliability, and usability of DSSAI.

Based on the results of the testing, it can be seen that all of the functions that exist in DSSAI have been built successfully, can run well and can meet the needs of end-user, especially the business and citizens that concern in the agribusiness investment. The testers really liked the interface of DSSAI, and the system can run smooth without any failure or error.

In addition, the simulation results of the system also being compared with the opinion and the experience from some economic experts. It can be seen that the simulation results of the system were quite suitable with the opinion and the experience of economic experts. For example, when users simulate to raise the value added tax for agroindustry sector, it will increase the value of Domestic Income of Non-Governmental from several household categories.

6 Discussion and Future Works

On the other CGE processing software, the user should really know the architecture inside the software, particularly in terms of file storage. Users must know where the input data files must be saved. The input data files must also be stored with specific extensions. In addition to the input data files, some other files also need to be prepared to run the simulation. However, in DSSAI, users simply dealing with a single interface. Users do not have to mess with the system architecture. User simply interacts with the interface and run it all quickly and easily. Even, novice user who has never used similar software, would be able to use DSSAI directly and easily.

In this research, it is shown that DSSAI can be created in a web-based technology, but the GAMS, as part of the system, not support simultaneous execution yet.

So, it still need further research about how to perform the CGE model execution simultaneously.

DSSAI would be meaningless if it is not supported by good and clear governance. Therefore, DSSAI must become an integrated part of the government organization or government institution that associated with the investment policy. A set of legal aspects can be prepared to support DSSAI, so that DSSAI can actually being used as a decision support in agribusiness investment.

7 Conclusion

DSSAI has successfully being established as one of the e-Government service to assist user in deciding the investment in agribusiness. DSSAI can perform complex CGE calculations, easy to use and has an attractive interface. The results of the DSSAI simulation are suitable with the opinion and the experience of economic experts, especially with the expert in agribusiness investment.

References

1. Heeks, R.: Implementing and Managing eGovernment An International Text. SAGE Publications, London (2006)
2. Mohora, M.C., Bayar, A.: Computable General Equilibrium Models for The Central and Eastern European EU Member States: A Survey. Romanian Journal of Economic Forecasting 1, 26–44 (2007)
3. Diao, X., Yeldan, E., Roe, T.L.: A Simple Dynamic Applied General Equilibrium Model of a Small Open Economy: Transitional Dynamics and Trade Policy. Journal of Economic Development 23(1), 77–101 (2008)
4. Abdelkhalek, T., Dufour, J.M.: Confidence Regions for Calibrated Parameters in Computable General Equilibrium Models. Annales D'economie et de Statistique (81), 1–31 (2006)
5. Martin, R.H.: Tourism Events: Impact on Imports. International Journal of Event Management Research 3(1), 15–28 (2007)
6. Waters, E.C., Holland, D.W., Weber, B.A.: Economic Impacts of a Property Tax Limitation: A Computable General Equilibrium Analysis of Oregon's Measure 5. Land Economics 73(1), 72–89 (1997)
7. PT. Primakelola Agribisnis Agroindustri. Final Report of The Formulation of Agro Industry Sector Investment Incentives Models (2009) (unpublished)
8. Galitz, W.O.: The Essential Guide to User Interface Design An Introduction to GUI Design Principles and Techniques. Wiley Publishing, Inc., Indiana (2007)
9. Pressman, R.S.: Software Engineering A Practitioner's Approach. Mc Graw-Hill, USA (2001)
10. Ramadhan, A., Sensuse, D.I.: e-Livestock as a New Paradigm in e-Government. In: Ramadhan, A., Sensuse, D.I. (eds.) Proceedings of the 3rd International Conference on Electrical Engineering and Informatics, vol. 1, pp. 1–4. IEEE Press (2011)

Web Security Testing Approaches: Comparison Framework

Fakhreldin T. Alssir and Moataz Ahmed

Abstract. Web applications security testing is becoming a highly challenging task. A number of approaches have been proposed to deal with such a challenge. However, up to date criteria that could be used to aid practitioners in selecting appropriate approaches suitable for their particular effort do not exist. In this paper we present a set of attributes to serve as criteria for classifying and comparing these approaches and provide such aid to practitioners. The set of attributes is also meant to guide researchers interested in proposing new security testing approaches. The paper discusses a number of representative approaches against the criteria.

1 Introduction

Web applications are becoming more complex. As more and more services and information are made available over the Internet and intranets, Web sites have become extraordinarily complex, while their security is often crucial to the success of businesses and organizations. Although traditional software testing is already a notoriously hard, time-consuming and expensive process, Web-site security testing presents even greater challenges. The most reliable method to ensure the quality of a piece of software is to proof the correct-ness of source code, e.g. through formal verification [28, 29]. Unfortunately, this approach is time consuming for a whole system. Only crucial parts of a system can be verified this way. In practice, only the uses of test cases are a way to improve the quality of a software-intensive system [18]. Software testing can be considered to have two aims. The primary aim is to prevent bugs from being introduced into code - prevention being the best medicine. The second is to discover those unprevented bugs, i.e. to indicate their symptoms and allow the infection to be cured. Software Testing is defined as the process of executing a program with the intent of finding errors [1]. Hence, a pair of input and its expected output, which is called a test case, is said to be successful if it succeeds to uncover errors, and not vice versa.

Fakhreldin T. Alssir · Moataz Ahmed
King Fahd University of Petroleum and Minerals, Dhahran 31261, Saudi Arabia, P.O. Box: 8652
email: fakhry@kfupm.edu.sa, moatza@kfupm.edu.sa

F.L. Gaol et al. (Eds.): Proc. of the 2011 2nd International Congress on CACS, AISC 144, pp. 163–169.
springerlink.com © Springer-Verlag Berlin Heidelberg 2012

2 Web Security Testing

Generally security testing is the process to determine that an information system protects data and maintains functionality as intended. The objective of security testing is to verify the effectiveness of the overall Web system defenses against undesired access of unauthorized users, as well as their capability to preserve system resources from improper uses, and to grant the access to authorized users to authorized services and resources. Taken into consideration this objective we can highlight two different types of security tests regarding to web applications:

Static Security Analysis: This type of test is kind of white box testing because the source code of the application analyze and inspected to find any possible security defects [8].

Dynamic Security Test: This category of test aims to find vulnerabilities by sending malicious requests, and investigating replies, mainly used when the source code is not available so we can consider it as sort of Black Test, in this case security testers are looking to the application from the attacker's point of view[12].To get the best results from the security testing and being more confidence about the web application safety, combination of both static and dynamic testing is recommended .The Open Web Application Security Project (OWASP) [9] defined the top 2 web application security risks for 2010 as:

Injection: such as SQL, OS, and LDAP injection, occur when untrusted data is sent to an interpreter as part of a command or query.

Cross-Site Scripting (XSS): XSS allows attackers to execute scripts in the victim's browser which can hijack user sessions, deface web sites, or redirect the user to malicious sites.

3 Comparison Framework

Shahriar et al. [31] presented a set of criteria's to compare automated security testing approaches , the selected approaches are from different types of applications utilities programs, network daemons, web scanners and web applications. Our proposed compression framework this specific for web applications , and we consider different criteria's that are not address by the compression framework [31], like the test case generation algorithm used and is the final result of the work is test data or test cases. In this paper we propose six criteria to compare web applications security testing works.

Covered Attacks: This is criterion is very important for selecting the approach to test for specific types of web applications security attacks.

Test case Generation Algorithm: This criterion describe the algorithm or the method used for generating test case, which is give an idea about the methods and algorithms used in automated security test cases generation.

White Box or Black Box (W/B Box): This criterion answer the question of will we need the web application source code or not during testing process.

Test Case or Test Data (TC/TD): This attribute determines the different output of the security testing work, is it generate test data or test case?

Source of Test Cases: This criterion identifies what artifacts of an application or environment are used for generating test cases. These include source code of the web applications, attacks databases, session data, mutation operators and perturbation operators.

Tool and Automation: One of the most important criteria to differentiate any work from another is how much automation is supported. we categories the approaches into three types which are fully automated, semi automates in a way that the tool doesn't cover the whole process and finally manual type which have no tool developed to assist the process at all.

4   Approaches Comparison

In this section the approaches are analyzed and brief description about each one is given also   we present the comparison between security testing approaches based on our set of attributes.   The list of considered approaches in our study is exhaustive, we gave attention to those works we considered significant and more recent as regards the subject under discussion.

Based on the above analysis of web application security testing approach, the primary observations of this study can be summarized as follows:

-The most tackled security vulnerabilities for web applications are cross site scripting (XSS), SQL injection (SQLIJ) and Buffer Overflow.
- The approaches tackled one attack are more accurate in term of number of reveling attacks comparing to the works claim that they are able to detect more than one attack.
-Most of the approaches are white box based, in which source code is needed to perform the testing process.
-Less work in the line of using heuristics search algorithms like genetic algorithm, hill climbing and simulated annealing, to search for adequate test cases.
-Most of the reviewed works are using kind of attacks databases.
-Not all web security vulnerabilities are tacked from software testing perspective, for example insecure direct object references, cross-site request forgery and security misconfiguration.

Table 1 Approaches Comparison

Approach	Attacks	Generation Algorithm	Tool/ Automat- ion	W/B Box	TD/ TC	Source of Test cases
[11] 2010	XSS SQLIJ	Perturbation based Algorithm.	Fully automated	White	TC	Perturbing regular ex- pressions.
[25] 2010	XSS	Genetic Algorithm.	Fully automated	White	TC	URL
[13] 2009	XSS SQLIJ	N/A	Manually.	White	TD	Source code
[17] 2009	XSS	N/A, they use attacks Database.	Semi- automated.	White	TC	Attacks Pool
[10] 2008	XSS SQLIJ	It combines concrete and symbolic execu- tion to covers paths.	Fully automated	White	TD	Source code and attacks database.
[23] 2008	XSS	N/A , test data derived from the recorded old user sessions.	Fully automated	White	TD	User session
[29] 2008	Buffer Overflow	N/A, they use attacks database.	Semi- automated	White	TC	Attacks Pool
[24] 2008	SQLIJ	N/A , they use attacks database	Semi- automated	White	TC	Attacks Pool
[28] 2007	SQLIJ	Attacks database is used to build URL requests	Fully automated	Black	TC	Http request
[20] 2007	SQLIJ	N/A, the work pre- sented model based framework to generate test cases.	Fully automated	White	TC	Source code
[26] 2006	XSS SQLIJ	Attacks database	Fully automated	Black	TC	Source code
[19] 2005	XSS SQLIJ Buffer Overflow	Attacks database.	Semi- automate.	White	TD	Source code
[16] 2005	XSS SQLIJ	Completion algorithm [16].	Fully automated	Black	TC	Fault database
[22] 2004	XSS SQLIJ	N/A,	Fully automated	Black	TC	Response pages
[27] 2004	XSS SQLIJ	N/A ,attacks database	Fully automated	Black	TD	Response pages

5   Conclusion

In this paper we presented an attribute-based framework to allow classifying and comparing approaches for web application security testing. The results of the study provides practitioners with an overview of prominent work in the literature and offer help with regard to making decisions as which approach would be appropriate. Moreover, this study is meant to guide researchers interested in developing new approaches. Also we identify some open issues for future work. As a follow-up to the work presented in this study, the authors are currently working on analyzing the possibility of using heuristics search algorithms such as genetic algorithm for automatically generating test data for web application security testing.

Acknowledgment. The authors wish to acknowledge King Fahd University of Petroleum and Minerals (KFUPM) for utilizing the various facilities in carrying out this research.

References

[1] Whittaker, J.A.: What is software testing? And why is it so hard? IEEE Software 17(1), 70–79 (2000)

[2] Ahmed, M.A., Hermadi, I.: GA-based multiple paths test data generator. Computers & Operations Research 35(10) (2008)

[3] Myers, G.J.: The art of software testing. Wiley, New York (2004)

[4] Di Lucca, G.A., Fasolino, A.R.: Testing Web-based applications: The state of the art and future trends. Information and Software Technology 48(1) (2006)

[5] Ricca, F., Tonella, P.: Web testing: a roadmap for the empirical research. In: IEEE International Symposium, pp. 63–70 (2005)

[6] IEEE Std. 610.12-1990. Glossary of Software Engineering Terminology. In: Software Engineering Standard Collection. IEEE CS Press, Los Alamitos (1990)

[7] Nguyen, H.Q.: Testing Applications on the Web: Test Planning for Internet-Based Systems. John Wiley & Sons, Inc. (2000)

[8] Chess, B., McGraw, G.: Static analysis for security. In: Security & Privacy, vol. 2(6), pp. 76–79. IEEE (November-December 2004)

[9] The Open Web Application Security Project,
http://www.owasp.org/index.php/Category:OWASP_Top_Ten_Pro
ject

[10] Kieżun, A., Guo, P.J., Jayaraman, K., Ernst, M.D.: Automatic creation of SQL injection and cross-site scripting attacks. MIT Computer Science and Artificial Intelligence Laboratory technical report, Cambridge, MA (September 2008)

[11] Li, N., Xie, T., et al.: Perturbation-based user-input-validation testing of web applications. Journal of Systems and Software 83(11), 2263–2274 (2010)

[12] Stytz, M.R., Banks, S.B.: Dynamic software security testing. In: Security & Privacy, vol. 4(3), pp. 77–79. IEEE (2006)

[13] Tian, H., Xu, J., Lian, K., Zhang, Y.: Research on strong-association rule based web application vulnerability detection. Computer Science and Information Technology, 237–241 (2009)

[14] Akhawe, D., Barth, A., Lam, P.E., Mitchell, J., Song, D.: Towards a Formal Foundation of Web Security. In: Computer Security Foundations Symposium, pp. 290–304. IEEE (2010)

[15] Shi, H.-Z., Chen, B., Yu, L.: Analysis of Web Security Comprehensive Evaluation Tools. Networks Security Wireless Communications and Trusted Computing 1, 285–289 (2010)

[16] Huang, Y.-W., Tsai, C.-H., et al.: A testing framework for Web application security assessment. Computer Networks 48(5), 739–761 (2005)

[17] Shahriar, H., Zulkernine, M.: MUTEC: Mutation-based testing of Cross Site Scripting. Software Engineering for Secure Systems, 47–53 (2009)

[18] Kurshan, R.: Formal Verification in a Commercial Setting. In: Proceedings of the 34th Annual Conference on Design Automation, New York, vol. 00, pp. 258–262 (2007)

[19] Tappenden, A., Beatty, P., Miller, J., Geras, A., Smith, M.: Agile Security Testing of Web-based Systems via HTTPUnit. In: Proceedings of Agile Development Conference (ADC), Denver, Colorad, pp. 29–38 (2005)

[20] Salas, P., Krishnan, Ross, K.J.: Model-Based Security Vulnerability Testing. In: Proceedings of Australian Software Engineering Conference, Australia, pp. 284–296 (2007)

[21] Eaton, C., Memon, A.M.: Advances in Web Testing. In: Advances in Computers, vol. 75(Computer Performance Issues), pp. 281–306. Elsevier (2009)

[22] Offutt, J., Wu, Y., Du, X., Huang, H.: Bypass Testing of Web Applications. In: Proceedings of the 15th Symposium on Software Reliability Engineering, France, pp. 187–197 (2004)

[23] Mcallister, S., Kirda, E., Kruegel, C.: Leveraging User Interactions for In-Depth Testing of Web Applications. In: Proceedings of the 11th Symposium on Recent Advances in Intrusion Detection, Massachusetts, USA, pp. 191–210 (2008)

[24] Shahriar, H., Zulkernine, M.: MUSIC: Mutation-based SQL Injection Vulnerability Checking. In: Proceedings of the Eighth International Conference on Quality Software (QSIC2008), pp. 77–86. IEEE CS Press, London (2008)

[25] Avancini, A., Ceccato, M.: Towards security testing with taint analysis and genetic algorithms. In: Proceedings of the 2010 ICSE Workshop on Software Engineering for Secure Systems, pp. 65–71. ACM, Cape Town (2010)

[26] Kals, S., Krida, E., Kruegel, C., Jovanovic, N.: SecuBat: A Web Vulnerability Scanner. In: Proceedings of the 15th International Conference on World Wide Web, Edinburgh, Scotland, May 2006, pp. 247–256 (2006)

[27] Huang, Y.-W., Tsai, C.-H.: Non-detrimental Web application security scanning. In: 15th International Symposium on Software Reliability Engineering, ISSRE 2004, November 2-5, pp. 219–230 (2004)

[28] Kosuga, Y., Kono, K., Hanaoka, M., Hishiyama, M., Takahama, Y.: Sania: Syntactic and Semantic Analysis for Automated Testing against SQL Injection. In: Proceedings of the 23rd Annual Computer Security Applications Conference, Miami, December 2007, pp. 107–117 (2007)

[29] Shahriar, H., Zulkernine, M.: Mutation-based Testing of Buffer Overflow Vulnerabilities. To appear in the Proceedings of the Second International Workshop on Security in Software Engineering (IWSSE 2008), pp. 979–984. IEEE CS Press, Turku (2008)

[30] Gold, R.: HTTPUnit, http://httpunit.sourceforge.net/

[31] Shahriar, H., Zulkernine, M.: Automatic Testing of Program Security Vulnerabilities. In: 33rd Annual IEEE International Computer Software and Applications Conference, COMPSAC 2009, July 20-24, vol. 2, pp. 550–555 (2009)

[32] WAVE - Web Accessibility Evaluation Tool, http://wave.webaim.org/

An Evolutionary Model of Enterprise Involvement in OSS: Understanding the Dynamism in the Emerging Strategic Engineering Dimension of OSS

Toshihiko Yamakami

Abstract. The size of open source software (OSS) projects has grown to the scale of an entire platform. Complete coverage of the platform facilitates and increases the opportunity for a new paradigm to drive an OSS project, as the strategic direction of an IT company. The author presents case studies of enterprise involvement in OSS projects, such as MySQL, Eclipse, Android and Chrome OS. Then, the author examines the aspects that drive enterprise involvement in OSS. The author proposes a four-stage transition model from shifts of focus in regard to enterprise involvement in OSS. The author discusses the viewpoint of the strategic evolution of OSS from lessons learned in the past.

1 Introduction

After decades of evolution, OSS has reached a state that it can provide a complete platform for commercial solutions. Once a complete platform is provided, the openness of OSS fosters an open collaborative environment to harness a global, distributed development environment, utilizing transparently shared information assets.

This also enables an emerging new trend in OSS, a strategic evolution driven by a major IT company or a group of companies.

In this paper, the author discusses the evolution of strategic involvement in OSS by enterprises. Then, the author presents an evolutionary model of that strategic involvement by commercial enterprises.

Toshihiko Yamakami

ACCESS, Research and Business Development,

1-10-2 Nakase, Mihama-ku, Chiba-shi, Chiba-ken, Japan 261-0023

http://www.access-company.com

F.L. Gaol et al. (Eds.): Proc. of the 2011 2nd International Congress on CACS, AISC 144, pp. 171–179.
springerlink.com

2 Purpose and Related Work

2.1 Purpose of Research

The aim of this research is to identify and understand the increasing visibility of strategic involvement in OSS by enterprises, by examining evolutionary views of the relationship between OSS and strategies of enterprises.

2.2 Related Work

OSS was separated from the concept of free software in the late 1990's in order to revisit the commercial issues of using OSS. It is paradoxical to publish source code, the core competence of the software industry, so openly. Fitzgerald discussed the contradictions, paradoxes and tensions of OSS in [Fitzgerald and Agerfalk(2005)]. Raymond discussed open source from the business model perspective in this famous open source work series [Raymond(2000)].

OSS has continued to evolve. Watson presented the second generation of OSS, or professional OSS [Watson et al(2008)Watson, Boudreau, York, Greiner, and Donald Wynn] in contrast to the three types of first generation OSS: community OSS, sponsored OSS, and corporate distribution.

Letellier discussed the third generation of OSS [Letellier(2008)] from the perspective of its organizational structure.

The long-term factors of OSS have also attracted the attention of researchers. Subramaniam discussed success factors using longitudinal data on OSS projects [Subramaniam et al(2009)Subramaniam, Sen, and Nelson] and presented the impacts of different license types. Yu discussed time series analysis techniques to study the time dependence of open-source software activities using mailing lists, bug reports, and revision history [Yu et al(2009)Yu, Ramaswamy, Lenin, and Narasimhan] and presented diversity in cyclic-ness and in seasonal dependency.

As the size of OSS software has grown, organizational governance has emerged. Examples include the Eclipse Foundation [The Eclipse Foundation(2004)] and Apache Software Foundation [The Apache Software Foundation(1999)]. There are new industrial organizations emerging for industry-specific software: for example, mobile handset software-related foundations including the OHA [Open Handset Alliance(2007)].

Alexy presented the impacts of OSS into intra-organization aspects [Alexy and Henkel(2009)].

I presented multiple views of the generations of OSS in order to understand the diverse evolutions of OSS [Yamakami(2011)].

The originality of this paper lies in its discussion of relationship between OSS and involvement by enterprises.

3 Open Source-Related Landscape

3.1 Relationship between OSS and Enterprises in Retrospect

Augustin discussed generations of OSS dating back to 1974 from the viewpoint of stacks in OSS. Augustin presented a 5-generation view, depicted in Fig. 1. The first generation of OSS consists of games distributed via mailing lists. The second generation consists of tools for development environments. The third generation is the OS (operating system). The fourth generation consists of infrastructure elements such as databases and web server scripting languages. The fifth generation is applications. It is a generational analysis by domains or completion of computing stacks.

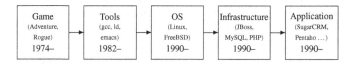

Fig. 1 Augustin's 5 generation-view

This presents the process by which OSS has enhanced its coverage of the building blocks of software. The size of OSS projects has reached a size that is comparable to any large-scale proprietary middleware or application. The quality of OSS has reached a level that large-scale commercial solutions can depend on.

Watson presented a view in terms of a business model. Watson discussed the emergence of professional OSS, as the second generation of OSS. The two generations of OSS are depicted in Fig. 2.

The second generation that emerges is professional OSS, in which companies contribute their assets to open source and explore a wide range of business models based around that. Full-time employees are engaged in OSS to leverage their business models.

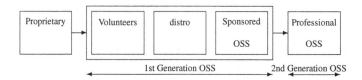

Fig. 2 Watson's 2 generation-view

The author has a sense that this increased interaction between OSS and business was similar to that between the Internet and business in the first half of the 1990s. In the case of the Internet, it was gradually recognized that a fusion with business was the way that would lead the Internet to its full potential. The author believes that a similar conclusion will be drawn in the case of OSS.

Fig. 3 3-generation view of diffusion using OSS.

OSS facilitates the unrestricted distribution of software code. This is used to leverage diffusion. The generations of diffusion using OSS are depicted in Fig. 3.

The first generation accommodates the diffusion of open standards. The implementation of open standards in OSS leverages the acceptance of a standard. This also fits OSS because open standards provide clear requirements, which eliminates overhead in OSS development with less ambiguity compared to other types of software.

The second generation leverages the diffusion of shared platforms. Splitting software into a shared platform and plug-ins provides efficient development of software as long as the architecture is properly designed. Eclipse is one example. When a platform is based on OSS, it provides transparency of governance, neutrality of delivery control, and public participation. It also facilitates open distribution of technical information.

The third generation leverages unbundling and unlocking. The digital economy accelerates the unbundling of digital content. Examples include PHP and database infrastructures.

As features are extended and communities grow with enhancements in the IT infrastructure, the volume of code simply continues to grow. This exposes OSS projects to the challenges of large-scale software development. In order to understand these challenges, the following generations are observed during the evolution of large-scale OSS, depicted in Fig. 4.

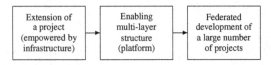

Fig. 4 3-generation view of large-scale software development.

In the first generation, a community grows and extends a number of small projects in the community. The IT infrastructure enables the management of larger-scale of software projects. Accumulation of community experience, project management experience, and the increased capabilities of development environments enable the development of large-scale software.

In the second generation, the idea of developing an entire software package within the community is abandoned. The architecture and ecosystem to enable further software development in relation to third-party software and corporations are developed. One example is the separation of platform and plug-in components. The shared platform is developed and maintained by the community. Each corporation can develop their own plug-in for its purposes, including business purposes. Examples include Eclipse.

In the third generation, a foundation is established to govern a large number of projects that are loosely connected. Each project is isolated in terms of functions and project management. The foundation has a higher level of orchestration. Examples include the Apache software foundation and the GNOME foundation.

As OSS has improved its presence and visibility in the software industry, it has increased opportunities for corporate involvement.

The generations of corporate involvement are depicted in Fig. 5.

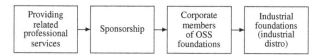

Fig. 5 4-generation view of corporate engagement.

In the first generation, companies provide related professional services. Examples include distro companies. A distro is a software package consisting of a specific software bundle, or even an entire operating system with included software, already compiled and configured. It is generally the closest thing to a turn key form of OSS. Each OSS component continues to evolve in beta releases. Therefore, it is necessary for some companies to provide professional skills to ensure a turn key OSS solution.

In the next generation, companies provide sponsorship. When Linux became mature, many large enterprises discovered the value of a free version of Unix and invested their full-time employees in the maintenance of OSS. Eclipse was created by IBM, however IBM opened this asset for the shared maintenance of OSS, because the IDE itself was not considered a core competency of their business.

In the third generation, OSS communities recognized the values of corporate involvement. Some foundations created corporate memberships, such as Advisory boards. This involvement created a revenue stream for OSS foundations for sustained growth. It enabled many large-scale corporations to secure a strategic position in OSS foundations.

In the fourth generation, corporate involvement lead to a foundation for specific industrial purposes. MeeGo and the Linux Foundation are two examples. This is something like a distro for a mobile handset middleware platform, including 100 million lines of OSS code.

These transition views are interesting and present a base for further discussion of the evolutionary development of corporate involvement in OSS.

3.2 Evolutionary View of Relationship between OSS and Enterprises

Examples of enterprise involvement in OSS are depicted in Table 1.

Table 1 Examples of enterprise involvement in OSS

Project	Enterprise	Description
MySQL	MySQL	Software owned by commercial companies, who seek to derive revenues from selling traditional software licenses, is made available under some OSS license. Dual licensing schemes enable the generation of revenue by the copyright-owner companies.
Eclipse	IBM	IBM invented Eclipse, a software development environment that consists of an integrated development environment (IDE) and extensible plug-ins. IBM coined an architecture where other players could extend the environment for their customized use, and released the software under an OSS license. IBM also helped create the Eclipse Foundation to act as a neutral driving force for the sustainable development and maintenance of the Eclipse platform.
Android	Google	Google purchased the Android company for its mobile software platform. Google initiated the Open Handset Alliance in order to strategically pursue the increased market share of Android in the mobile middleware market.
Chrome OS	Google	Google initiated the Chrome OS project to enable mainstream OS to use Chrome browser, in order to secure a leading position in embedded software for a wide range of digital appliances. Google created a press-release stating that it would be managed by OSS from scratch.

MySQL and Eclipse started in the 1990's. Android was initiated around 2003. Chrome OS was coined in 2010.

Driving global strategic endeavors requires openness because the costs to convince end users in global engagement are large. OSS provides this key openness from its origin.

Android became the leading middleware platform for mobile smartphones worldwide in late 2010. This exhibited the power of OSS projects as a business ecosystem enabler because the openness of Android and its third-party development friendliness are major drivers of Android success worldwide. The openness of the Android platform provides a basic scheme for governance by the Open Handset Alliance, the OSS-based foundation managing Android maintenance as mobile middleware.

An approach like Chrome OS is only possible through the shared understanding of social norms and conventions of OSS communities. Over the decade, the concept of OSS has matured and embodied the driving force of an ecosystem where many stakeholders are paying serious attention to this emerging platform that was built from scratch. It is based on a belief and expectation of stable OSS deliverables. This kind of belief can be built only through many successful OSS projects over decades.

These approaches are valid only when there is established trusts on both sides, depicted in Table 2.

The evolution of enterprise engagement can be viewed from two different aspects, depicted in Table 3.

The increasing importance of both aspects has continued to push enterprise engagement in many OSS projects.

Table 2 Required trusts as preconditions for industry-initiated strategy engineering in the domain of OSS

Side	Trust required
OSS community	Trust in Enterprise-engaged industrial OSS foundations. Trust in governance of enterprise in OSS-related activities.
Enterprise	Trust in quality and stability of OSS. Trust in governance and management of OSS communities.

Table 3 Two aspects of enterprise engagement in OSS in retrospect

Aspect	Description
OSS project as a building block for industrial solutions	The position of each OSS project or collection of OSS projects in an industrial solution.
OSS project as a business ecosystem enabler	The position of each OSS project or collection of OSS projects in a business ecosystem.

3.3 Factors Impacting Enterprise Involvement in OSS

Factors impacting enterprise involvement in OSS are shown in Table 4.

Table 4 Factors impacting enterprise involvement in OSS

Factor	Description
Quality	Stability and lack of bugs are important for commercial solutions.
Performance	Performance derived from OSS components is a major measure for commercial solutions.
Coverage	Coverage is the percentage of OSS components among the entire building blocks of a solution.
Completeness	Not just coverage, completeness is key for strategic involvement. Completeness facilitates a turn-key solution, which is important for selling the product to a wide range of customers.
Share Skills	Prevalence and availability of engineers with underlying technical know-how are important for long-term maintenance of large-scale solutions.

All of these measures have continued to grow in the last decades during OSS evolutionary history.

An evolutionary view of enterprise involvement in OSS is depicted in Fig. 6.

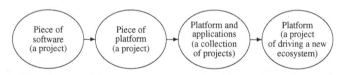

Fig. 6 Evolutionary view of enterprise involvement in OSS

4 Discussion

4.1 Advantages of the Proposed Approach

The proposed model presents a basic stage-awareness of the strategic involvement of enterprises in OSS. The awareness of these stages provides the players in OSS with the know-how to understand the strategic position by enterprise players. It also provides enterprise players an understanding of how they can utilize a collection of OSS projects in their global strategic engagements.

In past literature, the engagement of OSS has been captured from a viewpoint of voluntary sharing. In the era of million-line-code scales in OSS, this provides only a partial view of the engagement of enterprises in OSS.

The proposed model provides a perspective enabling enterprise engagement in large-scale OSS projects. This also helps large-scale OSS projects to be maintained in a healthy, economically-reasonable manner.

The proposed model provides a base for understanding the paradigm of open innovation. For example, 99 % of mobile handsets in Japan are 3G-enabled. Most of these 3G-enable mobile handsets are capable of handling video-phone features. However, the use rate of video-phone is very low even after a decade of diffusion. This indicates that the engineering challenges include not only the implementability or technical adoption, but also user acceptance and ecosystem acceptance.

4.2   Limitations

This is a descriptive study to identify the directions of enterprise engagement in OSS. The diversity of OSS projects makes any exhaustive research beyond the scope of this paper.

Quantitative verification of the evolution of OSS is beyond the scope of this paper. The comparisons are descriptive and lack any quantitative measures.

This is an exploratory work, therefore, in-depth analysis of the evolution of OSS remains for further research.

5   Conclusion

Even at an early stage, there was enterprise engagement in OSS projects. MySQL dated back to the middle of 1990's even when there was little acceptance of dual licenses in OSS.

It is known that Linux was driven by corporate employees such as IBM, Intel, HP and SGI towards commercial quality improvement. Corporate involvement was not welcome by some OSS projects, especially when OSS communities emphasized on absolutely-free sharing of project deliverables. Over the decades, OSS communities have gradually accepted the concept that the enterprise engagement provides more benefits than disadvantages. In parallel, many enterprises have captured the importance of OSS as the driving vehicle of open innovation.

As the amount of source code grows, more companies have started paying serious attention to OSS as part of their business ecosystem.

The author presents an analysis of the evolution of OSS from a viewpoint of enterprise engagement. The quality and completeness of OSS now provides a promising vehicle for driving the forces of strategic engineering, especially in global IT business. The author proposes a four-stage evolutionary model that explains the transitions of enterprise engagement in OSS for its strategic business development.

Today's OSS projects are important not only as software building blocks, but also as ecosystem building blocks. The era of open innovation promotes the need to understand OSS projects as a core part of business model engineering.

Trusts built upon many OSS projects over decades enable leading enterprise players to consider newly emerging OSS platforms as building blocks of global business ecosystems.

This aspect of OSS should be understood by many different stakeholders in OSS in order to minimize the conflicts between OSS communities and enterprise business model engineering.

References

[Alexy and Henkel(2009)] Alexy, O., Henkel, J.: Promoting the penguin: Intraorganizational implications of open innovation. Social Science Research Network (2009), http://papers.ssrn.com/sol3/papers.cfm?abstract_id=988363

[Fitzgerald and Agerfalk(2005)] Fitzgerald, B., Agerfalk, P.J.: The mysteries of open source software: Black and white and red all over? In: HICSS 2005: Proceedings of the Proceedings of the 38th Annual Hawaii International Conference on System Sciences, p. 196.1. IEEE Computer Society, Washington, DC, USA (2005), doi: http://dx.doi.org/10.1109/HICSS.2005.609

[Letellier(2008)] Letellier, F. (2008), Open source software: the role of nonprofits in federating business and innovation ecosystems (2008) (a submission for AFME), http://flet.netcipia.net/xwiki/bin/download/Main/publications-fr/GEM2008-FLetellier-SubmittedPaper.pdf

[Open Handset Alliance(2007)] Open Handset Alliance. Open Handset Alliance web page (2007), http://www.openhandsetalliance.com/

[Raymond(2000)] Raymond, E.S.: The magic cauldron (2000), http://www.catb.org/~esr/writings/cathedral-bazaar/magic-cauldron/

[Subramaniam et al(2009)Subramaniam, Sen, and Nelson] Subramaniam, C., Sen, R., Nelson, M.L.: Determinants of open source software project success: A longitudinal study. Decis. Support Syst. 46(2), 576–585 (2009), doi: http://dx.doi.org/10.1016/j.dss.2008.10.005

[The Apache Software Foundation(1999)] The Apache Software Foundation. The Apache Software Foundation web page (1999), http://www.apache.org/

[The Eclipse Foundation(2004)] The Eclipse Foundation. The Eclipse Foundation web page (2004), http://www.eclipse.org/

[Watson et al(2008)Watson, Boudreau, York, Greiner, and Donald Wynn] Watson, R.T., Boudreau, M.C., York, P.T., Greiner, E., Donald Wynn, J.: The business of open source. CACM 51(4), 41–46 (2008)

[Yamakami(2011)] Yamakami, T.: Generations of oss in evolutionary paths: Toward an understanding of where oss is heading. In: IEEE ICACT 2011, pp. 1599–1603. IEEE (2011)

[Yu et al(2009)Yu, Ramaswamy, Lenin, and Narasimhan] Yu, L., Ramaswamy, S., Lenin, R.B., Narasimhan, V.L.: Time series analysis of open-source software projects. In: ACM-SE 47: Proceedings of the 47th Annual Southeast Regional Conference, pp. 1–6. ACM, New York (2009), doi: http://doi.acm.org/10.1145/1566445.1566531

MI: An Information Support System for Decision Makers

Ohan R. Oumoudian and Ramzi A. Haraty

Abstract. Decision makers are key personnel working most of the time on critical duties. Information is the key to their success, and having the correct and up-to-date information is not an easy task; teamwork, where multiple parties have to work together is required. Though information is available everywhere in every action one may do during his/her daily life it has not yet been harvested as it should. Information in our approach is collected using a mobile device. Mobile devices are 90% of the time idle sitting in the pocket of its holder waiting for a call to take place or an SMS to be received. This 90% of the time is where this work focuses: monitoring GSM, Wi-Fi, GPS data and building a data warehouse. The data warehouse is the main source of information for decision makers, to succeed better in their daily tasks.

1 Introduction

Sources of information are multi-disciplinary; some information can be processed from the web, blogs, Facebook, Twitter, etc. Other types can be processed from data entry forms, whether simple opinion or a claim at a local restaurant to calling emergency numbers and placing complains. Surveillance data can also be processed for information retrieval. Surveillance data extends from local street cameras to surveillance camera in malls, individuals being watched, detectives, or even private eyes. Voice data can be captured and processed from people talking on the streets, to enterprise phone calls, to wiretapping and communication interference. With all this information, most people would agree that ignorance is bliss.

"Intelligence is the process of supporting the policy makers in making their decisions by providing them with the specific information they need" [1]. Information is the key feature of any law enforcement institution. Therefore, the more information is acquired and the more data is analyzed, the better decisions are taken and the safer are the streets and cities. Mobile devices are the source of information in this work. Mobile devices can provide different types of information, beginning with call log/SMS log to call recording, PDA activity monitoring, GSM

Ohan R. Oumoudian · Ramzi A. Haraty
Department of Computer Science and Mathematics
Lebanese American University
Beirut, Lebanon
e-mail: rharaty@lau.edu.lb

F.L. Gaol et al. (Eds.): Proc. of the 2011 2nd International Congress on CACS, AISC 144, pp. 181–189.
springerlink.com © Springer-Verlag Berlin Heidelberg 2012

network information, GPS satellite status, data communication over 2G network, 3G network, Bluetooth and Wi-Fi.

The rest of the paper is organized as follows: Section 2 provides related work. Section 3 provides the methodology of the work. Section 4 discusses mobile intelligence (MI) along with the technical details and database designs. Section 5 provides a real life scenario. And section 6 concludes the paper.

2   Related Work

In [2], the authors discuss the latest technology trends related to mobile internet - how mobile phones are becoming an everyday need. In 2008, 3.3 billion people-half of the world population - use mobile phones according to the international telecommunication union. They rely on 3G services such as:

- Mobile internet access for everyone with a mobile and a data plan.
- Mobile intranet and extranet access for enterprise users on the go, individuals who need to access office documents while at the client or outside the office.
- Customized infotainment.
- Multimedia messaging service between users, where large content of data over 3G network are sent.
- Location based services; services provided knowing the entourage of the user, like movie ad messaging of a theater nearby or resstaurant.

In [3], the authors present a tracking application integrated into a car targeted for car theft, teenage driving monitoring/speeding and vehicle tracking. The application can read the status of the car (parked, driven, locked) as well as more advanced sensor data if made available, also it can read the car alarm status if connected to the car alarm. The request for data is done in two ways: on demand and trigger based. The on-demand is controlled by a GSM phone that requests car status from the in-car system.

The authors of [4] introduce the GPS and GSM technology to the public transportation buses in the Punjab and Delhi cities of India. The system is made up of GSM modules to communicate, GPS devices to track, laser detectors to record passenger activities (going up on a bus or down) and a set of microcontrollers to establish communication between all sides. The approach is the integration of common new technology to simple aspects of the human life, the close monitoring and scheduling of public buses through the establishment of an operations room that track and audit the activity of public transportation.

The concentration of the authors of [5] is cellular tracking accuracy - how accurate can the output be by using the base cell location and signal strength.

The authors of [6] offer a roadmap from GERAN to 3G networks. The upgrade from GERAN to 3G passes from multiple phases starting from the core component all the way to the cells. The GSM will have to interface with UMTS, thus the protocols change. The way to go as proposed is either GERAN Iu or A/Gb. The first is a circuit switched network; thus, already taking the first step into 3G, since the GERAN Iu will traverse a 3G core network. The GERAN Iu upgrades to

UTRAN which makes the GERAN as if a fresh copy of the UTRAN 3G. Thus, the GERAN A/Gb was suggested and compared to the Iu. Both approaches have cost related considerations to be taken, it actually boils down to the operator to make the decision and see what is the more suitable in light of many considerations: cost, market, future services, etc.

The authors of [7] discuss location based services that are starting to emerge. Their proposed architecture for these services includes: target, position originator, location provider, service provider, content provider, and LBS user.

3 Proposed Methodology

Information is not only what one may find on the Internet, in printed media, or in visualized media. Information is born on sight or in thought or in action but rarely captured. "Rarely captured" because the amount of data generated compared to the amount of data collected is beyond comparison. Everything is information: daily actions (morning wake up, shower, breakfast) talks and conversations, as well as thoughts, ideas, dreams, situations. Any actions experienced by one of the human five senses can be considered as information; the human being itself is a giant data collector, data repository and data analyzer. The collection of data as per [1] can be classified in four categories: 1) the human based intelligence collection of data through trained personnel, 2) the imagery intelligence, 3) the open sources intelligence - these are publicly available, low-cost information sources, and 4) the signals intelligence.

3.1 GPS (Global Position System)

GPS system is the offspring of two military technologies: the American Navy's "TIMATION" program and the Air Force "Program 621B" [8]. Initially known as NAVSTAR GPS, the program took the best elements of TIMATION, used by the Navy for ships and submarine guidance and the Program 621B, used by the Air Force with four satellites constellation but unfortunately each served independently by ground-control stations. The NAVSTAR GPS was needed to replace the need of ground control stations as well as the TIMATION inability to provide position updates. The NAVSTAR consists of 24 satellites orbiting around the world providing all three locations axis, latitude, longitude and altitude. Initially designed for military use, this system has become more and more familiar in the domestic world.

3.2 GSM (Global System for Mobile Communications)

GSM is part of the 2G family of mobile communications. Mobile communications are the number one source of mobility. Users can call, SMS, access the Internet, anywhere, anytime as long as mobile coverage allows it. The main GSM network is made up of a cell (close to a bee-hive) structured network [9]. Mobile devices are connected to the GSM network using one main cell and six neighboring support cells. As the mobile user changes his/her location the main cell changes and the cell with the highest power becomes the main cell and the six others are ordered based on power strength. The GSM network has evolved from GPRS to the 2.5G with the presence of EDGE. The data rates have jumped from 2 to 4KBps to 32KBps (kilobytes per second) [10]. In mobile communication voice has always been priority over data. The ratio at the base station is 70/30. 70 for voice and 30 for data. Given the nature of the data packet switching, this ratio is increased for the benefit of voice. Thus, making it 80/20 even 90/10. This is about to change with the evolution of the mobile network. Voice is becoming cheap and data plans are being the main decision taking feature when users are to choose a mobile plan. UMTS, Universal Mobile Telecommunication Systems, is nowadays the new mobile standard or what is known as 3G [11]. The 3G network was created to support superfast connections compared to the current mobile network, from 236Kbps to 2Mbps and with HSUPA 7.2Mbps.

3.3 Wi-Fi

The wireless networks have grown large in the past decade, due to two simple facts: wire-less and mobile. More mobile devices are having the feature of Wi-Fi nowadays, especially in countries where 3G data plans are not available - actually where a3G network is not found [12].

3.4 GPS/GSM/Wi-Fi + Information

Wi-Fi cannot be localized given the huge number of wireless networks being created and removed every day. Wi-Fi unlike GSM does not require any legal papers to setup a Wi-Fi Internet in a certain area. For example a Wi-Fi hotspot W cannot be shown on a map because it is not registered while a cell id Y can be easily pointed out on a map. The GPS information is there to geotag. GSM network coverage is nationwide but the cells and power differ. What better way than geotagging the GSM cell and power at each GPS movement of a person on the move! Cell information is per serving cell, meaning the six neighboring cells information is not recorded. The Wi-Fi hotspots on the other side are as many as there can be; thus, Wi-Fi geo-tagging will geotag many Wi-Fi hotspots at a certain location latitude and longitude.

GPS coordinates, GSM cell id and power, Wi-Fi MAC address, name and power are three types of information that one walks through but never grabs. The MI project is a Wi-Fi, GSM, GPS information collector, Wi-Fi geo-tagger GSM information for future analysis.

4 Why MI?

The main motive behind the idea was to extract as much as possible information from a mobile device. Our work is, accordingly, what can be extracted from a mobile device. This includes: contacts, short messages, call logging, call conversations, media (pictures, music, tones, and videos), web browsing information, and geo-locations.

Primarily the information consists of the surrounding and not the activity of the mobile user itself (i.e., geo-locations). Geo-locations targets three components of the mobile device: the GSM/UTMS module, the GPS antenna and the wireless LAN (our work considers a modern mobile device as one that has all three services). The GSM module, mainly used for GSM/UMTS communications and GPRS/EDGE/CDMA/HSDPA for browsing, holds information of the main GSM cell as well as the signal strength and six neighboring cells. This information is subject to change due to the number of devices and voice calls taking place around the device. Wi-Fi hotspots are the easy internet access portal, available in any home, at the corner of every major city, in malls and servicing stations. Wi-Fi spots do not need licensing to work - just plug and play. Thus, the Wi-Fi stations of a country cannot be swept or calculated using a backbone device like in the case of the GSM network where the cells are in precise locations and their area of service is preconfigured. Last but not least, the GPS module is there to geotag the GSM network information as well as the Wi-Fi network information, creating four data layers: static GSM cell locations layer, dynamic GSM cell power layer, static Wi-Fi hotspot locations layer, and dynamic Wi-Fi hotspot signal coverage.

The Wi-Fi layer, shown in figure 1, portrays all the locations where Wi-Fi data are collected. This data will be processed into 30 to 50 meter buffers; thus, making it easier to visualize and query. Besides the above layers, the GSM and Wi-Fi information collected from the mobile device will be used as raw data, for information support system used by decision/policy makers.

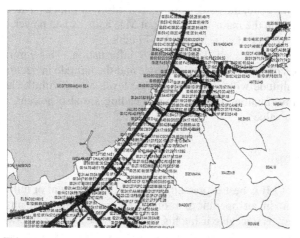

Fig. 1 The Wi-Fi layer.

MI collects upon movement the GSM cell data as well as all Wi-Fi hotspots in view. The data collected is appended to GPS coordinates and stored on the device. The device then synchronizes with the back office where the main data-warehouse is used for pre-processing and quick analysis. MI provides an intelligence database for decision makers. Consider a database with rich data about Wi-Fi hotspots, their MAC address, and their signal strength, about GSM cell ids and their signal, about GPS information speed, heading, and accuracy, all this data linked to a coordinate system, everything is there on a map to see and decide.

4.1 Technical MI

MI is a mobile application built using Windows Mobile 6 and Visual Studio 2008. The primary language of choice is VB.Net, and then we switched to C# since parts of the code required a wrapper for C++ libraries. The GPS library used is the default Windows Mobile Positioning Sample. The Wi-Fi and the GSM are C++ wrappers. Microsoft SQL Server is the main data collector. For the back office, the Enterprise 2008 Edition is used, and on the mobile devices the Compact Edition 3.5 Service pack 1 is used. The GIS data viewer is the ESRI. ArcGIS desktop 9.3.1 is used to display the data. The GIS data tests were done at a local GIS company.

4.2 Database Design

The database structure is made up of one main table (GCW) and three supporting tables (GPS, GSM, and Wi-Fi). GCW carries GPS data, cell information and Wi-Fi. The rest of the tables are supporting tables, holding additional information for the GCW record. The GCW and GPS records share only latitude and longitude, while the GPS table holds the speed, heading, altitude, GMT time, etc. The GCW and GSM have in common the Cellular ID and the Cellular ID power, and GSM holds the country code, the name base station information, etc. The GCW and Wi-Fi have in common the main Wi-Fi hotspot Mac address and power, the rest of the Wi-Fi in view are recorded inside the Wi-Fi table.

The GPS table holds all detailed GPS information. This information is the global positions on a 2D map, the longitude and the latitude in decimal degrees for accuracy, the altitude with respect to the sea level, along with the levels of precisions of this information. The speed and heading are also present, as well as the GMT date/time value.

5 An Example

The main use of MI is best expressed in the law enforcement or military field. The normal scenario in an operations room is to track outlaws. Consider the case where the outlaw is traceable using his mobile device and the outlaw is on the run. Given the sensitivity of the information, only law enforcement agencies can have access to a mobile device subscriber's current cell tower ID and power. Also, the information of cell towers and their area of service by power can only be provided for law enforcement companies. Now the information provided by the telecommunications company shows the locations of the cell towers, and the range of signal strength. The area covered is large; thus, making the decision hard. This is where the MI data comes in handy. The data will be drawn as intelligence over the maps layer. The tentative location of the fugitive will be highlighted given the amount of data and the accuracy with respect to time. The question to be asked is:

what data does intelligence provide for this cell information and power at the current time? Thus, the decision maker has more options to look into before taking his/her decision.

Another scenario is where a local hacker is doing an intrusion into a governmental site. Given the technology at hand in law enforcement companies as well as the military, one can know the set of routers the communication is being made through. Thus, at any moment the decision maker can compare this information to the data collected using MI and get the possible location of one of the routers.

As an example, the MI project for a certain city starts with data collection. The data is collected using handheld devices by agents patrolling the different areas of the city at different times of the day. While on the move, MI will record all the necessary GSM, GPS, Wi-Fi information and save it inside a local database on the PDA, using Microsoft SQL Compact Edition 3.5 SP1. When a certain agent returns to the field, the data is synched between the handheld device and the back office data warehouse using Microsoft SQL Server Enterprise 2008 web synchronization service. This latter can also be publicly available for sync on demand while agents are in the field. Once the data is made available, MI can start serving decision makers. In a mission that requires tracking a fugitive via his/her mobile phone, the decision makers will need to acquire up-to-date mobile phone information from the local mobile company operations room. The information received from the mobile company includes the current cell id and signal strength of the fugitive. Accordingly, the decision maker will see the main cell id location on the map and the area where the corresponding signal strength as shown in figure 2. This information is based on what the mobile company has provided as maps and signal strength area when the mobile company was established.

In figure 2 the cell id 1 and signal 40 is reflected by the -040 value next to the cell id tower, while cell id 2's signal strength is reflected by -060 and cell id

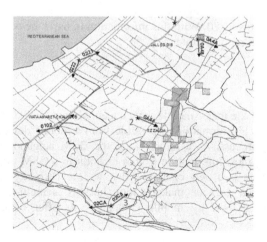

Fig. 2 Fugitive pursuit with GSM provider data.

3 by -070. The fugitive is talking from cell id 1 with signal strength -40. The information provided from the GSM network provider is poor and often old, not allowing the decision maker to implement a quick strategy to intercept the fugitive. Now the fugitive is on the move and has passed to cell id 2 (in figure 2, noted in red next to cell id). The area coverage is bigger but still no accurate data. The fugitive moves now to cell id 3 with and again the signal strength area is poor. The introduction of MI to the equation will give a scenario similar to figure 3.

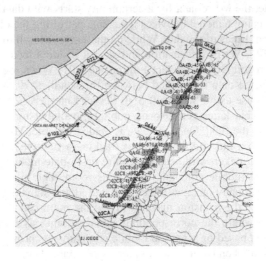

Fig. 3 Fugitive pursuit with GSM provider data and MI data.

MI has provided with the information found in hand the different location where a possible cell id 1 and power -40 can be, as well as cell id 2 and power -60 and cell id 3 and power -70. With this information in hand the decision maker can better target an interception strategy.

6 Conclusion

MI covers some of the major mobile device technologies such as Wi-Fi and GPS. MI extracts as much information as possible from a mobile device to help in the decision making process. As a future work, the GSM neighboring cells will be covered to get the most information from a location in a GSM network. Working with accelerometers is a new challenge that we are currently undertaking. MI was introduced as a decision maker's tool, but MI is multi-disciplinary - the information can be used by mobile companies to offer better services and plans.

References

1. Koltuksuz, A., Tekir, S.: Intelligence Analysis Modeling. In: Proceedings of the International Conference on Hybrid Information Technology, ICHIT 2006, Cheju Island, Korea, vol. 1 (November 2006)
2. Zhou, J., Ning, A., Deng, R., Meng, F.: Mobile Internet Application to Geographical Information Services. In: Proceedings of the International Conference on Multimedia Information Networking and Security, Hubei (November 2009)
3. Lita, I., Cioc, I., Visan, D.: A New Approach of Automobile Localization System Using GPS and GSM/GPRS Transmission. In: Proceedings of the 29th International Spring Seminar on Electronics Technology, St. Marienthal (May 2006)
4. Farooq, U., Ul Haq, T., Amar, M., Asad, M., Iqbal, I.: GPS-GSM Integration for Enhancing Public Transportation Management Services. In: Proceedings of the Second International Conference on Computer Engineering and Applications, Bali, Indonesia (March 2010)
5. Catrein, D., Hellebrandt, M., Mathar, M., Serrano, M.: Location Tracking of Mobiles: A Smart Filtering Method and its Use in Practice. In: Proceedings of the Vehicular Technology Conference (May 2004)
6. Proctor, T.: Evolution to 3G Services: Provision of 3C Services over GERAN (GSMIEDGE Radio Access Network) (2003)
7. Koutsiouris, V., Polychronopoulos, C., Vrechopoulos, A.: Developing 3G Location Based Services: The Case of an Innovative Entertainment Guide Application. In: Proceedings of the Sixth International Conference on the Management of Mobile Business (2007)
8. Parkinson, B., Gilbert, S.: Navstar:Global Positioning System - Ten Years Later. Proceedings of the IEEE 71(10), 1177–1187 (1983)
9. Pautet, M., Mouly, M.: GSM Protocol Architecture: Radio Sub-System Signallling. In: Proceedings of the Vehicular Technology Conference, St. Louis, MO, USA, pp. 326–332 (1991)
10. Gozalvez, J., Dunlop, J.: High-Speed Simulation of the GPRS Link Layer. In: The 11th IEEE International Symposium on Personal, Indoor and Mobile Radio Communications, London, UK (September 2000)
11. Shyy, D., Rohani, B.: Indoor Location Technique for 2G and 3G Cellular PCS Networks. In: Proceedings of the 25th Annual IEEE Conference on Local Computer Networks, Tampa, FL, USA (November 2000)
12. Taherkordi, A., Taleghan, M., Sharifi, M.: Dependability Considerations in Wireless Sensor Networks Applications. Journal of Networks 1(6) (2006)

Cost Optimization in Wide Area Network Design: An Evaluation of Created Algorithms

Tomasz Miksa, Leszek Koszalka*, Iwona Pozniak-Koszalka,
and Andrzej Kasprzak

Abstract. In the paper, a problem of cost optimization of Wide Area Network (WAN) is considered. For solving this problem two algorithms have been deigned and implemented: a heuristic algorithm called Marian and meta-heuristic algorithm called Kaflok. The paper presents the results of evaluation of these algorithms based on complex simulation experiments made using the designed experimentation system implemented in Java environment. The influence of distinct problem parameters on the total cost of WAN design and the comparison of efficiency of the algorithms are analyzed and discussed.

1 Introduction

Nowadays one can observe an increasing number of computer network users caused by the fact that more and more companies decides to connect their branches using private network [1]. Companies have different priorities concerning the parameters of the designed network. For some of them, the most important factor is reliability, for others the total cost of creation of the network is crucial.

In this paper, the minimization of the cost of WAN design is under consideration. Two algorithms for designing WAN are implemented. The first one called Marian is a new modification of the algorithm presented in [1]. The second algorithm called Kaflok is created by the authors of this paper basing on the evolutionary ideas of simulated annealing approach [2], [3].

The rest of the paper is organized as follows: in Section 2 the problem of WAN design is formulated. In Section 3 the two designed and implemented algorithms are shortly described. The results of the investigations made using multistage experiment design are analyzed and discussed in Section 4. Final remarks appear in Section 5.

Tomasz Miksa · Leszek Koszalka · Iwona Pozniak-Koszalka · Andrzej Kasprzak
Dept. of Systems and Computer Networks, Wroclaw University of Technology,
50-370 Wroclaw, Poland
e-mail: leszek.koszalka@pwr.wroc.pl

* Corresponding author.

F.L. Gaol et al. (Eds.): Proc. of the 2011 2nd International Congress on CACS, AISC 144, pp. 191–197.
springerlink.com

2 Problem Statement

The problem is to design a wide area network. The following assumptions are taken into consideration: (i) a monthly usage cost of WAN should be as low as possible, (ii) the average delay of the packet must be lower than a given value, (iii) each node must have at least N different connections with neighboring nodes.

The network can be represented (e.g. [2]) as a directed graph G consisting of set of vertexes V (network nodes) and set of arcs E (traffic between nodes). Thus, the considered problem may be stated as follows:

Given: set of vertexes V,
          distances $D(i,j)$ between node i and node j,
          traffic $E(i,j)$ between node i and node j,
          set of available bandwidths T,
          cost $C(i)$ of lease on bandwidth i from set T,
          maximum acceptable average delay of packet L,
          minimum number of neighboring nodes N_i.
To find: set of edges R of defined bandwidth,
          routing table for each node from V.
Such that: to minimize the total cost of network defined by the equation:

$$K = \sum_{i,j=0} D(i,j) \cdot C(R(i,j))$$ (1)

where $R(i,j)$ denotes the bandwidth between nodes i and j. The calculated cost concerns one month period. It was assumed, that: (i) the size of packet is 500 bytes, (ii) the available bandwidths are of 64000, 128000, 192000, 256000, 512000, and 1024000 bytes per second, (iii) the distances are measured in kilometers. The prices of leasing the connections were calculated according to the rules of the greatest Polish internet service provider named Telekomunikacja Polska [4]. It was also assumed, that the cost of 64000 b/s connection is equal to one unit.

3 Algorithms

Algorithm Marian. The algorithm consists of two phases: initial phase and iterative phase. The *initial phase* begins with creation of complete graph consisting of edges with maximum bandwidth and all vertexes from V. Next, the average delay of packet in the network is calculated. If it exceeds the maximum value, then the problem is unsolvable. If it fulfils such a condition then the procedure continues. The *iterative phase* begins with finding routing, transfers for each edge and calculating the value of effectiveness according to the equation (2).

$$\alpha = (T(R(i,j)) - B^R(i,j) + B^R(j,i)$$ (2)

where $T(R(i,j))$ is the current bandwidth between nodes i and j, $B^R(i,j)$ is the traffic from node i to node j assuming routing, $B^R(j,i)$ is the traffic from node j to node i assuming routing. Calculation of routing consists in a creation of a table for each node. The table consists of two rows. In the first row all other nodes are listed and in the second row the corresponding output interfaces are placed. The known Dijkstra algorithm (e.g. [5]), is applied to calculate the routing for each node. Next, the edge with the lowest effectiveness is chosen. The bandwith of this edge is reduced to the next allowed value only if, as a result of reduction, the node possesses still at least N connections (see Fig. 1) and the average delay of packet does not exceed the maximum allowed value. The average delay of packet in network T is expressed by the equation (3):

$$T = \frac{1}{\gamma} \sum_{\substack{i,j=0 \\ R(i,j) \neq 0}}^{E} \frac{B^R(i,j)+B^R(j,i)}{T(B(i,j))-(B^R(i,j)+B^R(j,i))} \tag{3}$$

where $B^R(i,j) + B^R(j,i)$ is the average quantity of bytes in the channel (sum of transfers in each direction), $T(B(i,j))$ is the bandwidth of a channel, and γ is the quantity of packets in network (during one unit of time – a second) defined by (4):

$$\gamma = \sum_{\substack{i,j=0 \\ i \neq j}}^{E} r_{ij} \tag{4}$$

In (4), r_{ij} means the quantity of packets in network between nodes i and j. Otherwise, the sequence of actions is repeated for the next edge according to that with the second lowest effectiveness. If all edges have been just considered and reduction of any of them is not possible, then the current graph is treated as a final solution. If the reduction has been possible, then the next iteration is being performed.

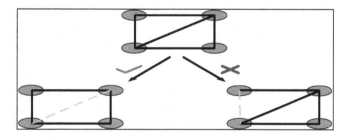

Fig. 1 Reduction of edges for N=2 : allowed (on the left) and prohibited (on the right).

Algorithm Kaflok. The algorithm is based on the ideas used in simulated anneal-ing. Some probabilistic mechanisms are implemented. When looking for the best route the neighborhood of an edge is generated at random and a change of its bandwidth is made at random, either. A new solution is checked for: (i) existence of N_I neighboring nodes, (ii) maximum average delay condition. The solution is accepted depending on the temperature in the current iteration. The algorithm can generate many different solutions for same input data. Fig. 2 presents an example with two different solutions for the same input data. The algorithm has several inner parameters which can be adjusted by the user. The most important is the probability of acceptance of a solution, denoted by P, defined by a chain of equations :

$$P = e^{\frac{-\Delta}{T_i}} \qquad T_{i+1} = \frac{T_i}{1+\lambda_i T_i} \qquad \lambda = \frac{T_p - N \cdot T_k}{T_p \cdot N \cdot T_k} \qquad (5)$$

where Δ is the cost difference between current and previous solution, T_i is the temperature at i iteration. The parameter lambda is computed in any iteration using T_p as the initial temperature and N as the number of all iterations.

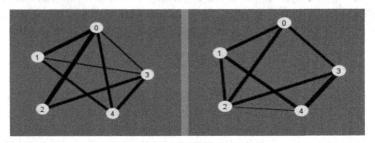

Fig. 2 Two different solutions for the same input data (for algorithm Kaflok).

4 Investigations

The investigations have been made using the designed simulator, called WANDesigner - window application implemented in Java with a use of Eclipse development environment [6]. The following research theses were stated:

- The results produced by the algorithm Kaflok are in average less efficient than results given by the algorithm Marian (Experiment 1).
- The increment of quantity of packets in the network with constant number of nodes results in the rise of the network cost (Experiment 2).
- The decrement of maximum allowed delay of packet increases the total cost of the network (Experiment 3).

In Experiment 4, we tried to determine which of the algorithms gives better results in a network with constant number of nodes and maximum transfer rate varying between 200 and 5000 B/s. Experiment designs based on ideas given in [7].

Experiment 1. It was carried out with input parameters: Start temp. = 10500, Iterations = 10000, Minimum average delay = 0.7, Max transfer = 25000. During the experiment the location of nodes did not change. The number of nodes was varying within following values: 4, 8, 16, and 32. The results are shown in Fig. 3.

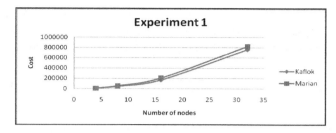

Fig. 3 Cost in relation to the number of nodes in Experiment 1.

Experiment 2. For the algorithm Marian the input data was set as follows: Maximum average delay = 0.7, Quantity of nodes = 10. Four networks with different nodes locations were tested. The varied parameter was the maximum transfer rate for the edge. Transfer rates were drawn between 0 and max transfer. The average cost was calculated as an arithmetical average of costs obtained for each trial with the same max transfer. The results are shown in Fig. 4.

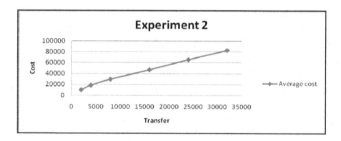

Fig. 4 Cost in relation to the maximum transfers in Experiment 2.

Experiment 3. For the algorithm Marian the input data was set as: Number of nodes = 10, Maximum transfer for edge = 5000. The same (as in Experiment 2) four networks were tested. The varied parameter was the maximum average delay. Transfer rates were drawn between 0 and 5000. The results are shown in Fig. 5.

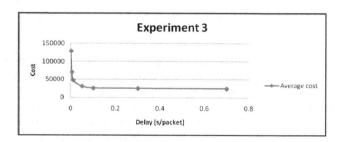

Fig. 5 Cost in relation to the delay in Experiment 3.

Experiment 4. During this experiment the input data was set as: Max average delay = 0.7, Quantity of nodes = 10. For each value of transfer, a different topology of network was generated. For Kaflok an average value was computed.

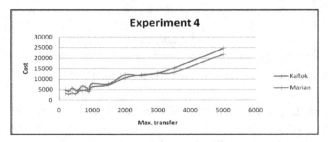

Fig. 6 Cost in relation to the maximum transfer in Experiment 4.

5 Conclusion

Having conducted these four experiments we can claim that results have met our expectations, but not without surprises. Experiment 1 did not confirm rightness of our prediction that Kaflok in average provides worse solutions (see Fig. 3, where Marian reached always better values than Kaflok). Basing on this observation we had decided to use only Marian algorithm in Experiment 2 and 3. However, a weak point of Marian algorithm is even reduction of bandwidth of all edges in the graph. Experiment 2 confirmed that the increment of quantity of packets in the network results in the rise of the network cost, assuming that there are a constant number of nodes. Fig. 4 shows the linear dependency of cost increment towards delay. Already conducted experiments encouraged us to claim, that it is more profitable to use connections with higher bandwidths with a use of less nodes, rather than more nodes with lower bandwidths. Experiment 3 justified our thesis, that decrement of maximum allowed delay of packet increases cost of the network. Improvement of delay from 0.7 to 0.1 s/packet was not followed by significant cost increase. However, further improvement of delay results in dynamic increment of the cost. Experiment 4 showed that for assumed configuration one can distinguish three ranges of transfer values: where transfer varies between 0 and 2000 b/s, the algorithm Marian provides better solutions; for subset (2000 – 3000 b/s) both algorithms provide comparable results but for values higher than 3000 b/s the algorithm Kaflok leads to better solutions.

Taking into consideration the way the algorithm Marian works, we notice that its main drawback is lack of possibility to increase the quality of connection. If it was possible to let the algorithm improve the bandwidth, the quality of solution could be better. However, it would result in increment of computational complexity that could lead us to the algorithm that checks every possible solution. This approach could be very inefficient for networks with a big number of nodes. Following [8] and [9] we plan in the nearer future to apply genetic ideas to our problem.

References

1. Kasprzak, A.: Packet Switching Wide Area Networks. WPWR, Wroclaw (1999)
2. Kirkpatrick, S., Gelatti, C.D., Vecchi, M.P.: Optimization by Simulated Annealing. Science, New Series 220(4598), 671–680
3. Laarhoven, J.M., Emile, H., Aarts, L.: Simulated Annealing: Theory and Applications. Springer, Berlin (1987)
4. Tarriff Catalogue of Polish Telecommunications Services S.A., http://www.tp.pl/b/binaries/PL/358251/_1.11.09_366849505.pdf
5. Youssef, H., Sait, S.M.: Iterative Computer Algorithms with Applications in Engineering, Washington (1997)
6. Eckel, B.: Thinking in Java. Helion, 4th edn. (2005) ISBN: 83-246-0111-2
7. Koszalka, L., Lisowski, D., Pozniak-Koszalka, I.: Comparison of Allocation Algorithms for Mesh Structured Networks with Using Multistage Simulation. In: Gavrilova, M.L., Gervasi, O., Kumar, V., Tan, C.J.K., Taniar, D., Laganá, A., Mun, Y., Choo, H. (eds.) ICCSA 2006. LNCS, vol. 3984, pp. 58–67. Springer, Heidelberg (2006)
8. Lenarski, K., Kasprzak, A., Pozniak-Koszalka, I.: Comparison of Heuristic Methods Applied to Optimization of Computer Networks. In: 7th ICN, IARIA (2007)
9. Glover, F., Kochenberger, G.A.: Handbook of Metaheuristics. Springer (2002)

Application of Domain Knowledge in Relational Schema Integration with Uncertainty

Wen Bin Hu[*], Hong Zhang, and Si Di Zhang

Abstract. Schema integration is the activity of providing a unified representation of multiple data sources. The core problems in schema integration are: schema matching and schema merging. There are uncertain problems in schema matching and schema merging. To solve the uncertain problems of relational schema integration, Domain Knowledge Application Model (DKAM) is proposed as a component of Uncertain Relational Schema Integration Model (URSIM). An autonomic computing approach is adopted in DKAM. Semantic integration approach and D-S evidence combination approach are applied in URSIM. A new method is proposed to calculate reliability of global integrated schema in the paper. Experimental results show that URSIM is feasible and DKAM is valuable and advanced. In contrast with current methods for schema integration with uncertainty, URSIM is efficient and the time complexity is reduced.

1 Introduction

Schema integration is the activity of providing a unified representation of multiple data sources. The core problems in schema integration are: schema matching [1] and schema merging [2]. Schema matching is a fundamental operation in the manipulation of schema information that takes two schemas as input and produces a mapping between elements of the two schemas that correspond semantically to

Wen Bin Hu
Nanjing University of Science and Technology, Nanjing, China
and

Huaihai Institute of Technology, Lianyungang, China
e-mail: hwb1008@163.com

Hong Zhang
Nanjing University of Science and Technology, Nanjing, China
e-mail: zhhong@mail.njust.edu.cn

Si Di Zhang
SINOPEC Jiangsu Oil Exploration Corporation, Yangzhou, China
e-mail: 296654979@qq.com

[*] This work was supported by a grant from the National Natural Science Foundation of China (No.60903027)

each other. Schema merging uses the discovered mappings to merge non-homologous schema into integrated schemas. The work in [4] has shown that if all semantic mappings are known, schema merging can be performed semi-automatically. In general it is impossible and difficult to fully identify the correct semantic mappings automatically. Manual schema integration is usually time-consuming, and it may be unfeasible, especially with a large database. Therefore, managing uncertainty has been recognized as the next issue on the research agenda in the realm of data integration [4].

Uncertainty is an intrinsic feature of automatic and semi-automatic schema integration processes, and it is inevitable because the semantics of schema objects cannot be fully derived from data and meta-data information [5]. Although many solutions have been proposed to reduce uncertainty, we risk losing relevant information and producing misleading results if we do not explicitly express uncertainty and keep it up to the end of the integration process [6]. Current techniques [7, 8] try to identify a single mapping for each pair of objects, which surely could be wrong. Top-k mappings [9] can be used to increase the recall of a schema integration process. However, we also notice that cutting low-probability mappings may result in loss of correct information.

An approach of uncertain schema integration with a multi-matcher is presented in [5], and uncertainty is kept up to the merging step after the matching phase. Data mappings are not covered in this paper, [10] and the extended version of this paper [11] focus on probabilistic schema matching. These papers do not provide a complete view of uncertain data integration processes, but focus on the implementation of probabilistic classifiers (matchers) that have not been covered in detail in [5].

Domain knowledge about semantic mappings and schema merging is essential to producing an integrated schema. Schema integration based on six semantic relationships in [5] and domain knowledge, and extended application of D-S evidence combination approach are studied in the paper. URSIM as a new model is provided. Reliability of every phase is combined by a new approach in it. Autonomic application of domain knowledge can reduce the complexity of schema matching and integration and is lucubrated.

2   Uncertain Factors in Schema Integration

The paper focuses on relational schema (r-schema) because most of the data are stored in relational database. In many applications, Uncertainty can be resulted from multiple reasons in schema integration. First, not all users are equipped with operation logic and can manage precise schema mapping expertly. Second, precise mappings are not created because some fields are not apprehended insufficiently. It is impossible to create and maintain all precise mappings in the integration of a large scale data [12].

Accuracy of integration can be increased when related information of schema is used in the process of schema integration. The characteristics of schemas of independently developed sources may be summarized two sorts: *Name heterogeneity* and *Structural heterogeneity*. There are three structural conflicts: type conflicts; key conflicts; domain conflicts.

3   Integration Model of Uncertain r-Schema

3.1   *Uncertain Relational Schema Integration Model*

An uncertain relational schema integration model (URSIM) based on domain knowledge is composed of seven submodels. They are initial matching reasoning submodel (IMRM), relational semantic reasoning submodel (RSRM), attribute matching reasoning submodel (AMRM), global schema generating submodel (GSGM), domain knowledge application submodel (DKAM), matcher choice submodel (MCM) and match type specify strategy (MTss). URSIM is illustrated in Figure 1. Domain knowledge is applied to every submodel in URSIM. The integration flow is illustrated in Figure 2. S_1 and S_2 are the input and global schema aggregate (GSA) is the output. KDB is a knowledge database. IMM is the initial matching matrix. SMA is the semantic matching aggregate. IISA is the initial integration schema aggregate. AMA is the attribute matching aggregate.

Fig. 1 $S_1 = \{O_{11}, O_{12}, \cdots, O_{1M}\}$ and $S_2 = \{O_{21}, O_{22}, \cdots, O_{2N}\}$ are two schema object sets in data sources needed in integration, respectively.

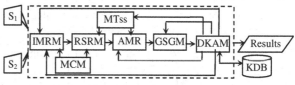

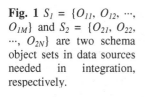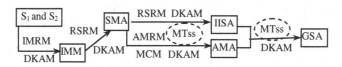

Fig. 2 Flow process of schema integration

3.2   *Definition of Uncertain Schema Integration*

In [12], one-to-one matching relationship between schemas is defined by probabilistic schema mapping (PSM) and the uncertainty of table-to-table is only expressed in schema level. Relative definition of uncertain is proposed as follows:

Definition 1 (*Uncertain Matching Relationship (UMR)*). Let S_1 and S_2 be relational schema. An uncertain matching relationship is a quad (O_{1i}, O_{2j}, D, m), where $O_{1i} \in S_1$, $O_{2j} \in S_2$, $i=[1,|S_1|]$, $j=[1, |S_2|]$, $D=\{\tilde{=}, \tilde{\subset}, \tilde{\supset}, \tilde{\cap}, \tilde{\not\subset}, \tilde{\approx}\}$ is a mutually exclusive semantic relationship set [9], and m is a Basic Probability Assignment (BPA) function [13].

Definition 2 (*Uncertain Schema Matching (USM)*). Let S_1 and S_2 be relational schema. An uncertain schema matching is a triple (S_1, S_2, UM), where S_1 and S_2 is a schema object set respectively, $UM=\{umr_1, umr_2, \cdots, umr_n\}$ is a set of UMR, n $\in [1, |S_1| \cdot |S_2|]$, and $umr_i \in UMR$.

Definition 3 (*Global Schema* (GS)). Global schema is a set that is composed of optional global schema with confidence degree. GS is generated by GSGM which is the component of URSIM.

Definition 4 (*Uncertain Schema Integration* (*USI*)). Let S_1 and S_2 be relational schema. An uncertain schema integration is a triple (*UM, SM, TR*), where *UM* is a set of *USM*, $SM=\{< sm_i, \ f \ (GSi_i) >| \ i= 1,2,\cdots, \ n\}$ is a integrated schema set, $f(GSi_i)$ is confidence degree of an optional global schema, and *TR* is a conversion rule set from *USM* to *SM*.

4   Autonomic Application Model of Domain Knowledge

4.1   Expression of Domain Knowledge

Domain knowledge about semantic mapping is essential to the creation of integrated schema. Domain knowledge is the information of domain range about schema objects in schema integration, including domain information, domain words, relationship information between integrated objects, etc.

Domain knowledge is partitioned into three sorts:

1. *Information knowledge*. Information of the domain range of schema objects is the information knowledge in integration.
2. *Task knowledge*. The interrelated information of knowledge reasoning is task knowledge.
3. *Policy knowledge*. Instruction knowledge and optimization knowledge are used to drive policy knowledge based on the tasks in reasoning phases.

There is uncertainty in three sorts of knowledge. Reliability of knowledge can be used to denote belief of uncertain knowledge.

Definition 5 (*Uncertain domain knowledge* (*UDK*)). Uncertain domain knowledge is a triple (*K, F, KR*), where *K* is a set $\{k_1, k_2, \cdots, k_n\}$ that is a knowledge aggregate about words, semantics and instances, *F* is a set $\{f_1, f_2, \cdots, f_n \}$ that is a function aggregate used to calculate reliability, and *KR* is a set $\{kr_{12}, kr_{13}, \cdots, kr_{(n-1)n} | \ kr_{ij} = (k_i, k_j)\}$ which is a relationship aggregate between different knowledge.

Reliability of knowledge rules in knowledge database is specified by domain experts. Uncertain knowledge rule is expressed by production rule as follows:

IF E THEN $H = \{h_1, h_2, \cdots, h_n\}$ $CF= \{c_1, c_2, \cdots, c_n\}$

In *CF*, c_i is a reliability of knowledge rules h_i.

Expression and reasoning of uncertain knowledge based on relational model are implemented efficiently [14].

4.2   Domain Knowledge Application Model

Domain knowledge application model (DKAM) is a task-driven multiple automatic submodel in URSIM. DKAM can update knowledge database according to the changing of multidimensional information. DKAM is composed of domain knowledge submodel (DKM), reasoning submodel (RM), task submodel (TM) and rule mapping submodel (RMM).

Domain knowledge in knowledge database is input of DKAM. A framework of DKAM is illustrated in Figure 3. In RM, DWM is domain words matcher, IM is instance matcher, TM is type matcher, SM is semantics matcher, RUC is rules update component, and MEC is matcher extending component. Domain knowledge inputted is operated by RMM according to the task in TM after clustering. Needed processors and rules are chosen according to the task in RM. DKAM apply its output to other submodels of URSIM in order to improve integration efficiency and validity.

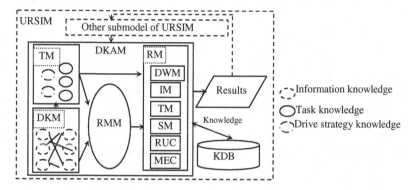

Fig. 3 A framework of DKAM

4.3   Autonomic Application of Domain Knowledge

Application of domain knowledge is a complicated problem in URSIM. DKAM is designed to manage itself according to the high-level guidance from tasks of other submodels in URSIM. DKAM is viewed as an autonomic application component of URSIM. DKM and TM are the autonomic elements based on knowledge model. RM and RMM are the autonomic elements based on mathematical model and knowledge model.

Definition 6 (*Autonomic Element (AE)*). AE is a triple (AM, MR, Sensors/Effectors), where AM is an autonomic manager, and MR is a managed resource or other managed autonomic elements.

Three sorts of work mechanism of AM are adopted based on different problem domains in DKAM. They are listed as follows:

(1) $1 \rightarrow 2a \rightarrow 4$
(2) $1 \rightarrow 2b \rightarrow 3 \rightarrow 4$
(3) $1 \rightarrow 2c \rightarrow 3 \rightarrow 4$

Work mechanism of AM includes four important steps [15]: 1, self-awareness/context-awareness, which is the gist for making decision; 2, decision-making based on strategies, which is the core component of AM; 3, planning based on objects, which is optional; 4.action/planning execution.

Work mechanism (1) is adopted in DKM and TM, while work mechanism (2) and (3) are adopted in RMM and RM, respectively.

5   Global Schema Creation Submodel

Reliability in every submodel can be combined by D-S evidence combination approach. Global schema (GS) can be generated in global schema generating submodel (GSGM). $IISA_i$ is in combination with the matching results of attributes of schema object in order to create optional global schema. A GS is calculated as:

$$F(GS_i)=\omega_1 f(IISA_i)+\omega_2 f(att) \tag{4}$$

Where $\omega_1 = \dfrac{f(IISA_i)}{f(IISA_i)+f(att)}, \omega_2 = \dfrac{f(att)}{f(IISA_i)+f(att)}$, $f(IISA_i)$ is the reliability of $IISA_i$, and $f(att)$ is the reliability of attribute matching in IISA.

Several global schemas with reliability are created after attribute matching and are operated as input by IISA.

6   Example and Analysis of Time Complexity

6.1   Example

As a motivating example, consider the schemas S_1 and S_2. S_1 models a data source of undergraduate students. Undergraduates register (reg) in courses that are taught (tch) by staff members. S_2 models a data source of postgraduate students, who may as well register in fourth-year undergraduate courses optionally to refresh their knowledge or to familiarize themselves with new subjects.

Detail information of the two schemas is illustrated as follows:

S_1: Student (PID, pname, sex, class, native, birthday, specialityid, dep_id);
Course (speciality, course, TID);
Staff (TID, tname, sex, department, birthday);
S_2: Student (UID, name, sex, classtype, native_p, speciality, department);
Course (course_id, speciality, TID);
Staff (TID, tname, sex, department, birthday, workdate).

Part of matching rules and part of reliability of semantic relationship are illustrated in the Table1.

The two global schemas GS_1 and GS_2 are created with reliability f_1 and f_2, respectively. The global schema is expressed as follows:

Student (ID, name, sex, class, native, birthday, speciality, dep);
Course (TID, speciality, course);
Staff (TID, name, sex, department, birthday, workdate).

Semantics between schema objects are not the same though forms of the two GS are identical. Explanation of semantics is not discussed in this paper.

Table1 Matching rules of semantic relation and confidence degree

Premise E_i (initial matching relationship)	Conclusion B_i (semantic relationship)	CF	$f(E_i)$	$f(B_i)$
$< S_1.\text{stu}, S_2.\text{stu} >$	$\{ \tilde{\cap}, \tilde{\not\nearrow} \}$	(0.2,0.7)	$f(E_1)=0.9$	$f(B_1)=0.88$
$<S_1.\text{course}, S_2.\text{course}>$	$\{ \tilde{\simeq} \}$	(0.9)	$f(E_2)=0.8$	$f(B_2)=0.956$
$<S_1.\text{staff}, S_2.\text{staff} >$	$\{ \tilde{\simeq}, \tilde{\subset} \}$	(0.5,0.3)	$f(E_3)=1$	$f(B_3)=0.86$
$< S_1.\text{reg}, S_2.\text{reg} >$	$\{ \tilde{\cap}, \tilde{\nearrow} \}$	(0.2,0.6)	$f(E_4)=0.9$	$f(B_4)=0.832$
$< S_1.\text{tch}, S_2.\text{tch} >$	$\{ \tilde{\simeq}, \tilde{\subset} \}$	(0.35,0.6)	$f(E_5)=1$	$f(B_5)=0.97$

6.2   Analysis of Complexity

Time complexity of IMM is $O(mn)$ in IMRM, where m is number of raw and n is number of column in IMM. The number of matching created in IMRM is decreased by nearly 50% by our approach. Time complexity of basic D-S evidence combination approach is $O(2^{kn})$, while time complexity of reformative combination approach [13] adopted in this paper is $O(k2^{2n}+k)$. Complexity of attribute matching is $O(ln)$, where l and n is number of attribute of two schema objects, respectively. Time Complexity is decreased to a certain extent through analysis.

7   Conclusion

In this paper, a new framework URSIM and a valuable application submodel DKAM are presented. Compared with current frameworks, our framework manages the inherent uncertainty in (semi-) automatic schema matching and schema integration with uncertainty, and supports six kinds of semantic relationships between schema objects.In particular, we have proved that the reliability of all schema integration phases can be combined by reformative D-S evidence combination approach [13], and the autonomic application of domain knowledge can improve the quality and performance of schema integration, enabling schema integration to be operated easily. Moreover, time complexity of schema integration is reduced.

However, there are still many open problems to be discussed as briefly listed in [6]. Meanwhile, uncertainty management of schema matching between relational schema and XML schema is not studied. In future work, we still need to experimentally verify the efficiency and effectiveness of our method. Further research should also be done to the complexity matching of attributes, creating of knowledge reasoning model, and acquiring of initial reliability.

References

1. Rahm, E., Bernstein, P.: A survey of approaches to automatic schema matching. VLDB Journal 10, 334–335 (2001)
2. Bernstein, P.: Applying model management to classical meta data problems. In: Proc. CIDR, pp. 209–220 (2003)
3. Bernstein, P., Pottinger, R.A.: Merging models based on given correspondences. In: Proc. 29th VLDB Conference, Berlin (2003)
4. Madhavan, J., Bernstein, P.A., Domingos, P., Halevy, A.Y.: Representing and reasoning about mappings between domain models. In: Proc. 18th NC on AAAI/IAAI, pp. 80–86 (2002)
5. Magnani, M., Rizopoulos, N., McBrien, P., Cucci, F.: Schema integration based on uncertain semantic mappings. In: Delcambre, L.M.L., Kop, C., Mayr, H.C., Mylopoulos, J., Pastor, Ó. (eds.) ER 2005. LNCS, vol. 3716, pp. 31–46. Springer, Heidelberg (2005)
6. Magnani, M., Montesi, D.: Uncertainty in data integration: current approaches and open problems. In: VLDB Workshop on Management of Uncertain Data, pp. 18–32 (2007)
7. Madhavan, J., Bernstein, P., Rahm, E.: Generic schema matching with Cupid. In: Proc. 27th VLDB Conference, pp. 49–58 (2001)
8. Melnik, S., Garcia-Molina, H., Rahm, E.: Similarity flooding: A versatile graph matching algorithm and its application to schema matching. In: ICDE, pp. 117–128 (2002)
9. Gal, A.: Managing Uncertainty in Schema Matching with top-K Schema Mappings. In: Spaccapietra, S., Aberer, K., Cudré-Mauroux, P. (eds.) Journal on Data Semantics VI. LNCS, vol. 4090, pp. 90–114. Springer, Heidelberg (2006)
10. Nottelmann, H., Straccia, U.: Information retrieval and machine learning for probabilistic schema matching. In: Proc., Manage., vol. 43(3), pp. 552–576 (2007)
11. Nottelmann, H., Straccia, U.: splmap: A probabilistic approach to schema matching. In: ECIR, pp. 81–95 (2005)
12. Dong, X.L., Halevy, A.Y., Yu, C.: Data integration with uncertainty. The VLDB Journal 18(2), 469–500 (2009)
13. Xia, W.J., Zhu, L.H., Tao, T.R.: Evidence combination approach based on uncertainty measure. Journal of Computer Application 29(8), 2257–2260 (2009)
14. Huang, F.Y., Feng, Y.Q., Wang, L., Lu, P.Y.: Relation-model-based Indefinite Knowledge Representation and Inference Technology and Its Application in KMS. Journal of NanJing University of Science and Technology 30(5), 653–658 (2006)
15. Liao, B.-S., Li, S.-J., Yao, Y., Gao, J.: Conceptual Model and Realization Methods of Autonomic Computing. Journal of Software 19(4), 779–802 (2008)

Multi-pose Face Recognition Using Fusion of Scale Invariant Features

I Gede Pasek Suta Wijaya, Keiichi Uchimura, and Gou Koutaki

Abstract. This paper presents a new multi-pose face recognition approach using fusion of scale invariant features (FSIF). The FSIF is a face descriptor representing 3D face images features which is created by fusing some scale invariant features extracted by scale invariant features transforms (SIFT) from several different poses of 2D face images. The main aim of this method is to avoid using 3D scanner for estimating any pose variations of a face image but it still have reasonable achievement compare to 3D-based face recognition method for multi-pose face recognition. The experimental results show the proposed method is sufficiently to overcame large face variability due to face pose variations.

1 Introduction

The comprehensive state of art of face recognition and current two-dimensional face recognition algorithms has been presented in Ref. [1], the recent techniques for 3D face recognition has been reviewed in Ref. [2], and the PCA-based algorithm for 3D face recognition has been developed in Ref. [3]. In addition, some statistical approaches such as Cook's Gaussian mixture-based Iterative Close Point algorithm, and Lee's Extended Gaussian Image model have been proposed, as presented in Ref. [4]. Furthermore, 3D face recognition based multi-features and multi-features fusion of face images were proposed in Ref. [4]. However, face recognition still have several challenges in terms of pose and illumination variations which are known as most difficult problems in the 2D face recognition. The 3D scanning techniques have

I Gede Pasek Suta Wijaya
Electrical Engineering Department, Mataram University, Jl. Majapahit 62 Mataram Indonesia
e-mail: gdepasek@navi.cs.kumamoto-u.ac.jp

I Gede Pasek Suta Wijaya · Keiichi Uchimura · Gou Koutaki
Computer Science and Electrical Engineering Department GSST, Kumamoto University, Kurokami 2-39-1, Kumamoto Japan
e-mail: [uchimura,koutaki]@cs.kumamoto-u.ac.jp

F.L. Gaol et al. (Eds.): Proc. of the 2011 2nd International Congress on CACS, AISC 144, pp. 207–213.
springerlink.com   © Springer-Verlag Berlin Heidelberg 2012

been proposed to overcome the mentioned problems in face recognition. However, the 3D scanning techniques require 3D scanner/camera which is an expensive tool for data registration.

In this paper, we propose an alternative technique for multi-pose face recognition using fusion of scale invariant features (FSIF) based face descriptor which is built from 2D face images. The main aim of this method is to avoid using the 3D scanner for estimating any pose variations of a face image with still having reasonable achievement compare to recent 3D-based face recognition for multi-pose face images, as presented in Ref. [4]. It can be realized because some pose variations of a person contains the sub-3D face features which can be extracted by scale invariant features transforms (SIFT) algorithms. From the extracted face features, the FSIF-based face descriptor can be built using fusion technique. The detail explanation of the SIFT and its implementation can be found in Refs. [5, 6].

2   FSIF-Based Face Descriptor

The algorithm for creating FSIF-based face descriptor is presented in Fig. 1(a), which works as follows:

1. Suppose, we have there face images (denoted by P_1, P_2, and P_3) as input set. From the selected images, the scale invariant features of each face images are extracted using SIFT algorithm from Ref. [5].
2. Removing the redundant scale invariant features of P_1 and P_2, and of P_3 and P_2, step by step using intersection ($\cap$) and subtraction operation. The intersection operation is implemented to find out the location of the redundant features. Then, the redundant features are removed by subtracting the original features with the detected redundant features.

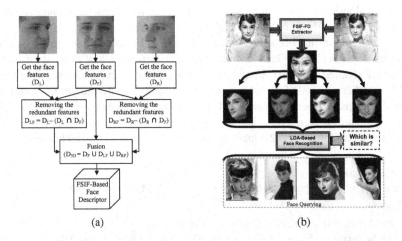

(a)   (b)

Fig. 1 The block diagram of: creating FSIF-based face descriptor (a) and proposed face recognition (b).

3. Finally, fusing all of the non-redundant features into a descriptor using the union ($\cup$) operation.

In this case, the FSIF-based face descriptor is a two dimensional data which represents some pose variations. The benefit of this representation is simple/compact and requires less memory space compare to the real 3D data. In addition, the more 2D-face images are included for creating of the FSIF-based face descriptor the more rich face descriptors will be gotten. Consequently, if the more rich face descriptors are used for recognition, the higher recognition rate will be achieved.

3 The Implementation for Face Recognition

To implement the FSIF-based face descriptor for multi-pose face recognition, we employ two techniques: SIFT features matching and Linear Discriminant Analysis (LDA) based classifier.

The first technique classifies directly the probe FSIF-based face descriptor with the registered FSIF-based face descriptor using SIFT matching algorithm which determines the number of matching features points using the matching algorithm in Ref. [6]. The classification criterion is defined based on number of matching features points. The face descriptor which has the largest number of matching features points is concludes as the best likeness.

Secondly, the FSIF is integrated with LDA based classifier for face recognition. The detail explanation of LDA-based classifier can be found in Refs. [3, 7]. The block diagram of the FSIF-LDA-based face recognition can be seen on Fig. 1 (b), which works as follows:

1. The input images are extracted into a FSIF-based face descriptors representing several pose variations features. In this case, each class of registered face is represented using a descriptor.
2. From input FSIF set, the LDA-based features cluster is trained to obtain the optimum projection matrix (W_{LDA}) which satisfies the following criterion.

$$J(W) = arg\,max \frac{|W^T S_b W|}{|W^T S_w W|} \qquad (1)$$

Where S_w and S_b are within class scatter and between class scatter of the training set, respectively. The optimum W_{LDA} is obtained by eigen analysis. Next, the registered FSIF-based face descriptors can be projected for reducing the dimensional size using the $D_i^p = W_{LDA}^T D_i$, where D_i^p is i-th projected FSIF-based face descriptor, D_i is i-th input FSIF-based face descriptor, $i = 1 \cdots N$, and N is total classes. The size of D_i^p is much less then D_i. Finally, the optimum W_{LDA} and the D^p are saved in the database which are used in the recognition process.
3. In the matching process, the query FSIF-based face descriptor (D_q) is projected into D_q^p as done on training process and then the D_q^p is matched with each registered D_i^p using the procedure in the first technique classification.

4 Experimental and Results

In order to know the performance of the proposed method, several experiments using data from challenges databases: the ITS-Lab.[7], ORL[8], CVL[9], and GTV databases[10] were carried out using PC with specification: Core-Duo Processor 1.7 GHz and 2 GB RAM. The example of face pose variations from the mentioned databases is shown in Fig. 2.

The first experiment was carried out to investigate the effect of FSIF-based face descriptor on the recognition rate. In this test, we selected first there images ($P_1 - P_3$ of face pose variations as shown in Fig. 2) for creating FSIF-based face descriptor (training set) and the remaining face images were selected for testing set. In addition, the achievements of our proposed method were compared with the previous methods: PCA and LDA based face recognition[?, 7] called as base-line methods. In this case, we implemented the DCT-based holistic features (HF) instead of raw face images on base-line methods. The PCA and LDA with HF have been reported to provide reasonable recognition rate compared to that of without HF, as described in Refs. [7]. The experimental result shows the FSIF-based face descriptor is sufficiently for representing any face pose variations which is proved by providing higher recognition rate than that of base-line method, as shown in Table 1(Left). Furthermore, the proposed method provides sufficient improvement for ITS-Lab and ORL databases and significant improvement for CVL and GTV databases containing large face pose variations (see Fig. 2) than that of base-line methods. It means the proposed FSIF-based face descriptor can be used for solving the multi-pose difficulty on face recognition.

The second experiment was performed to know the performance of the integration of FSIF-based face descriptor with the dimensional reduction technique (PCA and LDA) for face recognition. Although the descriptor dimensional size decrease significantly (25% of the original size), it still provides higher and much higher recognition rate than that of without LDA (#3) and base-line method (#1 and #2), as shown in Table 1 (Right). It can be achieved, because the PCA and LDA not only use for dimensional reduction but also have ability to transform the original data into more separable data. In other words, the PCA and LDA have ability to separate

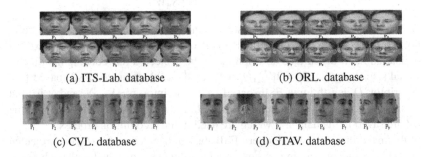

(a) ITS-Lab. database   (b) ORL. database

(c) CVL. database   (d) GTAV. database

Fig. 2 Example of face pose variations of single subject.

Table 1 Left: the comparison of recognition rate of our proposed method with that of the baseline methods (PCA and LDA) and Right: performance of the combination of 3D-descriptor with PCA and LDA compared vs. baseline methods.

Database	Recognition Rate (%)			Improvement (%)		Database	Recognition Rate (%)		Improvement (%)		
	#1	#2	#3	#1 vs #3	#2 vs. #3		#4	#5	#1 vs. #4	#2 vs. #5	#3 vs.# 5
ITS	85.54	97.00	97.42	11.88	0.41	ITS	95.56	98.24	10.02	1.24	0.8264
ORL	93.00	94.50	97.25	4.25	2.75	ORL	95.00	97.25	2.00	2.75	0.0000
CVL	65.29	66.67	70.80	5.51	4.14	CVL	69.42	72.93	4.14	6.27	2.1300
GTAV	49.24	52.44	64.90	15.66	12.46	GTAV	55.47	66.33	6.23	13.89	1.4300

Note HF+PCA: #1, HF+ LDA: #2, and FSIF-FD: #3.   Note FSIF-FD +PCA: #4 and FSIF-FD + LDA: #5.

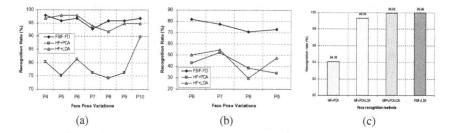

Fig. 3 The recognition rate vs. face pose variations on ITS database (a), on GTAV database (b), and the recognition our method vs. that of recent 3D-methods (c).

the overlapping between class data of the training set. In addition, the impacts of dimensional reduction are not only decrease the memory space requirement (from 200 KB to 50 KB per class) but also the recognition processing time (from 1.5 to 0.56 second).

The third experiment was performed to evaluate the stability of the proposed method against face pose variations. It was performed in the ITS-Lab. database representing small pose variations data, and GTAV database representing large pose variations data. From the ITS-Lab. and GTAV databases three images ($P_1 - P_3$ of Fig. 2(a)) and five images ($P_1 - P_5$ of Fig. 2(d)) per class were selected for creating FSIF-based face descriptor, respectively. The result shows that proposed method gives better and more stable recognition rate than that of the based-line method, as plotted in Fig. 3 (a and b). This result supports the previous experimental results which are presented in Table 1. In addition, the significant recognition rate improvement is given by GTAV face database containing large face variability due to pose variations. It means the proposed FSIF-based face descriptor created from 2D dimensional images is sufficiently for representing the large variability face images due to pose variations.

The last experiment was carried out to compare our proposed method with the recent 3D-face recognitions which work based on combination multi-features (MF) and multi-feature fusion (MFF) of 3D face image with PCA and PCALDA (see Ref. [4]). In this test, we compare three best variants of those methods called as MF+PCA, MF+PCALDA and MFF+PCALDA. The experiment was done in the

ITS-Lab face database version 1 which contains 40 classes and each class consists of 10 face pose variations which were acquired by Konica Minolta 3D-camera series VIVID 900. The testing parameters were setup the same as done in Ref. [4]: five images of each class were chosen for training set and the remaining images were selected for testing and the features size was 25% of the original size.

The experimental result shows that our proposed method can recognize almost all of the testing data (see Fig. 3(c)). It means our proposed method is an alternative solution for multi-pose face recognition with having reasonable achievements compared to 3D-based face recognition. Even though the recognition of FSIF+LDA is not much different with the MFF+PCALDA but our proposed method does not require 3D camera sensors for making the multi-features at all. In other words, our proposed method is the cheapest 2D-based face recognition approach which can be implemented for real time multi-pose face recognition with 2D web-camera as the image capturing.

5 Conclusion and Future Works

A multi-pose face recognition approach based on FSIF-based face descriptor is an alternative solution for building multi-pose face recognition without requiring the 3D sensor. The proposed method gives good recognition rate, provides robust achievements to face pose variations, and provides better recognition rate over recent 3D-face recognition methods. In future, this research will be continued to the visualize the FSIF-based face descriptor into 3D face descriptor and to modify the FSIF-based face descriptor extraction algorithm which can work faster in both building and matching the descriptor than that of current implementation.

Acknowledgement. I would like to send my appreciation and my great thank to Innovation Project of Kumamoto University for many support to this research, and also to owner of ORL, CVL, GTAV databases.

References

1. Zhao, W., Chellappa, R., Phillips, P.J., et al.: Face Recognition: A Literature Survey. ACM Computing Surveys 35(4), 399–458 (2003)
2. Chang, K.I., Bowyre, K.W., Flyn, P.J.: An Evaluation of Multi-modal 2D+3D Face Biometrics. IEEE Transactions on PAMI 27, 619–624 (2005)
3. Chen, W., Meng, J.-E., Wu, S.: PCA and LDA in DCT Domain. Pattern Recognition Letter 26, 2474–2482 (2005)
4. Cuicui, Z., Uchimura, K., Zhang, C., et al.: 3D Face Recognition Using Multi-level Multi-feature Fusion. In: Proceedings of the 4th Pacific-Rim Symposium on Image and Video Technology (PSIVT 2010), Singapore, pp. 21–26 (2010)
5. Lowe, D.G.: Distinctive Image Features from Scale-Invariant Keypoints. International Journal of Computer Vision 60(2), 91–110 (2004)
6. Lowe, D.G.: Object Recognition from Local Scale-Invariant Features. In: Proceedings of the International Conference on Computer Vision, Corfu (1999)

7. Wijaya, I.G.P.S., Uchimura, K., Hu, Z.: Improving the PDLDA Based Face Recognition Using Lighting Compensation. In: Proceedings of Workshop of Image Electronics and Visual Computing 2010, Nice France, CDROM (2010)
8. Samaria, F., Harter, A.: Parametrization of a stochastic model for human face identification. In: The 2nd IEEE Workshop on Applications of Computer Vision, Sarasota Florida, pp. 138–142 (1994)
9. http://lrv.fri.uni-lj.si/facedb.html
10. http://gps-tsc.upc.es/GTAV/ResearchAreas/UPCFaceDatabase/GTAVFaceDatabase.htm

Development of a Cost Predicting Model for Maintenance of University Buildings

Chang-Sian Li and Sy-Jye Guo

Abstract. If a university building's performance fails to conform to what is required or anticipated, the engineer-of-record for the facility can be charged with negligence and may bear liability for damages arising from that failure. An engineer's negligence is assessed by arranging the engineer's actions relative to the standard maintenance cost and budget of the profession. This paper briefly describes the meaning and application of the standard of maintenance costs and budgets, and addresses the meaning of maintenance cost and budgeting with regard to engineering. To examine this issue, this paper presents a case study on the operation maintenance phase of 4 university buildings on the campus of National Taiwan University. Using historical data on maintenance and repair over a 42-year period, a cost prediction model using the life-cycle cost (LCC) was determined using three different methods: (1) simple linear regression (SLR); (2) multiple regression (MR); and finally (3) a back propagation artificial neural network (BPN). The research results showed that the BPN model had keen estimation ability. This paper implemented the BPN model in a case study to analyze the problems of maintenance costs and budgeting for university buildings. The paper helps to set a legitimate standard for arranging repair maintenance costs, and proposes a plan and standard for the repair maintenance strategy of structures.

1 Introduction

The life cycle of a university building is similar to the human life cycle. Affected by age, university buildings may suffer physical deterioration, leading to both functional obsolescence and external obsolescence, which then affect the building's usability. Therefore, long-term planning and maintenance management are quite important. Taking the operation stage of the life cycle of existing university buildings in Taiwan as an example, maintenance and renovation costs may increase with a building's age, but there are no annual maintenance and renovation

Chang-Sian Li
Department of Civil Engineering, National Taiwan University, Taipei, Taiwan
email: d95521006@ntu.edu.tw

Sy-Jye Guo
Department of Civil Engineering, National Taiwan University, Taipei, Taiwan
email: sjguo@ntu.edu.tw

F.L. Gaol et al. (Eds.): Proc. of the 2011 2nd International Congress on CACS, AISC 144, pp. 215–221.
springerlink.com © Springer-Verlag Berlin Heidelberg 2012

costs that can be used as a reference for creating a budget. There are numerous university buildings, but with inadequate financial funds, removing and re-building them is difficult. In recent years, financial assistance from the Ministry of Education, which is responsible for universities in Taiwan, has been gradually re-duced. Subsidies for universities have also been decreasing steadily, leading to an imbalance of supply and demand in the allocation of maintenance and renovation budgets for university buildings. With the shortage of funds, the gradual obsoles-cence and deterioration of the facilities, and the continued increases in mainten-ance and renovation costs, the proper allocation of limited maintenance and renovation budgets is of great significance. This study had the following objectives:

Build a cost prediction model where life cycle cost (LCC) is determined using three different methodologies: (1) simple linear regression (SLR); (2) multiple re-gression (MR); and finally (3) a back propagation artificial neural network (BPN). Implement the BPN model in a case study to analyze the problems of maintenance costs and budgeting for university buildings. The advantages and limitations of each method are discussed with the findings of the case study.

2 Model Implementation

A. Build a Cost Prediction Model

The building age, number of floors, number of classrooms, and the elevators were independent variables of the prediction model. This study arranged four indepen-dent variables to build different prediction models, found the relationship between independent variable and dependent variables, and then calculated the accumu-lated maintenance costs of university buildings over time (Table 1).

Table 1 Prediction Models

Model Variables	Model I	Model II	Model III	Model IV
Variables	Building Age, Number of Floors, Number of Classrooms, Elevators	Building Age, Number of Floors, Elevators	Building Age, Number of Floors, Number of Classrooms	Building Age, Number of Floors
Independent Variables	Accumulation of Maintenance Costs			

B. Model A: SLR for the Cost Prediction Model

SLR for a cost prediction model was built using historical data on maintenance and repair for university buildings over a 42-year period, in which the equation

=2.6388x^2+35.25x-26.678, and R^2=0.9008. Research results showed that the maintenance and renovation costs estimated by the SLR model although could reflect the linear relationship between building age and the accumulation of maintenance costs over the years. The SLR model proved that the maintenance and renovation costs of university buildings develop linearly and may be of a fluctuation type. But SLR is unable to explain fluctuation type and reflect other important parameters. The SLR model had a poor estimation effect and explanation capability.

C.   Model B: MR for the Cost Prediction Model

MR for the cost prediction model was built using historical data on maintenance and repair for university buildings over a 42-year period. This study predicted the maintenance costs and related data, and then compared these calculations with the actual maintenance costs for the buildings. Research results showed had two situations that the date for the 4 university buildings. (1) Derived from MR for General Courses Building and Freshman building will result in negative cost prediction for building ages under 2 years (Fig.1. and Fig.4.). (2) Derived from MR for Common Courses Building and General Purpose building will result in negative cost prediction for building ages under 5 years (Fig.2. and Fig.3.). The results for the first two years and first five years were negative. Research results showed differed from the actual maintenance cost. Therefore, the prediction yielded from the MR models differed greatly from the actual results. Research results showed that the maintenance and renovation costs estimated by the MR model increased linearly and could not reflect the relationship between them and the actual maintenance and renovation costs of university buildings. The MR model also proved that the maintenance and renovation costs of university buildings develop non-linearly and may be of fluctuation type. The MR model had a poor estimation effect and explanation capability.

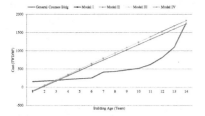 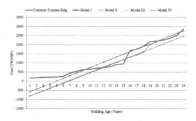

Fig. 1 MR for the General Courses Bldg. (1USD=30TWD)

Fig. 2 MR for the Common Courses Bldg. (1USD=30TWD)

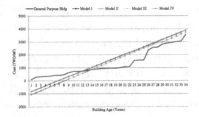 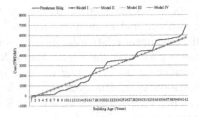

Fig. 3 MR for the General Purpose Bldg. (1USD=30TWD)

Fig. 4 MR for the Freshman Bldg. (1USD=30TWD)

D.   Model C: BPN for the Cost Prediction Model

Renovation and component variables from the four buildings were used for the BPN. The basic structure of the network included hidden layers, an input layer and an output layer. Setting different numbers of hidden layers, nodes, loops, learning speeds and momentum factors in the training process yielded different results. This study obtained the parameters with a better learning effect and estimation effect through trial and error. BPN for the cost prediction model was built using historical data on maintenance and repair for university buildings over a 42-year period. The yearly maintenance records for the 4 university buildings in the 42 – year data are 114. A total of 76 renovation samples were used for training and the other 38 renovation samples were used for testing. Renovation information referred to actual yearly maintenance and renovations, and was divided into three types: periodic maintenance, non-periodic repairs, and demand change costs. The model was then applied to the maintenance and renovation cost prediction for the General Courses building (Fig.5.), the Common Courses building (Fig.6.), the General Purpose building (Fig.7.) and the Freshman building (Fig.8.). Research results showed that the maintenance and renovation costs estimated by the BPN model had distribution trends that were similar to the actual maintenance and re-novation costs of university buildings, and they showed fluctuation. The BPN model proved that the maintenance and renovation costs of university buildings develop non-linearly, and it indicated high correlation. The BPN model had a good estimation effect and explanation ability.

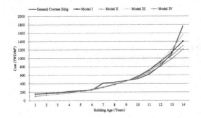 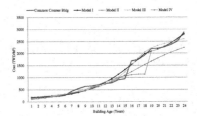

Fig. 5 BPN for the General Courses Bldg. (1USD=30TWD)

Fig. 6 BPN for the Common Courses Bldg. (1USD=30TWD)

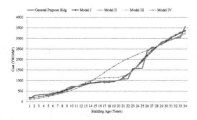
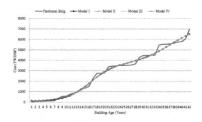

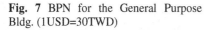

Fig. 7 BPN for the General Purpose Bldg. (1USD=30TWD)

Fig. 8 BPN for the Freshman Bldg. (1USD=30TWD)

E.   Prediction Result Error

Table 2 shows the calculated error by the Root Mean Square Error between the MR prediction model and the actual maintenance cost of university buildings. The results were compared with the result from the BPN prediction model to select the optimization model. From the predicted maintenance error, the maintenance and renovation costs were predicted. The BPN prediction model was more accurate than the MR prediction model in predicting cost.

Table 2 Prediction Result Error (UNIT: USD/M^2)

Methodology	Model Bldg.	Model I	Model II	Model III	Model IV
MR	General Courses	15.5	14.07	15.47	15.6
	Common Courses	10.23	13.57	10.27	12.7
	General Purpose	20.57	20.23	20.5	24.17
	Freshman	14.73	16.2	14.8	14.8
BPN	General Courses	3.63	4.7	2.87	4.4
	Common Courses	4.1	7.87	3.77	9.1
	General Purpose	3.8	7.83	3.93	12.8
	Freshman	7.27	6.97	7.17	7.47

3   Budget Arrangements for Maintenance and Renovation Costs

This study predicted the accumulation of maintenance costs over time using BPN Model II; if a building was aged 30 years, every five years would be a cycle and every five floors would be a level. Whether elevators had been installed was also

Table 3 Budget Arrangement Base for University Building Maintenance Arrangement (UNIT: USD/M^2)

Floors Age	Elevator	1-5 Years	6-10 Years	11-15 Years	16-20 Years	21-25 Years	26-30 Years	30+ Years
Fewer than 5 Floors	With Elevator	1.67	2.4	6.07	9.07	4.5	4.4	6.67
More than 5 Floors		3.27	5.67	9.63	6.4	6.83	8.93	6.33
Fewer than 5 Floors	Without Elevator	0.7	1.13	2.57	6.27	7.5	3.4	3.57

noted. Table 3 shows the predicted maintenance and renovation results for university buildings according to different conditions.

4 Summary and Conclusions

Development of a budget arrangement benchmark: The budget arrangement for maintenance and renovation was developed by the BPN prediction model. The prediction of cost budget arrangements according to the building age (each five year period was a cycle), number of floors (every five floors was a layer) and elevators (whether to set elevators) was used in the budget planning for maintenance and renovation. The study results showed that the BPN model achieved a successful application of different building ages for university buildings. An empirical formula was built on historical data for maintenance and repair, so as to learn and deduce the maintenance and renovation costs of university buildings.

Comparative analysis of the prediction model for maintenance and renovation cost: This paper successfully built a cost prediction model using the life-cycle cost of maintenance costs and budgets for university buildings. Research results showed that the BPN model was superior to both the MR model and the SLR model. The BPN model was built on historical data on maintenance and repair, which solved the problem of maintenance and renovation costs being subject to fluctuation with building age. The BPN model solves problems that cannot be addressed by either the SLR model or the MR model. The BPN model accepts logistic and numerical independent variables as well as those of orderly classification and disorderly classification, has a strong model learning and estimation ability, and can be applied to the problems of function mapping, the prediction of number sequences, sample classification and non-linear development. Examples of this include maintenance and renovation costs for buildings. Recommendation of arrangement benchmark: Research results indicated the peak distribution and periodicity for the maintenance and renovation of university buildings. If a building has one to five floors (without an elevator), its first peak of renovation will be around 25 years of age. If a building has one to five floors (with an elevator), its first peak of renovation will be around 20 years of age and its second renovation peak will come when it is 35 years of age. For a building with more than five floors (with an elevator), its first renovation peak will be around 15 years of age and its second renovation peak will occur when it is 30 years of age.

References

[1] ASCE Task Committee on Application of Artificial Neural Networks in Hydrology, Artificial Neural Networks in Hydrology, Part I & II: Preliminary Concepts. Journal of Hydrology Engineering, ASCE 5(2), 115–137 (2000)
[2] Farran, M.: Comparative Analysis of Life-Cycle Costing for Rehabilitating Infrastructure Systems. Journal of Performance of Constructed Facilities, ASCE 23(5), 320–326 (2009)
[3] Kim, G.K.: Development of a life cycle cost estimate system for structures of light rail transit infrastructure. Automation in Construction 19(3), 308–325 (2010)
[4] Ottoman, G.R.: Budgeting for Facility Maintenance and Repair. I & II: Methods and Models & Multicriteria Process for Model Selection. Journal of Management in Engineering, ASCE 15(4), 71–95 (1999)

Functional Treatment of Bilingual Alignment and Its Application to Semantic Processing

Yoshihiko Nitta

Abstract. In this paper we propose a functional treatment of bilingual alignment together with some new mathematical functions to describe sentential semantics transparently. We propose that every sentence is composed of two categories: kernel sentence and meta-sentence. Meta-sentence can be understood as a sentence structuring operator, while kernel sentence is a simple structured, mono-predicate sentence. Kernel sentence has obvious translation of canonical form, while meta-sentence represents logical-semantic structure of sentence which takes kernel sentence(s) as its dominating variable(s). From meta-sentences we can draw a lot of useful semantic information. We also show typical examples of meta-sentences obtained from typical Japanese short sentence known as "haiku".

Keywords: Bilingual Alignment, Meta-sentence, Kernel Sentence, Translation, Semantic Processing, Haiku.

1   Introduction

We propose some new view of bilingual alignment together with some functions operating on alignment. These new functions intend to treat various aspects of language information processing such as translation, filtering, condensation, summarization, keyword extraction, etc. in some simple mechanical view.

Currently it is widely recognized that bilingual alignments are effective means for constructing various machine translation[7,8] and foreign language education. So far word-by-word or phrase-by phrase alignments are focused and various automatic constructing techniques have been investigated. (For example see: [2,4,13]. A yet another interesting treatment of bilingual alignment can be seen in [5].)

In this paper, however, we treat bilingual alignment is a kind of mathematical system constructing from translation function Tran, kernel sentence K, and meta-sentence M( ).

If we denote Sj as source language sentence, Se as target (=translated) sentence and Tran as translator or transformation from Sj to Se, then we can write:

$$\text{Tran}(Sj) = Se$$

F.L. Gaol et al. (Eds.): Proc. of the 2011 2nd International Congress on CACS, AISC 144, pp. 223–229.
springerlink.com   © Springer-Verlag Berlin Heidelberg 2012

If Sj and Se belong to same language, Tran means paraphrase or interpretation.

We propose that each sentence S is composed of two categories: kernel sentence(s) K and meta-sentence M (). Meta-sentence can be understood as a sentence structuring operator. Thus we can write

$$S=M(K),\quad Tran(S)=Tran(M(K))=Tran(M)\,(Tran(K)).$$

Roughly speaking, kernel sentence K represents simple structured, mono-predicate sentence which has obvious translation or interpretation of canonical form, while meta-sentence M() represents logico-semantic structure of sentence or pragmatic superficial structure of sentence, which takes K as its dominating variable.

In this treatment, bilingual alignment can be understood as meta-functional mapping Tran between two meta-sentence Mj() and Me().

$$Tran:\ Mj()\rightarrow\ Me().$$

From Meta-sentences we can describe a lot of useful semantic and pragmatic knowledge about sentence formation. We show typical examples of meta-sentences obtained from typical Japanese sentences which contain various ellipses together with English translation.

2 Meta-Sentence and Bilingual Alignment

First of all we claim that every sentence is composed of two categories: kernel sentence K and meta-sentence. M() Let us denote K as kernel sentence and M() as meta-sentence. Meta-sentence behave like a function over kernel sentence, thus we can write:

$$S = M(K)$$

Here S denotes a sentence. If we would emphasize that sentence S is written by language j or language e, then we would write:

$$Sj = Mj\,(Kj)$$

or

$$Se = Me\,(Ke)$$

where, Mj means the meta-sentence in language j, and Kj means kernel sentence in language j. (j can be replaced by e.)

If we denote the translation between language j and language e as Tran,
Tran: Sj → Se
Then at the same time,
Tran: Mj()→ Me()

(Mj (Kj), Me (Ke)) can be interpreted as the bilingual alignment between language j and e, whose concrete example will be shown later.

(1) Explanation about Kernel Sentence K

The "kernel" designates the kernel of sentential meaning and the kernel of translation or interpretation. Roughly speaking the kernel sentence has a simple predicate-and-argument structure, which has one verb and dominating noun(s). Kernel sentence can be a form of no verb(-phrase) , that is of noun(-phrase) only.

The kernel sentence has usually an obvious translation of canonical form and has high possibility of pseudo symmetrical translation property.

If sentence K has the symmetrical translation property, then

$$\text{Tran}(K) = K' \quad \text{and} \quad \text{Tran}^{-1}(K') = K$$

In case of pseudo- symmetrical translation property,

$$\text{Tran}^{-1}(K') \fallingdotseq K$$

The antonym of "symmetrical translation property" is "directional translation property", that is $\text{Tran}^{-1}(\text{Tran}(S)) \neq S$.

If sentence S has symmetrical translation property, then this sentence is easily translatable and has obvious translation. Thus these kinds of sentences are expected to form some baseline for overall translation or interpretation.

(2) Explanation about Meta-Sentence M()

The meta-sentence is a kind of hyper-sentence that operates over kernel sentence(s). This operation is viewed as sentence constructing operation, which makes actual surface sentence from kernel sentence(s). The essential part of meta-sentence depends on function word in Japanese; and depends on preposition and connection word in English.

3   Translation of Meta-Sentences

First let us show concrete examples.

- Surface sentences:
  S1: Keiji toiu Shokugyou gara kare-wa metuki-ga surudoi..
      (刑事という職業がら彼は目つきが鋭い。)
  S2: Basho-gara-mo wakimae-zu kare-wa oogoe-wo dasu.
      (場所がらもわきまえず彼は大声を出す。)
  S3: Being a detective, he has a sharp eye.
  S4: His talks in a loud voice regardless of the occasion.
- Kernel sentences:
  K1 = "Kare-wa metuki-ga surudoi"
        ("彼は目つきが鋭い")
  K2 = "kare-wa kane-niwa binkan de-aru"
        ("彼は大声を出す")

K3= "He has a sharp eye."
K4="He talks in a loud voice."

- Meta-sentences:

M1(N,Sent) = N gara(がら)   Sent

M2(N,Sent) = N   gara-mo wakimae-zu (がらもわきまえず) Sent

M3(N,Sent) = Being N, Sent

M4(N,Sent) = Regardless of N, Sent

Here N and Sent are meta-symbols (syntactic variables) as usual.

Later we will use simplified expressions such as follows,

M1 = $\sim$ gara ...

M3 = Being $\sim$, ...

Eventually translation is the functional mapping Tran( ) over kernel sentence K and/or meta-sentence M( ). Bilingual alignments are the outcomes from Tran( ).

Tran: K1 $\rightarrow$ K3   or   Tran(K1) = K3

Tran: M1( )$\rightarrow$ M3( )   or   Tran(M1( )) = M3( )

M1(N, K1)= M1("Keiji to-iu Shokugyou", "Kare-wa metuki-ga surudoi")

   ="Keiji to-iu Shokugyou gara   Kare-wa metuki-ga surudoi"

Tran(M1) (Tran("Keiji to-iu Shokugyou"), Tran("Kare-wa metuki-ga surudoi"))

   = M3(Tran("Keiji to-iu Shokugyou"), Tran("Kare-wa metuki-ga surudoi"))

   = M3("a detective", "he has a sharp eye. "

   = "Being a detective, he has a sharp eye."

For each meta-sentence M( ), its mapped image Tran(M()) is essentially depend on the value of its dominant variable(s) K. The typical evidence of this dependency is illustrated in the following examples.

In the above N="Keiji to-iu Shokugyou" and K= "Kare-wa metuki-ga surudoi".

Now if we put N="Sensei toiu shokugyou" ("先生という職業")

and K = "Watashi-wa okonai-wo tsutsusinde-iru."("私は行いを慎んでいる")

Then,

Tran (M1(N, K)) $\neq$ M3(Tran(N), Tran(K))

but

Tran (M1(N, K)) = M5(Tran(N), Tran(K))

or

Tran: M1(N, K) $\rightarrow$ M5(Tran(N), Tran(K)

Here M5(N, K) = As N, K

Finally,

M5 (Tran(N), Tran(K)) = "As Tran(N), Tran(K)."

                     = "As a teacher, I try to be careful of my behavior.".

4   Collected Examples of Meta-Sentences

First we have to note that the directionality of translation.

   $\text{Tran}^{-1}\,(\text{Tran}(S)\,)\neq S$ for almost all translation Tran and sentence S.

The notion of directionality or unsymmetrical property has been studied pragmatically in translation theory [6, 9,10] in opposition to the notion of natural translation.

Thus translation of meta-sentence M( ) and bilingual alignment inevitably have directionality or unsymmetrical property, $\text{Tran}^{-1}\,(\text{Tran}(M(\,))\,)\neq M(\,)$

This is the main reason of overall difficulty of translation. Here "j" stands for Japanese language and "e" stands for English language.

In order to collect meta-sentences useful for j-to-e directional translation, we can use some effective trigger word or phrases as follows.

   ・ ～shite...suru [～して...する]
   ・ ～de-aru-ga-yue-ni ...de-aru（da）[～であるがゆえに ...である（だ）]
   ・ ～shita-node... [～したので...]
   ・ ～da-kara... [～だから...]
   ・ ～(N1)ya～(N2)wa...(VP)sase-te～(Adv) [～や～は...させてで～]
   ・ ～(VP)ya～(N1)ga...(VP)shite-mo～(N2) wa～(N3) [～や～が...しても～は～]
   ・ ～(N1)ga～(N2)wo...(VP)shitaru～(N3)kana [～が～を...したる～かな]
   ・ ～(N1)ya～(Adv) ...(VP) ～(N2) [～や～...する～]

(The latter four meta-sentences are taken from "haiku", typical simplified Japanese sentence that has various interesting ellipses.)

These kinds of trigger-word oriented meta-sentences are able to be obtained relatively easily from well-designed bilingual corpus by using regular expression based-text mining technique. The character class switching from Kanji to Hiragana gives good clue to effective mining [11].

Typical Japanese meta-sentebces have generally the structure as follows:

   M (K1, K2) =   "K1   shite   K2   shita"
          =   "K1   shite   K2   ni-natta"
          = "K1   [da]kara   K2shita"

Here   K1 = kernel sentence representing cause or reason,
          K2 = consequence, result, event
   ・ The translated meta-sentence has generally the style as follows:
   Me(Tran(K2), Tran(K1))   =   Sent2 +Sent1
                    =   Sent2 + Conjunction + Sent1
                    =   Sent2 + VPing1
                    =   Sent2 + Preposition + VPing1
                    =   Sent2 + to + VPinfinitive1
   Here,   Sent2 = Tran(K2),   VP1 = Tran(K1).

(1) K1∈origin, cause)   K2∈effect、 reward

Mj(K1,K2)= "Kare-wa shingou-wo mushi-**shite** omoi batsu-wo uke-**ta**."

[彼は信号無視を　して　大怪我をした。]

Me(tran(K2), tran(K1)) = He was severely injured **for** runn**ing** a red light.

Me( ) = Sent2 + for + VP**ing**1

(2) K1∈involuntary event occurrences   K2∈cause, reason

Mj(K1,K2)= "Wa-ga ware-wo oki-wasure-taru atusa-kana."

[我（わ）が我を置き忘れたる暑さかな。]

Me(tran(K2), tran(K1)) = Because of the heat, I forget where to put myself.

Me( ) = Because of N2 + Sent1

If bilingual alignment is equipped with satellite circumstance information, the meta-sentence construction will be much more powerful. A somewhat similar idea can be seen in [1]. Currently we are putting the satellite information on Adv-position.

5   Concluding Remark

We have proposed the functional view of bilingual alignment and demonstrate its power and effectiveness in semantic processing together with beginner level translation aid. Most vital feature exists in the notion of meta-sentence which is a kind of operator for surface sentence formation. Meta-sentence is working over the domain of kernel sentences. The kernel sentence behaves like a fundamental element in language translation or interpretation, which has canonical form of translation correspondence in target language and also has bi-directional (symmetrical) translation property. Thus kernel sentence has simple structure and is easily translatable to other linguistic form, which forms the baseline of overall translation.

Currently meta-sentences have been extracted and collected semi-automatically using naive pattern matching programs based on regular expression[3, 12].

Our next step is to construct a more powerful program to extract and manipulate meta-sentences.

References

1. Bentivogli, L., Pianta, E.: Exploiting Parallel Texts in the Creation of Multilingual Semantically Annotated Resources: the MultiSemiCor Corpus. Natural Language En-gineering 11(3), 247–261 (2005)
2. Church, K., Dagan, I., Gale, W., Fung, P., Satish, B., Helfman, J.: Aligning Parallel Texts: Do Methods Developed for English French Generalize to Asian Languages? In: Proceedings of the Pacific Asia Conference on Formal and Computational Linguistics (1993)
3. Kinyon, A.: A Language-Independent Shallow-Parser Compiler. In: Proc. 39th ACL Ann. Meeting (European Chapter), pp. 322–329 (2001)

4. Macklovitch, E., Marie-Louise, H.: Line 'em up: Advances in Alignment Technology and Their Impact on Translation Support Tools. In: AMTA, pp. 145–156 (1996)
5. Mihalcea, R., Simard, M.: Parallel Texts. Natural Language Engineering 11(3), 239–246 (2005)
6. Munday, J.: Introducing Translation Studies. Taylor & Francis (2009)
7. Nitta, Y.: Idiosyncratic Gap: A Tough Problem to Machine Translation. In: Proc. Comp. Linguistics, COLING 1986, ACL (Assoc. Comp. Ling.) (1986)
8. Nitta, Y.: Problems of Machine Translation: From a Viewpoint of Logical Semantics. Economic Review of Nihon University, Nihon University, Tokyo 72(2), 23–42 (2002)
9. Nitta, Y.: The Utility and Problem of Insufficient Machine Translation. Economic Review of Nihon University 80(4), 1–54 (2001)
10. Pim, A.: Exploring Translation Theories. Routledge, Taylor & Francis (2010)
11. Saraki, M., Nitta, Y.: The Semantic Classification of Verb Conjunction in the "Shite" Form. In: Proceedings of Spring IECEI Conference, IECEI, Japan (2005)
12. Saraki, M., Nitta, Y. (ed.): Regular Expression and Text Mining, Second Printing, Akashi-Shoten, 312 p (2008) (in Japanese)
13. William, A.G., Church, K.W.: A Program for Aligning Sentences in Bilingual Corpora. Computational Linguistics 19(3), 75–102 (1993)

Digital Image Stabilization Using a Functional Neural Fuzzy Network

Chi-Feng Wu, Cheng-Jian Lin, Yu-Jia Shiue, and Chi-Yung Lee

Abstract. This study proposed a real-time video stabilization method to eliminate the unwanted shakes, preserve the intended panning of camera, and improve the stability of the captured video sequence. The proposed method uses a functional neuro-fuzzy network to learn the phenomena of different shakes and then it chooses adequate compensation weight for two different methods to calculate the compensated motion vector. Experimental results show that the proposed method has superior performance than other motion compensation methods.

1 Introduction

Video capture devices have been used extensively in applications such as consumer electronics, security surveillance and driver assistance systems (DAS). The visual quality of the captured image sequences is degraded by the unwanted camera shakes. In order to improve the system performance, motion compensation is used to eliminate the impact and improve the stability of image. Motion compensation system can be divided into electronic image stabilization (EIS), optical image stabilization (OIS), and digital image stabilization(DIS). DIS uses digital image processing techniques to remove the shake of image. The system does not require additional motion-sensing device, and it can distinguish between intended panning and pure shake. DIS is generally composed by three major units: local motion estimation, global motion vector estimation and motion compensation. Global motion estimation is to estimate the global motion vector (GMV) between two

Chi-Feng Wu
Department of Information Management and Communication, Wenzao Ursuline College of Languages, Kaohsiung City, Taiwan, R.O.C.
e-mail: cfwu@mail.wtuc.edu.tw

Cheng-Jian Lin, Yu-Jia Shiue
Department of Computer Science and Information Engineering, National Chin-Yi University of Technology, Taichung County, Taiwan, R.O.C.
e-mail: cjlin@ncut.edu.tw

Chi-Yung Lee
Department of Computer Science and Information Engineering, Nankai University of Technology, Nantou, Taiwan 542, R.O.C.
e-mail: cylee@nkc.edu.tw

F.L. Gaol et al. (Eds.): Proc. of the 2011 2nd International Congress on CACS, AISC 144, pp. 231–237.
springerlink.com © Springer-Verlag Berlin Heidelberg 2012

successive frames of the acquired video sequence. Based on GMV, the motion compensation unit produces a shift vector for current frame, which is used to eliminate the annoying camera shake while preserving the intended camera panning.

Nine fixed observation feature blocks, distributed equally over the frame, are selected, and full search method is applied to get their local motion vectors (LMV) [1]. To increase the reliability of the LMV, feature block with large content changes compared with co-located block in previous frame is weep out. The method of [2] uses representative point matching (RPM) to reduce the amount of computation, and an inverse triangle method to examine the reliability of the derived LMVs. Instead of block matching, many methods compare the characteristic values, like edges and feature points, to get GMV [3-6]. For motion compensation, Kalman filter is a frequently used method [3], it can filter out the abnormal vibration, but retain the lens movement. For shock absorption, the concept of cumulative motion vectors is proposed in [7]. However, this method cannot work well in constant motion condition, and improved method with one integrator is proposed in [2]. In this study, we focus on the problem of motion compensation, and a solution based on functional neuro-fuzzy network is proposed.

2   The Proposed Video Stabilization Algorithm

At first, the frame is divided into 4 blocks of equal size. Although full-search method can find the motion vector close to the movement of image, but its computational complexity is too high to fit the requirement of real-time systems. To reduce the computational complexity, we use the representative point matching (RPM) as the search method. RPM method divides the block into numbers of sub-blocks, and takes center point of each sub-block as its representative point. Then, the amount in the calculation of MAD_r will drop to S, where S is the amount of sub-blocks, and the formula is:

$$MAD_r(p,q) = \sum_{k=1}^{S} (I(t-1, x_k, y_k), I(t, x_k + p, y_k + q)) \tag{1}$$

where (x_k, y_k) is the coordinates of the representative point of sub-block k, t is the frame number, and I is the intensity of the pixel. Then, the vector (p, q) with minimum MAD_r is the desired motion vector.

To get the GMV of current frame, we select the four LMVs, mean vector of LMVs, GMV of previous frame, and zero motion vector as candidates of GMV. Then, nine fixed observation blocks with size 16×16 pixels are preselected, which are distributed equally over a frame. MAD_p of all pairs for each observation block and each GMV candidate are calculated, and the formula is:

$$MAD_p(u,v) = \sum_{k=1}^{16} \sum_{l=1}^{16} |I(t-1, x_i + k, y_i + l) - I(t, x_i + u + k, y_i + v + l)| \tag{2}$$

where (x_i, y_i) is the coordinates of pixel at left-top corner for the i-th observation block, t is the frame number, and vector (u, v) is a candidate of GMV. In addition to camera movement, there may be some other object move into and out of the screen. These moving objects may just move to the location of observation block, causing the calculated MAD_p is abnormal. And the derived motion vector may reflect the movement of objects in the picture, not the global movement of background. To avoid choosing the wrong GMV, we will use the ballot to get the GMV. Each observation block ranks the GMV candidates according to their MAD_p values. The candidate with minimum sum of rank will be the GMV of current frame. If two or more candidates have the same sum of rank, comparison of average MAD_p value will be made.

For motion compensation, AMV is a cumulative motion vector of current frame relative to the first reference frame [7]. AMV use decay factor k to achieve the purpose of shake absorption, and following is AMV's formula:

$$AMV(t) = kAMV(t-1) + (\alpha GMV(t) + (1-\alpha)GMV(t-1)) \tag{3}$$

where k is a constant, $0<k<1$, and the greater value of k, the greater the ability of shock absorption. However, when the camera moves at constant speed, a larger k will make AMV value larger. Due to fixed search range of motion estimation, the ability of shake absorption will be decreased. Moreover, compensated trajectory of AMV will be late than the actual trajectory. Hsu add an integrator to improve the delay of constant speed moving [2]. The formula is as follows:

$$\begin{aligned} CMV(t) &= k \times CMV(t-1) + GMV(t) - \beta \times CMV\_I(t-1) \\ CMV\_I(t) &= CMV\_I(t-1) + CMV(t) \\ CMV(t) &= clipper(CMV(t)) \\ &= \frac{1}{2}(|CMV(t)+l| - |CMV(t)+l|) \end{aligned} \tag{4}$$

where the constant k is used to make panning of the screen can play more smoothly, constant β is used to filter out abnormal vibration, and l is the search boundary. But there are overshooting phenomena of trajectory in this method. These two methods have their own advantages and disadvantages when used in different situations. Hsu combined these two compensation methods with fuzzy reasoning [8]. In this study, we use a functional neuro-fuzzy network (FNFN) to learn the characteristics of AMV and CMV in different shakes. The inputs of FNFN are the vector difference, and short-term smoothness difference between AMV and CMV defined in [8]. The formula of short-term smoothness is as follows:

$$SI(t) = \frac{1}{p} \sum_{i=t-p-1}^{t} |GMV(i) - GMV(i-1)| \tag{5}$$

where p is size of window. The output of FNFN is adequate compensation weight $\gamma(t)$ for different methods, and the formula of compensated motion vector is modified as follows:

$$\begin{aligned} FMV(t) &= k \times FMV(t-1) + \gamma(t) \times ((\alpha GMV(t) + (1-\alpha)GMV(t-1))) \\ &\quad + (1-\gamma(t)) \times (GMV(t) - \beta \times CMV\_I(t-1)) \\ FMV(t) &= clipper(FMV(t)) \end{aligned} \tag{6}$$

3   Functional Neuro-Fuzzy Network (FNFN)

This subsection describes the FNFN model [9], which uses a nonlinear combination of input variables (FLNN). Each fuzzy rule corresponds to a sub-FLNN, comprising a functional link. Fig. 1 presents the structure of the proposed FNFN model.

The FNFN model realizes a fuzzy if-then rule in the following form.

Rule-j

$$\text{IF } x_1 \text{ is } A_{1j} \text{ and } x_2 \text{ is } A_{2j} \dots \text{ and } x_i \text{ is } A_{ij} \dots \text{ and } x_N \text{ is } A_{Nj}$$

$$\text{THEN } \hat{y}_j = \sum_{k=1}^{M} w_{kj} \phi_k \tag{7}$$

$$= w_{1j}\phi_1 + w_{2j}\phi_2 + \dots + w_{Mj}\phi_M$$

where x_i and $\hat{y}_j$ are the input and local output variables, respectively; A_{ij} is the linguistic term of the precondition part with Gaussian membership function; N is the number of input variables; w_{kj} is the link weight of the local output; ϕ_k is the basis trigonometric function of input variables; M is the number of basis function, and Rule-j is the jth fuzzy rule.

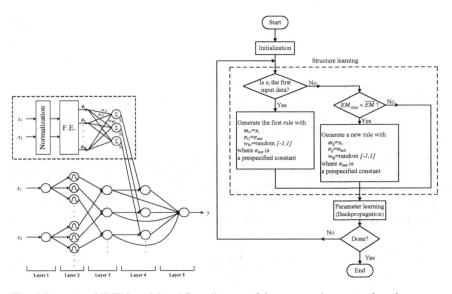

Fig. 1 Structure of FNFN model and flow diagram of the structure/parameter learning

The structure learning illustrated in Fig. 1 is used to determine whether a new rule should be extracted from the training data and to find out the number of fuzzy sets in the universal of discourse of each input variable. One cluster in the input space corresponds to one potential fuzzy logic rule, in which m_{ij} and σ_{ij} represent the mean and variance of that cluster, respectively. Entropy values between data points and current membership functions are calculated to determine whether or not to add a new rule. For computational efficiency, the entropy measure can be calculated using the firing strength $u_{ij}^{(2)}$ of FNFN as follow:

$$EM_j = -\sum_{i=1}^{N} D_{ij} \log_2 D_{ij} \tag{8}$$

where $D_{ij} = \exp\left(u_{ij}^{(2)-1}\right)$ and $EM_j \in [0,1]$. This measure used to generate a new fuzzy rule and new functional link bases for new incoming data is described as follows. The maximum entropy measure

$$EM_{max} = \max_{1 \leq j \leq R_{(T)}} EM_j \tag{9}$$

is determined, where $R(t)$ is the number of existing rules at time t. If $EM_{max} \leq \overline{EM}$, then a new rule is generated, where $\overline{EM} \in [0,1]$ is a pre-specified threshold that decays during the learning process.

After the structure learning phase, the network enters the parameter learning phase to adjust the parameters of the network optimally based on the same current training data. The goal to minimize the cost function E is defined as

$$E(t) = \frac{1}{2}[y(t) - y^d(t)]^2 = \frac{1}{2}e^2(t) \tag{10}$$

where $y^d(t)$ is the desired output and $y(t)$ is the model output for each discrete time t.

When the BP learning algorithm is adopted, the weighting vector of the FNFN model is adjusted such that the error defined in Eq. (10) is less than the desired threshold value after a given number of training cycles. The well-known BP learning algorithm may be written briefly as

$$W(t+1) = W(t) + \Delta W(t) = W(t) + \left(-\eta \frac{\partial E(t)}{\partial W(t)}\right) \tag{11}$$

where η and W represent the learning rate and the tuning parameters of the FNFN model, respectively.

4   Experimental Results

After learning phase, only 11 rules were generated in the FNFN model. We selected three video sequences with 320×240 resolution for testing, which consist of different movement. The contents of videos are that the boat of coastguard drives with constant speed, baby who crawls around the room, and a kart drives within fixed region. The performance of smoothing is evaluated by smoothness index (SI) proposed in [2], which is a average value of motion vector variation between frames. The SI is given by

$$SI = \frac{1}{N-1}\sum_{t=2}^{N}|GMV(t)-GMV(t-1)| \qquad (12)$$

where t is the frame number, N is the number of total frame. The frame rate of the experiment is 20 frame/sec, and the search range is limited to ± 40 point.

Table 1 SI of Each Compensation Methods

Video#	Total no. of frame	SI				
		Original	Paik	Hsu	Fuzzy	FNFN
1	299	0.3154	0.1007	0.1745	0.1611	0.0604
2	607	2.6810	0.6083	0.4678	0.6314	0.3421
3	3007	4.1304	1.3598	1.1174	1.7679	1.1174

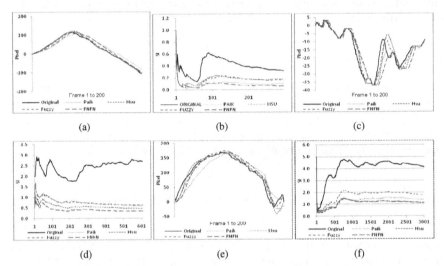

Fig. 2 Original and compensated motion trajectories in horizontal direction and SI of each frame with respect to video 1, video 2 and video 3.

Table 1 shows the SI comparison of FNFN method and other CMV generation methods. The result show that Paik and Hsu method work well in some case. However, FNFN is close with the result of best method, or has superior performance than other methods. Fig. 2(a)(c)(e) depicts the motion trajectories of original image and compensated motion trajectories. From Fig. 2(a)(e), it can be apparently seen that the trajectories of proposed method is smoother than other method, more close to original trajectory than Paik's method, and has no overshooting problems. The SI diagram shown in Fig. 2(b)(d)(f) prove that the compensated video of proposed method has less vibration through the whole video.

5   Conclusions

In this study, we proposed a robust and efficient digital video stabilization method. The method is based on functional neural fuzzy network (FNFN). Experimental results indicate that the proposed method can provides a smoother trajectory of video which is taken with panning and vibration of camcorder.

References

1. Chen, C.H., Kuo, Y.L., Chen, T.Y., Chen, J.R.: Real-time video stabilization based on motion compensation. In: Fourth Inter. Conf. on Innovative Computing, Information and Control, pp. 1495–1498 (2009)
2. Hsu, S.C., Liang, S.F., Fan, K.W., Lin, C.T.: A robust in-car digital image stabilization technique. IEEE Trans. on System, Man, and Cyber. Part C: Applications and Reviews 37(2), 234–247 (2007)
3. Wang, C., Kim, J.H., Byun, K.Y., Ni, J., Ko, S.J.: Robust digital image stabilization using kalman filter. IEEE Trans. on Consumer Electronics 55(1), 6–13 (2009)
4. Cai, J., Walker, R.: Robust video stabilization algorithm using feature point selection and delta optical flow. IET Comput. Vis. 3(4), 176–188 (2009)
5. Yang, J., Schonfeld, D., Mohamed, M.: Robust video stabilization based on particle filter tracking of projected camera motion. IEEE Trans. on Circuits and Systems for Video Technology 19(7) (July 2009)
6. Liang, Y.M., Tyan, H.R., Chang, S.L., Liao, H.Y.M., Chen, S.W.: Video stabilization for a camcorder mounted on a moving vehicle. IEEE Trans. on Vehicular Technology 53(6), 1636–1648 (2004)
7. Paik, J.K., Park, Y.C., Kim, D.W.: An adaptive motion decision system for digital image stabilizer based on edge pattern matching. IEEE Trans. on Consum. Electron. 38(3), 607–616 (1992)
8. Hsu, S.C., Lin, C.T.: Fuzzy inference applied to digital image stabilization techniques. Image and Recognition 13(3), 55–66 (2007)
9. Lin, C.J., Liu, Y.C., Lee, C.Y.: An efficient neural fuzzy network based on immune particle swarm optimization for prediction and control applications. Int. Journal of Innovative Computing, Information and Control 4(7), 1711–1721 (2008)

A Combined Assessment Method on the Credit Risk of Enterprise Group Based on the Logistic Model and Neural Networks[*]

Xiao Min, Liu Wenrui, Xu Chao, and Zhou Zongfang

Abstract. As the credit risk of enterprise group is closely related to its financial state, in this paper, we chose the financial indicators and characteristic index of the enterprise groups used in literature [1] and applied the combined method based on the Logistic model and Neural Networks to assess the credit risk of Chinese listed enterprise group. Then we chose correlative Chinese listed enterprise group over the 2004-2008 period as study sample and conducted empirical research by using the combined method. Finally, we made a comparison among the evaluation results of three methods.

1 Introduction

The development of the enterprise groups can improve the optimal combination of production factors and the rational allocation of resources, so the enterprise groups play a vital role in the economic development. But with the intensification of the global financial crisis, many well-known multinational enterprise groups went bankrupt, such as Lehman Brothers and Sanlu Group. So how to assess the credit risk of enterprise group now becomes a main problem that the whole financial community concern about. As the enterprise group has the features such as cross-industry business, complex relationship and so on, it is difficult to assess its credit risk.

At present, there are not many research on the credit risk of enterprise group in both domestic and overseas and most of them study the causes and how to deal with it from the view of legal and institutional or from the theoretical level[2]~[6]. In literature [7], through construct multi-hierarchy T-S fuzzy model to evaluate the credit risk of parent and subsidiary company in an enterprise group, and discuss the process of default contagion. So far, there is few research on credit risk of enterprise group from the empirical aspect.

In this paper, we apply the combined method based on the Logistic model and Neural Networks to assess the credit risk of Chinese listed enterprise group.

Xiao Min · Liu Wenrui · Xu Chao · Zhou Zongfang
School of Management and Economics, UESTC, Chengdu, China
e-mail: xcwdl@126.com

[*] This research has been supported by National Natural Science Foundation of China (No. 70971015)

F.L. Gaol et al. (Eds.): Proc. of the 2011 2nd International Congress on CACS, AISC 144, pp. 239–245.
springerlink.com
© Springer-Verlag Berlin Heidelberg 2012

The main contents are as follows: first of all, we choose the financial indicators in literature [1], add in characteristic index of the enterprise group (gross related party transactions, frequency of transactions, asset-liability ratio, liability ratio of cash flow, ROE, cost profit rate, inventory turnover, asset turnover rate, growth rate of IMP, growth rate of total assets, industry fragmentation, denoted by X_{11}, X_{12}, X_{23}, X_{24}, X_{31}, X_{34}, X_{41}, X_{43}, X_{52}, X_{54}, X_{62}) ,and set up an evaluation index system for assessing the credit risk of enterprise group. Second, we choose correlative Chinese listed enterprise group over the 2004-2008 period as study sample and conduct empirical research by using the combined method and compare the evaluation results of three methods. Finally, make a summary of the full-text.

2   Combined Assessment Method on Assessing the Credit Risk of Enterprise Group

A.   The Construction Principle of Combined Assessment Model

From the construction principles of the Logistic and neural network we can know that each method has its characteristics, each assessment model also has its own features. As a statistical method, Logistic regression does not need to meet the requirements of the statistical assumptions such as normal distribution, so it can be widely applied in reality. At the same time, the method has a high degree of interpretation of the model and strong robustness. It can be not only for classification, but also to generate linear score card. However, Logistic regression lacks of identification with high accuracy which directly affects the application of this method. Neural network as a non-statistical method has many advantages compared with traditional statistical methods, such as a higher classification accuracy and a stronger self-learning ability, adaptive ability and the ability to handle nonlinear problems. Neural network also has its shortcomings, like the robustness of learning and memory in the network is not very high and dose not has a high degree of interpretation of the model.

In order to overcome these defects of the two methods and take advantage of both methods better, according to the idea of combined model in literature [7] and [8], we propose a combined method based on the Logistic model and Neural Networks to assess the credit risk of Chinese listed enterprise group. First, we use Logistic regression to establish a credit evaluation model, then take the score values generated by model into the neural network model together with other indicators as the explanatory of the model which based on neural network.

Define the probabilities of default generated by Logistic regression model are $p_1, p_2, ..., p_h$, which will be explanatory variables of the final combined assessment model. The input of neural network are $x_1, x_2, ..., x_m$. The output of neural network are $y_1, y_2, ..., y_h$. Assume that the n assessed value of variable Y_i are $\hat{Y}_{1_i}, \hat{Y}_{2_i}, \hat{Y}_{3_i}, ..., \hat{Y}_{n_i}$, so the combined assessment model is $\hat{Y}_{a_i} = f(\hat{Y}_{1_i}, \hat{Y}_{2_i}, \hat{Y}_{3_i}, ..., \hat{Y}_{n_i})$ and the scores of the classification are $a_1, a_2, ..., a_h$.

According to the construction of the combined assessment model, the assessed values generated by Logistic regression model are $p_1, p_2, ..., p_h$ and the output of neural network are $y_1, y_2, ..., y_h$. Both of them can measure the classification of samples. Define the scores of classification produced by combination of them are $a_1, a_2, ..., a_h$. $(a_1, a_2, ..., a_h) = f(p_1, p_2, ..., p_h, y_1, y_2, ..., y_h)$ For the neural network model we have $Y_\Omega = (y_1, y_2, ..., y_h) = F(X_I) = F(x_1, x_2, ..., x_m)$, so the combined model can be transformed to $(a_1, a_2, ..., a_h) = h(p_1, p_2, ..., p_h, x_1, x_2, ..., x_m)$. The schematic of the combined assessment model is shown as below:

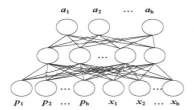

Fig. 1 The schematic of neural network with the probability of default

For the neural network, input is $X_{I_k} = (p_{1_k}, p_{2_k}, ..., p_{h_k}, x_{1_k}, x_{2_k}, ..., x_{m_k})$ and output is $Y_{O_K} = (a^*_{1k}, a^*_{2k}; \cdots, a^*_{hk})$. a_j represents the distance between the value obtained by samples and different classifications. The greater value is, the smaller possibility that the sample belongs to Class j is

$$a^*_{jk} = \begin{cases} 0 & \text{sample belong to class } j \\ 1 & \text{others} \end{cases}.$$

B.   The Construction of Combined Assessment Model

The combined assessment model uses the same training samples and the same algorithm. We take the probability of default calculated from Logistic regression into the original neural network to construct the combined assessment model.

Determination of Nodes of Different Layers
(1) Determination of nodes of different layers
The input nodes should correspond to the number of indicators of credit discrimination. This model includes the score value obtained from Logistic regression and 11 initial indexes after the index reduction. Therefore, there are 12 secondary indicators in the combined assessment model, the input nodes $n=12$.

As the neural network built above, the enterprise groups can be classified as default and non-default. As a result, the corresponding output value are (1, 0), (0,1). The nodes of output $m=2$.

(2) Determination of nodes of hidden layer

In this paper, after conducting a number of calculations of nodes of hidden layer we find that when the hidden nodes are 20, the rate of miscarriage of justice reach the least.

Training of Combined Assessment Model

We choose 117 correlative Chinese listed enterprise groups over the Shanghai and Shenzhen A-share as study samples, 82 of which are training samples and the other 35 are test samples. According to the relevant financial data in CCER database to dimensionless the index. The training samples and the test samples are the same with what in Logistic model and neural network.

According to the general requirements for neural network and the needs of the model, the samples use the common type of Sigmoid Logarithmic function. Set the expected distortion $E=1*e^{-0.06}$, learning rate is 0.01, the maximum number of learning is 3000 and the system shows the curves of the changes in the training error every 10-steps. After determining the parameters, we use Matlab software to train the network model. After 47 iterations the actual error $\varepsilon=7.02733\times e^{-0.07}$ meet the requirements of target. The operation is as follows:

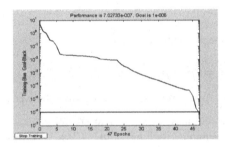

Fig. 2 The training plans of neural network with the probability of default

C.   The Test for the Combined Assessment Model

We choose the data from 35 enterprise groups as the test sample and there is 26 normal credit enterprises, which is the non-ST enterprises, and also 9 credit default enterprises which is ST enterprises. Put test sample into the combined assessment model that have been trained, we have the test result as follows:

Table 1 Test Result for the Combined Assessment Model

		Forecast value		
		Whether is ST		
Observed value		0	1	Accuracy (%)
Whether is ST	0	25	1	96.15
	1	8	1	87.5
Total accuracy				94.29

(0 = Normal credit ; 1 = a credit default)

It can be known form table 1 that one mistake happened among the total 26 when the normal credit companies are assessed, so the accuracy is 96.15%; also one mistake happened among the total 9 when the credit default companies are assessed, so the accuracy is 87.5%. The total accuracy is 94.29%. The classification results of both training sample and test sample are as follow:

Table 2 Classification Result

Types of model	training sample (82, non-ST 61, ST 21)			test sample (35, non-ST 26, ST 9)		
	Accuracy (%)	The first error ratio (%)	The second error ratio (%)	Accuracy (%)	The first error ratio (%)	The second error ratio (%)
Neural network	97.14	12.5(1)	0.(0)	94.29	12.5(1)	3.85(1)

From table 2, the distinguishing normal credit enterprises from all are precise, and the accuracy is 96.15%; the accuracy of distinguishing credit default enterprises from all is 87.5%. Comparing with other two single models, there is a large improvement. And the total accuracy is 94.29%.

In the assessment of 9 credit default companies in the teas sample, there are 2 credit default companies mistakenly judged as normal business of credit by this model. We define this kind of error as the first error, so the first error ratio is 22.22%. And there is only one normal credit company mistakenly judged as credit default company among the 26 normal credit companies, which we define as the second error, so the second error ratio is 3.85%.

Whether is the training sample or the test sample, although the total accuracy is 91.43% the first error ratio is much larger than the second error ratio, which reflects that the model still has large zoom to improve, so how to raise the accuracy of forecasting is still our research work in next step.

D. Comparison among Three Models

We compare the assessment results among Logistic model, neural network and the combined assessment model with the same sample as follows:

Table 3 Comparison of Accuracy among Three Models

Types of models	Training Sample			Test Sample		
	Total Accuracy (%)	The first error ratio (%)	The second error ratio (%)	Total Accuracy (%)	The first error ratio (%)	The second error ratio (%)
Logistic	86.6	38.1	4.9	85.71	44.4	3.85
neural network	96.80	22.22	1.64	91.43	22.22	3.85
combined assessment	97.14	12.5	0	94.29	12.5	3.85

(1) The accuracy rate of neural network model, 91.43%, is higher than that of Logistic regression model, which is only 85.71%.

(2) The accuracy rate of the combined assessment model is higher than both neural network model and Logistic regression model.

(3) The combined assessment model always keeps the first error ratio in a low level, showing a better classification ability to the credit default enterprises.

From the empirical results we can know that combined assessment model in all aspects is much better than both neural network model and Logistic regression model, especially with regard to the classification of credit default enterprises, which makes a reasonable explanation for the evolution from traditional statistical model to the intelligent method applied to credit risk assessment of enterprise groups.

3   Conclusion

We evaluated the credit risk of Chinese listing enterprise group by a combined assessment method with Logistic regression model and neural network model, in which we added both financial information and the characteristic index of enterprise group. We put the result generated by Logistic regression model, together with other index, as the explanatory variables in neural network model and constructed the assessment model of credit risk of Chinese listing enterprise group. Then we got the accuracy of this model according to the test sample. The conclusion is that neural network model is better than Logistic regression model; the combined assessment model is better than any single model. Especially for the error rate which may indicate the real credit losses of creditors such as commercial banks, the combined assessment model shows its obvious advantages.

References

[1] Liu, W., Xiao, M., Zhou, Z.: Empirical Study on the Credit Risk of Enterprise Group Using BP Neural Networks. An Academic Edition of ManaMaga 10, 37–43 (2010)

[2] Altman, E., Resti, A., Sironi, A.: Default recovery rates in credit risk modeling: a review of the literature and empirical evidence. Economic Notes 33(2), 183–208 (2004)

[3] Berger, A.N., De Young, R.: Problem loans and cost efficiency in commercial banks. Journal of Banking and Finance 21, 849–870 (1997)

[4] Liu, D., Wang, X.: On the Construction of System of the Credit Risk Management of Banks upon Group Customers. Journal of Changsha University of Science & Technology (Social Science) 2, 67–70 (2009)

[5] Chen, L., Zhou, Z.: The Research on Measure Default Correlation of Related Corporations Controlled by an Enterprise Group. Chinese Journal of Management Science 18(05), 159–164 (2010)

[6] Zhou, Z.: The theories and methods on emerging technology enterprise's credit risk evaluation. Science Press, Beijing (2010)

[7] Wang, C., Wan, H., Zhang, W.: Application of Combining Forecasts in Credit Risk Assessment in Banks. Journal of Industrial Engineering/Engineering Management 13(1), 5–9 (1999)

[8] Lee, T.-S., Chiu, C.-C., Lu, C.-J., Chen, I.-F.: Credit scoring using hybrid neural discriminant technique. Expert System with Applications 23(3), 245–254 (2002)

Research on Factors Influencing European Option Price by Using Hybrid Neural Network

Zhang Hong-yan

Abstract. According to European option pricing models such as the Black-Scholes model, option price is determined by five factors which are underlying asset price, striking price, interest rate, maturity, and volatility. Gross open interest reflects the activity of the transaction traded by the buyers and sellers in option markets. Several different kinds of gross open interest are used as the input variables of the neural networks in this work. The experimental results show that the hybrid neural network model using a kind of average gross open interest as the input variable performs better than the other models put forward in this article. It may hints that the activity of an option market could be one of the factors influencing European option price.

1 Introduction

According to the Black-Scholes model[1] , option price depends on five factors such as spot price of the underlying asset, striking price, interest rate, maturity, and volatility. Are there other factors that could lead to the change of an European option price? There is little discussion in academe. But as Mcmillan [2] says, one often uses trading volume as the index of a stock option. It may hints that the activity of an option market could influence option price .

Gross open interest reflects the activity of a real option market. Nonparametric techniques such as artificial neural networks don't necessarily involve directly any financial theory. In order to study whether the activity of a real market is relative to an European option price, several kinds of average gross open interest are applied to the input variables of the neural networks including pure neural networks and hybrid neural networks in this paper.

2 Option Pricing

2.1 Option and Black-Scholes Model

Until today, the Black-Scholes model[3] (the Black-Scholes model is simplified by the B-S model in the following section) is the most dominant model to derive

Zhang Hong-yan
Wang Yanan Institute for Studies in Economics, Xiamen University, Xiamen, China
email: zhanghongyan72@gmail.com

F.L. Gaol et al. (Eds.): Proc. of the 2011 2nd International Congress on CACS, AISC 144, pp. 247–252.
springerlink.com   © Springer-Verlag Berlin Heidelberg 2012

European option prices in the world financial markets. The original Black-Scholes formula has five inputs, which are the stock price S, the strike K, the option maturity τ, the interest rate r, and the volatility σ. The call price is derived as:

$$C = S\phi(d_1) - Ke^{-r\tau}\phi(d_2) \tag{1}$$

where

$$d_1 = \frac{\ln(S/K) + \tau(r + \sigma^2/2)}{\sigma\sqrt{\tau}} \tag{2}$$

$$d_2 = d_1 - \sigma\sqrt{\tau} = \frac{\ln(S/K) + \tau(r - \sigma^2/2)}{\sigma\sqrt{\tau}} \tag{3}$$

Where C : call price; $\phi()$: cumulative normal distribution; S : spot price of the underlying asset; r : continuously compounded risk-free interest rate; K : strike; $\ln()$: natural logarithm function; τ : maturity; σ : volatility.

2.2   Architecture of Hybrid Neural Network

The basis of the hybrid neural network applied in this work is the same as the hybrid neural network used in Boek[4] . The activity function of the hidden layer of the hybrid neural network is tansig, and the activity function of the output layer is logsig. The mode of sample studying is batch disposal. The training algorithm of the hybrid neural network is the modified self-adaptive Levenberg-Marquardt algorithm[5].

2.3   Architecture of BP Neural Network

The architecture of the BP neural network adopted in this work is the same as the architecture adopted by Yao[6]. The activity function of the hidden layer of the BP neural network is tansig. The activity function of the output layer of the BP neural network is logsig. The mode of sample studying is batch disposal. The training algorithm of the BP neural network is the modified self-adaptive $Levenberg - Marquardt$ algorithm[5].

2.4   Input Variables of Neural Networks

According to the B-S model, there are five factors such as S, X, $T-t$, r, σ which can influence European option prices. S, X, $T-t$ have already been known among of them. Scholars have drawn different conclusions on whether the interest rate r is used as an input variable of a neural network. For example, Boek [4] holds r as an input variable of a neural network. At the same time, Yao [6] consider that you cannot make any improvement by taking r as the input variable in the research on the Asian option market Nikkey 225. Considering the historical data has only 13 days and the variation of the interest rate is very little, we do not take r into account as an input variable of the neural networks in this work.

Gross open interest reflects the activity of the transaction traded by the buyers and sellers in an option market. Just think of an example, as an important news comes, the market usually fluctuates violently , both buyers and sellers could buy or sell some option product to obtain profit, the price of the underlying asset would not have smart change at the same time. The option price could change with the activity of the financial market in this case. In order to find whether the option price could have a link with the activity of the financial market , two kinds of gross open interest are taken into account as the input variables of the neural networks in this paper.

According to our experience, traders care more about expiration date rather than strike price. For example, one day in April 2004, traders heard that a major event which would affect markets violently would take place on June 1, 2004. The sum of the gross open interest that have all different types of exercise price of an option would change and the gross open interest that has specific strike of the option would not change perhaps. Moreover, the quantity of the strike of the option traded in varied trading day is quite different. For example, there are 28 kinds of strike of Hong Kong Index call option which is traded on May 17, 2004 and the expiration date of which is September 2004 . Meanwhile, there are 29 kinds of strike of Hong Kong Index call option which is traded on May 18, 2004 and the expiration date of which is September 2004. In order to facilitate comparison, the following expression of quotient is applied in this work:

$$avgoI1_{i,j} = \frac{\sum_{i=1}^{n} grossoI_{i,j-1}}{n} \qquad (4)$$

Where $avgoI1_{i,j}$: Average gross open interest of ith option traded on jth trading day; $grossoI_{i,j-1}$: Gross open interest of the option traded on j-1th trading day whose maturity and strike are the same as the ith option traded on jth trading day; n: Quantity of the option traded on jth trading day whose maturity is the same as the ith option traded on j-1th trading day and whose strike is not the same as the ith option traded on jth trading day.

Just think about another kind of average gross open interest. Traders heard about some news in some day. For example, on May 17,2004, traders heard that a major event which would affect the market heavily would take place on June 1, 2004. Traders would not only interested in the specific option whose maturity is June 2004, but also would buy or sell the option products whose maturity are across all(June 2004, July 2004, September 2004, December 2004, March 2005, June 2005,December 2005). The sum of the gross open interest of the option whose maturity are across all would change accordingly. Furthermore, just consider that the quantity of the strike of the option traded in different trading day may be different, another kind of average gross open interest is applied in this work:

$$avgol2_{i,j} = \frac{\sum_{i=1}^{n_1} grossol_{i,j-1}}{n_1} \tag{5}$$

Where $avgol2_{i,j}$: Average gross open interest of ith option traded on jth trading day; $grossol_{i,j-1}$: Gross open interest of the option traded on j-1th trading day whose maturity and strike are the same as ith option traded on jth trading day; n_1: Quantity of the option whose maturity and strike are across all.

In summary, there are six kinds of neural network in this paper. It is NN1(BP neural network, Input variables: (S(underlying asset price), X(strike), T-t(maturity)); NN2(BP neural network, Input variables: S(underlying asset price), X(strike), T-t(maturity), $avgol1_{i,j}$); NN3(BP neural network, Input variables: S(underlying asset price), X(strike), T-t(maturity), $avgol2_{i,j}$); NN4(Hybrid neural network, Input variables: S(underlying asset price), X(strike), T-t(maturity)); NN5(Hybrid neural network, Input variables: S(underlying asset price), X(strike), T-t(maturity), $avgol1_{i,j}$); NN6(Hybrid neural network, Input variables: S(underlying asset price), X(strike), T-t(maturity), $avgol2_{i,j}$). The output variable of these six kinds of neural network are C.

The criterion used for evaluating these approaches is mean absolute percentage error (MAPE):

$$MAPE = \frac{1}{n}\sum_{i=1}^{n}\left|\frac{y-\hat{y}}{y}\right| \tag{6}$$

Where $\hat{y}$: Output value of option pricing models; y: Target value; n: Number of samples.

3   Experimental Results

The dataset used in this work is Hong Kong Index call option dataset. Hong Kong derivative market is one of the biggest derivative markets in the world. The research on Hong Kong financial derivative market is of universal significance.

The output of the above neural networks is a re-integration of the generating output in an anticipatory way by two sub-models. The training data and validation data are selected from the call option traded in 16 trading days from May 4,2004 to May 25, 2004, and the forecasting value are the call options prices of 17th trading day on May 27,2004. There are eight kinds of call options whose maturity is May 2004, June 2004, July 2004, September 2004,December 2004,March 2005,June 2005,December 2005 respectively.

The call option which is satisfied following conditions is eliminated from the training and validation data set in this paper:

1)   If the call option traded on some day whose expiration date is the same as the call option traded on the last trading day has no corresponding strike, for example, the call option traded on May 4th,2004 , whose strike is 9400 and expiration date is July 2004, should be eliminated from the option data set, for the sake that the call option traded on May 3rd,2004 whose strike is 9400 and expiration date is July 2004 does not exist in the call option data set.

2)   The call option whose price is less than 10 points is eliminated.

In satisfying such above conditions, the call options traded from May 4th,2004 to May 25th,2004 include 2672 data. All these 2672 data are used as training data set. The call options traded on May 27th, 2004 include 175 data, and these 175 data are used as testing data set. Forecasting results are different due to different hidden neuron nodes of the neural network. After validated and compared reiteratively, the architecture of NN1 we applied in this work is 3X12X1, the architecture of NN2 we applied in this work is 4X9X1, the architecture of NN3 we applied in this work is 4X9X1, the architecture of NN4 we applied in this work is 3X32X1, the architecture of NN5 we applied in this work is 4X16X1, the architecture of NN6 we applied in this work is 4X5X1. When we adopt the above architectures , the forecasting results are the best comparing with the results using same neural network with different number of hidden neuron nodes. And these forecasting results using neural network models and the B-S model are shown in table 1:

Table 1 Forecasting results of all kinds of models

Different models	NN1 model	NN2 model	NN3 model	NN4 model	NN5 model	NN6 model	B-S model
MAPE	0.0792	0.0562	0.0465	0.0752	0.0547	0.0373	0.1573

First, it can be shown in Table 1, all the forecasting results based on the neural network models are better than the classical B-S model. This is because that the assumption under which the B-S model follows often violates real markets and neural networks can reflect the real price of underlying asset better in the near future. It can be demonstrated on Hong Kong Index call option market in this paper. Secondly, The forecasting results of NN2 model and NN3 model are better than the forecasting result of NN1 model. As the sum of the gross open interest of the option across all kinds of expiration date is taken as an input variable of Neural Networks, we make a big improvement in Forecasting accuracy. This suggests that the sum of the gross open interest could change a lot when the market meets some news that can affect market heavily. At the same time, we can forecast the option price traded in the next trading day effectively by the change of the gross open interest of the option data traded in the last trading day. Thirdly, all the hybrid neural network models perform better than the BP neural network models with the same input variables used in the neural networks in forecasting. And the hybrid neural networks converge faster than the BP neural networks. Fourthly, NN6 model performs the best in all the neural network models. i.e. as the architecture of hybrid neural network is applied and the input variables of the neural network are S, X, T-t, $avgol2_{i,j}$, it performs the best in all the models.

4 Conclusion

Traders often change market strategy that they buy or sell option products. They hope that they can predict the option price of the next stage based on the present option price more accurately, and this is often the key to obtain profit for option traders. We predict the option price of the next trading day by the research on the short-term historical market data of the Hong Kong Index call option, and we get better results in the empirical research on the Hong Kong Index call option market.

There are much difference between the option price of a real market and the B-S model price due that the assumptions followed by the B-S model are very strict. Gross open interest reflects the activity of the transaction traded by buyers and sellers in option markets. Two kinds of gross open interest are used as the input variables of the hybrid neural networks in this work. The experimental results show that the hybrid neural network model using a kind of average gross open interest as the input variable performs more accurate than the other models put forward in this article. It may suggests that the activity of option markets may become one of the factors influencing European option price.

References

[1] Black, F., Scholes, M.: The pricing of options and corporate liabilities. J. Political Econ. 81, 637–659 (1973)
[2] Mcmillan, L.G.: Mcmillan on options, 2nd edn. John Wiley & Sons, New York (2004)
[3] Hull, J.C.: Options, Futures and Other Derivatives, 3rd edn. Prentice Hall, Inc. (1997)
[4] Boek, C., Lajbcygier, P.: A hybrid neural network approach to the pricing of options. IEEE Neural Network in Financial Engineering (1995)
[5] Hagan, M.T., Menhaj, M.B.: Training Feedforward Networks with the Marquardt Algorithm. IEEE Transactions on Neural Netwoks 6, 989–993 (1994)
[6] Yao, J.T.: Option Price Forecasting using Neural Networks. Omega 28, 455–466 (2000)

Tibetan Processing Key Technology Research for Intelligent Mobile Phone Based on Android

Nyima Trashi, Yong Tso, Qun Nuo, Tsi Qu, and Duojie Renqian

Abstract. The author has been engaging in the research of Tibetan language information processing techniques, and has presided over the development of CDMA 450M network-based Tibetan mobile phone and vehicle phone, as well as smart mobile phone Tibetan software package based on Windows Mobile and Symbian. In 2011, required by China Mobile Communications Group Co., Ltd. Tibet Branch, Android 2.2 operating system based smart mobile phones was developed. Through in-depth study of Android operating system, the author firstly proposed Android 2.2-based Tibetan processing technology, and used own national patents to design a highly adaptive Tibetan keyboard layout, thereby achieving rapid Tibetan input established the basis of standard for Tibetan mobile phone keyboard layout. The technology is a kind of technological innovation, filling the blanks of related fields both at home and abroad.

Keywords: Android, Tibetan, Mobile Phone, Input/Output.

1 Introduction

With the development of mobile communication technology and wide use of mobile phones, the number of mobile phone user grows fast and the demand on Tibetan mobile phone keeps on rising. Therefore, Tibetan mobile phone software package research and development has a special significance in realizing mobile phone Tibetan processing improving the country's overall level of information, promoting socio-economic development of minority areas.

The implementation of Tibetan input/output is the core of mobile phone Tibetan processing technology. This paper firstly proposes a kind of Tibetan input/output technology based on Android 2.2 operating system, designs a strongly adaptable Tibetan keyboard layout, which achieves fast Tibetan input and fills the blanks of Tibetan processing in the open source operating system of smart mobile phone.

Nyima Trashi · Yong Tso · Qun Nuo · Tsi Qu · Duojie Renqian
Modern Education Technology Center, Tibet University, Tibet, China
e-mail: {nmzx,yc,q_nuo,cq,djrq}@utibet.edu.cn

F.L. Gaol et al. (Eds.): Proc. of the 2011 2nd International Congress on CACS, AISC 144, pp. 253–259.
springerlink.com © Springer-Verlag Berlin Heidelberg 2012

2   Android and Its Input/Output Technology

2.1   Android

Android was launched in November 2007 by Google, which was based on Linux kernel smart mobile phones. To promote Android, Google has established Open Handset Alliance with dozens of mobile phone companies. Alliance members include 34 leading companies in the area of technology and wireless application such as Motorola, HTC, Samsung, LG, China Mobile and China Telecom. Android has the following advantages: (a) full open source, reduce the platform's development costs; (b) Linux-based general-purpose operating system, with good hardware compatibility; (c) rich database resources, support multiple functions; (d) great support to the third party development. Currently, java is the main development language supported by Android, and Android Tibetan input/output technology is achieved in the common version 2.2.

2.2   Android Input/Output Technology System Structure

Android system provides input method framework after version 1.5 and the input method framework is shown as Fig.1. This framework provides InputMethodManagerService components to manage various input methods installed in system. Android system regards Input Method as the Service type program and the input method must inherit InputMethodService class. InputMethodService represents a specific input method process, which is responsible for initial, setting up and destruction of input method. InputConnection is responsible for applications and input method data exchange. At present, the input method framework is complete in Android system and Tibetan input/output can be achieved through inheriting InputMethodService, and implemented definition of methods, input method input logic, specific algorithm engine, soft keyboard layout and candidate display in InputMethodService class.

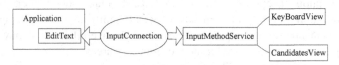

Fig. 1 Android System Input Technology System Structure

3   Implementation of Android Tibetan Input/Output

3.1   Implementation Principle

Android Tibetan input method consists of input method, soft keyboard and candidate window of Tibetan word, of which input method service plays a major role and is the implementation of android. Input method service. Input Method Service.

In input method implementation process, the following methods are the core part of input methods:

```
public class InputService extends InputMethodService
{
     public void onCreate();
     public void onInitializeInterface();
     public View onCreateInputView();
     public View onCreateCandidatesView();
     public void onStartInput();
     public boolean onKey(int keyCode);
}
```

When the input method is loaded and started, the system firstly calls onCreate function to do some necessary initialization. Later, it calls onInitializeInterface function to initialize the input method visual interface. Then, the system calls on-CreateInputView and onCreateCandidatesView functions to return to soft keyboard view and candidate window view. When user uses input method, the system calls onStartInput to remind user to prepare input, and also display a soft keyboard returned by input services onCreateInputView function. After receiving the key information, system calls onKey function to notify input services.

3.2   Input/Output Process

Android Tibetan input/output implementation process is shown in Fig.2:

    a. Input method received the information of soft keyboard;

    b. Determine whether the keyboard keys are control keys;

    c. If they are not control keys, search Tibetan characters in the input method word library, send the results to data output module, and go to step (f);

    d. If it is the control key in input state, search Tibetan characters in input method word library after paging and send results to data output module, go to step (f);

    e. If it is the control key in non-input state, directly send to the data output module;

    f. Determine whether the characters to be output in data output module are Tibetan words;

    g. If they are Tibetan words, send to the target window being edited;

    h. If they are not Tibetan words, send control characters to the target window being edited;

    i. End.

Note: The above input process does not involve associated Tibetan (word) phase processing.

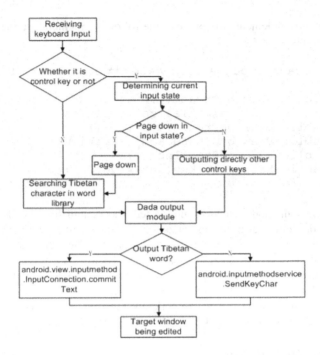

Fig. 2 Tibetan Input/Output Process

3.3 Coding Scheme

Tibetan coding uses "Information technology-Tibetan coded character sets for information interchange, Basic set" (hereinafter referred to as the "Basic Set") and "Information technology- Tibetan coded character set-Expansion A" (hereinafter referred to as "Expansion A") national standards. The "Basic Set" coded character set is located in Unicode(ISO/IEC 10646 BMP) U+0F00~U+0FBF; "Expansion A" coded character set is located in Unicode(ISO/IEC 10646 BMP) 0xF300~0xF8FF, that is, the PUA area.

"Information technology-Tibetan coded character set-Expansion B" (hereinafter referred to as "Expansion B") national standards collect 5702 Sanskrit Tibetan characters and all characters are encoded in a OF of GB 13000.1-1993, whose location is 0xF0000-0xF1645, and the code of each character consists of 3 bytes. Because the encoded characters in "Expansion B" are less used Sanskrit Tibetan characters, which are rarely used in handheld devices, there is no need for Tibetan input/output to support "Expansion B".

3.4 Soft Keyboard

Presently, Android 2.2 Tibetan input method is equipped with two soft keyboards, namely, Tibetan and Sanskrit Tibetan, as shown in Fig.3. We use XML file for the construction of Tibetan virtual keyboard and properties, such as:

```
<Key     android:codes="character     coding"     andro-
id:keyLable= "key name"/>
```

In soft keyboard, "key name" represents the letters and characters of Tibetan and Sanskrit Tibetan. "Character coding" means the corresponding Unicode code. Use the coded characters in system font through key pressing, then output and display. The prototype of soft keyboard is as follows:

```
public class TibetanKeyBoardView extends KeyboardView
  {
    public void onDraw(Canvas canvas);
  }
```

Soft keyboard is a subclass of KeyboardView and the most important subclass function is onDraw, used to provide the corresponding keyboard display style.

(a) Tibetan soft keyboard   (b) Sanskrit Tibetan soft keyboard

Fig. 3 Input Soft Keyboard Key Position Diagram

The Tibetan shown in Fig.4 means I'm a teacher in the school. According to the Tibetan coding scheme and keyboard layout design, the corresponding keyboard input code and internal code is as follows.

Tibetan input method keyboard input code: 10,77940,175680,3190,7130,6830

Tibetan word internal code: 0F440F0B,F5D90F560F0B,F36CF5910F0B, 0F51F35E0F0B,F3740F530F0B,F5A40F530F0D.

Fig. 4 Example of Tibetan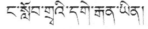

3.5 Font

Android 2.2 Tibetan input/output uses TrueType font, which achieves the loading of Tibetan font through original system document expansion. Specific methods are as follows:

In the directory of Android operating system /system/fonts, the font resources used by the system include DroidSansFallback.ttf, DroidSans-Bold.ttf,

DroidSans.ttf, DroidSansMono.ttf, DroidSans_Subset.ttf, etc, of which Droid-SansFallback.ttf is the most basic system font file and all fonts supported by Android are included in this file. To make Android 2.2 system correctly output the Tibetan, we apply the method of direct Tibetan coded character expansion into DroidSansFallback.ttf font file. Use FontCreator tool to expand Tibetan coded character to DroidSansFallback.ttf and use expanded font file to replace the original font file.

3.6 Screen Resolution Self-adaptation

Android mobile phone screen self-adaptation mainly considers screen resolution. At present, Android mobile phone resolution includes HVGA (320x480), QVGA(240x320) and WVGA (480x800). Android 2.2 Tibetan input/output studied by the author mainly considers HVGA based medium resolution model.

4   Instance of Input Method Application

In the environment of Android input/output system structure and its technology environment, the author has developed and achieved Android 2.2 Sunshine Tibetan input method. Fig.5 is an instance of Tibetan input method application in SAMSUNG P1000 smart mobile phone. The interface includes Soft Keyboard, Candidate Window and Text Window. It is known from the instance that the Android 2.2 mobile phone-based Tibetan input/output technology is feasible and effective.

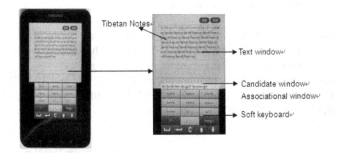

Fig. 5 Example of Input Applications

5   Conclusion

There is significant difference between Android Tibetan input/output technology development and development of Windows Mobile and Symbian Tibetan Input/out technology. Through in-depth study on Android input/output system structure and its technology environment, the author first proposes a kind of

Tibetan input/output method based on Android 2.2 smart mobile phones. To improve Tibetan input speed, association input function has been achieved. In order to make Tibetan keyboard layout adaptable to most mobile phones, a highly adaptive Tibetan keyboard layout is designed by the owned national patents, laying the basis for the standard of mobile phone Tibetan keyboard layout. The technology is a kind of technological innovation, filling the blanks of related fields both at home and abroad.

References

1. Zhong, M.: Android-based Smart Phone Platform Solutions. Engineering Master Thesis, Shangdong University
2. Xiong, G.: Android-based Smart Phone Design and Realization. Engineering Master Thesis, Wuhan University of Technology
3. Liu, F., Wang, Y., Tang, B., Wang, X., Wang, X.: Android-based smart Chinese Input method. Computer Engineering 36(7), 225–227 (2011)
4. Zhou, Y.: Mobile Phone Text Input Technology Research. Engineering Master Thesis, Chongqing University
5. Li, P.: Android System Plug-in Input Program Design. Computer Knowledge and Technology 5(35), 9979–9981 (2009)
6. Trashi, N., Shu, L.Z., Nuo, Q., Dun, P., Yang, C.Z., Tso, Y.: A Tibetan Mobile Phone based on CDMA System. In: Communication Systems and Networks, IASTED AsiaCSN 2007, Phuket, Thailand, April 2-4 (2007) ISBN CD:978-0-88986-658-4:24-28
7. Nyima, T., Li, Z.-S., Qun, N., Pu, D., Yong, T., Chen, A.-L.: A Method of Implementing Tibetan Processing for Mobile Phone. Journal of Sichuan University (Engineering Science Edition) 41(1), 162–167 (2009)
8. Nyima, T., Li, Z.-S., Yong, T., Qun, N., Pu, D.: Research on Key Issues of Implementing Computer Tibetan Fast Input. Journal of University of Electronic Science and Technology of China 38(1), 102–107 (2009)
9. Nyima, T., Li, Z.-S., Qun, N., Yong, T.: Computer Conversion Model of Tibetan Coding. In: International Conference on Communications, Circuits and Systems Proceedings. Communications, Networks and Signal Processing (2009) ISBN:978-1-4244-4887-6:562-565
10. Nyima, T., Cai, C.: Approach of Tibetan Input, Display and Messaging Implementation on Hand-held Electronic Terminal Device. China, Patent of Invention, Patent No.: ZL, 1 0037578.5 (2005)
11. Chen, Y., Yu, S.: Tibetan Information Processing: Past, Present, and Future. China Tibetology 4, 97–105 (2003)
12. Rui, J., Wu, J., Sun, Y.: Study on Implementing Tibetan Operating System Based on ISO/IEC10646. Journal of Chinese Information, Processing 19(5), 97–104 (2005)
13. Gao, D., Gong, Y.: The Optimal Arrangement of Tibetan Character onto Keyboard. Journal of Chinese Information, Processing 19(6), 92–97 (2005)
14. Lu, Y.: A Study of Layout and Input Method of A General Tibetan Computer Keyboard. Journal of Chinese Information Processing 20(2), 78–86 (2006)
15. Lin, S., Dong, Y., Wang, S., Li, T., Nyima, T., Pu, D.: A Tibetan Input Method Based on Syllable Word for Mobile Phone. In: ICESS 2005, Second International Conference on Embedded Software and System, pp. 16–18. Xian, China (2005)

Self-adaptive Clustering-Based Differential Evolution with New Composite Trial Vector Generation Strategies

Xiaoyan Yang and Gang Liu

Abstract. Differential evolution (DE) algorithms is a population-based algorithm like the genetic algorithms. But there are some problems in DE,such as slow and/or premature convergence. In this paper, a self-adaptive clustering-based differential evolution with new composite trial vector generation strategies (SaCoCDE) is proposed for the unconstrained global optimization problems. In SaCoCDE, the population is partitioned into k subsets by a clustering algorithm. And these cluster centers and the best vector in the current population are used to design the new differential evolution mutation operators. And these different mutation strategies with self-adaptive parameter settings can be appropriate during different stages of the evolution. This method utilizes the concept of the cluster neighborhood of each population member. The CEC2005 benchmark functions are employed for experimental verification. Experimental results indicate that CCDE is highly competitive compared to the state-of-the-art DE algorithms.

1 Introduction

Differential evolution (DE) algorithms, proposed by Storn and Price [1], compose an efficient type of evolutionary algorithm (EA) for the global optimization domain. DE is a population-based stochastic search technique. It uses mutation, crossover, and selection operators at each generation to move its population toward the global optimum. It has successfully been applied to diverse domains of science and engineering, such as [2, 3], and many others. However, DE may be trapped in a local optimum point and also be premature convergence in optimization. In recent years,

Xiaoyan Yang · Gang Liu
State Key Lab of Software Engineering, Computer School, Wuhan University, Wuhan PR China
e-mail: yxy0197@126.com, lg0061408@126.com

Xiaoyan Yang
School of Science, Hubei University of Technology, Wuhan PR China

F.L. Gaol et al. (Eds.): Proc. of the 2011 2nd International Congress on CACS, AISC 144, pp. 261–267.
springerlink.com   © Springer-Verlag Berlin Heidelberg 2012

a lot of researches have been proposed to improve the performance of DE. The adaptive DE is a research focus. These DE variants are listed in the literature [4,5,6].

Motivated by these findings, we propose a self-adaptive clustering-based differential evolution with new composite trial vector generation strategies (SaCoCDE). SaCoCDE is the hybridization of the one-step k-means clustering, the self-adaptive parameter settings and the composite trial vector generation strategies with DE. The new mutation operators are proposed in this paper. These mutation operators are inspired by [7] and the concept of the cluster. They are called the cluster neighborhood mutation model, the global mutation model and the random mutation model. In the cluster neighborhood mutation model the population is partitioned into k subsets by a clustering algorithm, and these cluster centers are used to yield the new mutation operators. The best vector created by three trial vector generation strategies is selected into the next generation population. The parameter control technique is based on the self-adaptive of two parameters (the scaling factor F and the crossover rate Cr), associated with the evolutionary process. And each mutation operator has the different the self-adaptive F and the same the self-adaptive Cr. The main goal here is to produce a flexible DE, in terms of control parameters and the composite trial vector generation strategies. The CEC2005 benchmark functions are employed for experimental verification. It compares the performance of SaCoCDE with several state-of-the-art DE variants. Experimental results indicate that our approach performs better, or at least comparably, in terms of the quality of the final solutions.

The remainder of this paper is organized as follows. In Section 2, we provide a brief outline of the DE algorithm and SaCoCDE is presented in detail in Section 3. Experimental results are reported in Section 4. Finally, Section 5 concludes this paper.

2   Differential Evolution

DE creates new candidate solutions by combining the parent individual and several other individuals of the same population. A candidate replaces the parent only if it has better fitness. There are two parameters: the scaling factor F and the crossover rate Cr in DE. In general conditions, the control parameters depend on the results of preliminary tuning. But, the time for finding these parameters is unacceptably long. Mutation is very important for DE. Some well-known mutation schemes are listed as follows:

$$DE/rand/1 : V_i = X_{r1} + F * (X_{r2} - X_{r3}) \tag{1}$$

$$DE/rand/2 : V_i = X_{r1} + F * (X_{r2} - X_{r3} + X_{r4} - X_{r5}) \tag{2}$$

$$DE/best/1 : V_i = X_{best} + F * (X_{r1} - X_{r2}) \tag{3}$$

where $r1, r2, r3, r4, r5 \in \{1, 2, \ldots, NP(populationsize)\}$ are randomly chosen integers, which are different from each other and also different from the running index i. X_{best} represents the best individual in the current generation. $F(> 0)$ is a scaling factor which controls the amplification of the differential vector.

3 Self-adaptive Clustering-Based Differential Evolution with New Composite Trial Vector Generation Strategies

The characteristics of the trial vector generation strategies and the control parameters of DE have a great impact on the performance of search. Based on the above considerations, one-step k-means clustering algorithm, the self-adaptive parameter settings and the composite trial vector generation strategies are introduced in DE. The proposed DE algorithm is named SaCoCDE. One-step k-means clustering algorithm is used to get the cluster centers of the current population. We propose three kinds of mutation models for DE. The first one is called the cluster neighborhood mutation model, where each vector is mutated using the cluster centers. The second one is called the global mutation model, where each vector is mutated using the DE/current-to-best/1. The third one is called the random mutation model, where each vector is mutated using the DE/rand/1. The best vector in 3 trial vectors is selected to replace the target vector. It is greedy selection scheme. And these mutation model have different the self-adaptive F, respectively.

3.1 The Self-adaptive Parameter Settings

In order to choose suitable control parameter values, we propose a self-adaptive approach for control parameters. This method is motivated by the literature [8]. The Cr values is taken in the range $[0.1, 0.9]$ and the F values is taken in the range $[0.6, 1]$. The self-adaptive Cr and F may be expressed as:

$$F_i = 0.5 + rand[0, 1] * 0.5 \tag{4}$$

$$Cr = 0.1 + rand[0, 1] * 0.8 \tag{5}$$

where $rand[0, 1]$ is random number from $[0, 1]$ and $i = 1, 2, 3$. And there are three self-adaptive F for three mutation operators and a self-adaptive Cr for all mutation operators.

Each member in the initial population is randomly associated parameter values taken from the respective range. The target vectors produce trial vectors using the assigned parameter values. If the generated trial vector produced is better than the target vector, the parameter values are retained with trial vector which becomes the target vector in the next generation. If the target vector is better than the trial vector, then the associated parameter values are randomly reinitialized from the respective range. This leads to an increased probability of production of offspring by the better the associated control parameters in the future generations.

3.2 The New Mutation Operators

In this work, one-step k-means clustering is used to get the cluster center of the current population. The one-step k-means clustering is chosen because of its simplicity and linear time complexity. K is the number of clusters and is generated from

$[2, \sqrt{NP}]$ randomly. It is a rule of thumb used by many investigators in the literature [9]. Other clustering approaches can also be employed as well.

For each member of the cluster, a local donor vector (the cluster neighborhood donor vector) is created by employing the cluster center and any two other vectors chosen from the population. Similarly, the global donor vector is created. The random donor vector is also created. The three models may be expressed as:

$$L_i = X_i + F1 * (C_i - X_i + C_{r1} - X_{r2} + C_{r2} - X_{r2}) \tag{6}$$

$$G_i = X_i + F2 * (X_{best} - X_i + X_{r3} - X_{r4}) \tag{7}$$

$$R_i = X_i + F3 * (X_{r5} - X_{r6}) \tag{8}$$

where C_i, C_{r1} and C_{r2} are the cluster center of the cluster of X_i, X_{r1} and X_{r2}, X_{best} is the best vector in the entire population at generation G. And $r1 \neq r2 \neq i, r3 \neq r4 \neq i,$ $r5 \neq r6 \neq i, r1, r2, r3, r4, r5, r6 \in [1, NP]$. $F1, F2, F3$ are self-adaptive.

It is worth pointing out that 3 new trial vector are generated in SaCoCDE. The equation (6), (7)and (8) generate 3 new trial vectors respectively. And then the best vector in 3 new trial vectors is selected to replace the target vector in the population. This method is effective to improve the search efficiency of DE. The advantages of cluster neighborhood do not need to assign the neighborhood size and do not need to consider population topology.

The SaCoCDE algorithm is described in Algorithm 1, where t is the iteration counter; $rndint[2, \sqrt{NP}]$ is a random integer from $[2, \sqrt{NP}]$. The deterministic method is not used to refine the cluster centers. It is suitable that the number of cluster centers is random for fast computing and is flexible.

Compared with the technique in [7,8], our approach has five main characteristics: (1) the one-step k-means clustering algorithm is used to create the cluster neighborhood and it is not need to define a neighborhood of radius. In [7], it need to define a neighborhood of radius; (2) Three kinds of mutation models for DE are different with the ones in [7,8]; (3) Each mutation operator has the different self-adaptive parameter F.(4) Three new trial vectors are generated respectively and the best vector in 3 new trial vectors is selected to replace the target vector if it is better than the target vector. In [7], only one trial vector is generated and it is not used the self-adaptive method. In [8], although there are 3 the mutation strategies, each vector uses only one of the mutation strategies in mutation step. It means that only one trial vector is generated. (5)The method to generate the self-adaptive parameters is different from one in [8].

Hence, the SaCoCDE algorithm can improve search efficiency and the ability to find the optimal solution. And SaCoCDE is simplicity, efficiency and flexibility.

4 Experimental Results and Discussions

In this section, the CEC 2005 benchmark suite which consists of 25 scalable benchmark functions [10] is employed for an experimental evaluation of the proposed algorithm. The number of variables of all the functions is scalable (up to 50), and

Algorithm 1. The SaCoCDE algorithm

Step1. Generate the initial population P and initial self-adaptive parameters. Evaluate each individual in P, $t = 1$.

Step2. While the halting criterion is not satisfied do

Step3. Randomly generate $k = rndint[2, \sqrt{NP}]$, one-step k-means clustering is used to create k cluster centers.

Step4. For i=1 to NP do

The equation (6), (7)and (8) generate 3 new trial vectors using the self-adaptive parameters associated respectively. The best vector in 3 new trial vectors is selected to replace the target vector in the population if it is better than the target vector.

Step5. Update step

$$F_i = \begin{cases} F_{i+1} & \text{if } f(U(best) > f(X_i)) \\ F_i & \text{otherwise} \end{cases} \quad j = 1,2,\ldots n$$

$$Cr_i = \begin{cases} Cr_{i+1} & \text{if } f(U(best) > f(X_i)) \\ Cr_i & \text{otherwise} \end{cases} \quad j = 1,2,\ldots n$$

In DE algorithms, U( ) is the crossed vector.

Step6.End for

Step7. t = t + 1.

Step8. End while

Step9. Output the best vector.

we have considered in this paper dimensions D= 30 for all the problems. In this paper, the unimodal functions F1-F5, the basic multimodal F6-F10, the expanded multimodal functions F13-F14 functions are used.

SaCoCDE was compared with two other state-of-the-art DE variants. They are EPSDE [8], DE-GPBX-α [11]. In our experiments, we used the same parameter settings for these methods as in their original papers. The number of function evaluations (NFEs) in all these methods was set to 300 000, as the same as in SaCoCDE. The population size in SaCoCDE was set to 100. Each algorithm was executed independently 30 times, to obtain an estimate of the mean solution error and its standard deviation.

The experimental results are given in Table 1. For unimodal functions F1-F5, we can see that SaCoCDE performs better than DE-GPBX-α, EPSDE on five(i.e.F1-F5), two(i.e.F4-F5) test functions, respectively. EPSDE is better than other algorithms on F2-F3. DE-GPBX-α cannot outperform SaCoCDE on any test function.

For basic multimodal functions F6-F10, SaCoCDE is significantly better than DE-GPBX-α, EPSDE on two(i.e.F6-F7), two(i.e.F6-F7) test functions, respectively. For F8, the results obtained by all algorithms are not very different. For basic multimodal functions F6-F10, the performance of these methods are same.

For expanded multimodal functions F13-F14, SaCoCDE exhibit better performance than other methods on F14. EPSDE is better than SaCoCDE on F13.

Table 1 Experimental Results of SaCoCDE, DE-GPBX-α, EPSDE Over 30 Independent Runs on 25 Test Functions of 30 Variables With 300 000 NFEs(Only List 12 Representative Functions).

F	SaCoCDE		DE-GPBX-α		EPSDE	
	Mean	Std dev	Mean	Std dev	Mean	Std dev
F1	00E+00	00E+00	2.84E-15	1.25E-14	0.00E+00	0.00E+00
F2	1.12E-01	6.16E-02	3.34E+01	5.17E+01	**4.23E-26**	4.07E-26
F3	1.51E+06	5.47E+05	8.70E+06	6.21E+06	**8.74E+05**	3.28E+06
F4	**5.45E+00**	2.52E+00	7.01E+02	7.26E+02	3.49E+02	2.23E+03
F5	**3.20E+02**	2.11E+02	2.03E+03	6.28E+02	1.40E+03	7.12E+02
F6	**5.17E-03**	4.97E-03	5.04E+01	4.34E+01	6.38E-01	1.49E+00
F7	**6.57E-04**	2.50E-03	4.70E+03	4.09E-13	1.77E-02	1.34E-02
F8	2.09E+01	5.56E-02	2.09E+01	5.54E-02	2.09E+01	5.81E-02
F9	4.15E+01	4.93E+00	8.12E+00	4.37E+00	**3.98E-02**	1.99E-01
F10	1.59E+02	1.18E+01	1.01E+02	3.95E+01	**5.36E+01**	3.03E+01
F13	8.50E+00	2.56E+00	1.28E+00	2.93E+00	**1.94E+00**	1.46E-01
F14	**1.20E+01**	3.77E-01	1.31E+01	2.28E-01	1.35E+01	2.09E-01

In summary, SaCoCDE performs better, or at least comparably, in terms of the quality of the final solutions on basic multimodal functions, expanded multimodal functions.

The evolution of the mean function error values derived from DE-GPBX-α, EPSDE and SaCoCDE versus the number of NFEs is plotted in Figure.1 for some typical test functions.

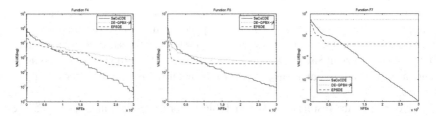

Fig. 1 Evolution of the mean function error values derived from SaCoCDE, DE-GPBX-α, EPSDE versus the number of FES on test functions.

From Figure.1, it can be seen that for the test functions SaCoCDE converges faster than the competitors. In view of the above discussion it can be concluded that the overall performance of SaCoCDE is better than the two competitors for the CEC2005 benchmark functions.

5 Conclusion

In this paper, we proposed the cluster neighborhood mutation operator, the global mutation operator and the random mutation operator, in an attempt to balance their effects. SaCoCDE employed three trial vector generation strategies and self-adaptive parameter settings. The structure of SaCoCDE is simple and it is easy to implement.

The experimental studies in this paper were carried out on 12 global numerical optimization problems used in the CEC2005 special session on real-parameter optimization. SaCoCDE was compared with two other state-of-the-art DE variants. The experimental results suggested that its overall performance was better than the two competitors.

In the future, we will generalize our work to other EAs for other hard optimization problems. The self-adaptive parameter settings will be further studied.

References

1. Price, K., Storn, R., Lampinen, J.: Differential Evolution-A Practical Approach to Global Optimization. Springer, Berlin (2005)
2. Chang, W.-D.: Two-dimensional fractional-order digital differentiator design by using differential evolution algorithm. Digital Signal Processing 19(4), 660–667 (2009)
3. Wang, L., Li, L.-P.: Fixed-Structure H Controller Synthesis Based on Differential Evolution with Level Comparison. IEEE Transactions on Evolutionary Computation 15(1), 120–129 (2011)
4. Qin, A.K., Huang, V.L., Suganthan, P.N.: Differential evolution algorithm with strategy adaptation for global numerical optimization. IEEE Transactions on Evolutionary Computation 13(2), 398–417 (2009)
5. Tvrdk, J.: Adaptation in differential evolution: A numerical comparison. Applied Soft Computing 9(3), 1149–1155 (2009)
6. Thangaraj, R., Pant, M., Abraham, A.: A simple adaptive differential evolution algorithm. In: 2009 World Congress on Nature and Biologically Inspired Computing, pp. 457–462 (2009)
7. Das, S., Abraham, A., Chakraborty, U.K., Konar, A.: Differential Evolution Using a Neighborhood-Based Mutation Operator. IEEE Transactions on Evolutionary Computation 13(3), 526–553 (2009)
8. Mallipeddi, R., Suganthan, P.N., Pan, Q.K., Tasgetiren, M.F.: Differential evolution algorithm with ensemble of parameters and mutation strategies. Appl. Soft Comput. 11(2), 1679–1696 (2011)
9. Sheng, W., Swift, S., Zhang, L., Liu, X.: A weighted sum validity function for clustering with a hybrid niching genetic algorithm. IEEE Transactions on Systems, Man, and Cybernetics, Part B: Cybernetics 35(6), 1156–1167 (2005)
10. Suganthan, P.N., Hansen, N., Liang, J.J., Deb, K., Chen, Y.P., Auger, A., Tiwari, S.: Problem definitions and evaluation criteria for the CEC 2005 special session on real-parameter optimization, Nanyang Technol. Univ., Singapore, and IIT Kanpur, Kanpur, India, KanGAL Rep.2005005 (May 2005)
11. Dorronsoro, B., Bouvry, P.: Improving Classical and Decentralized Differential Evolution with New Mutation Operator and Population Topologies. IEEE Transactions on Evolutionary Computation 15(1), 67–98 (2011)

Research on Structure Design of Health Information Management System for Women and Children

Hangkun Ling and Ning Ling

Abstract. Women and Children are the hope of human. On Millennium Development Goals [1], World Health Organization (WHO) was expressing the requirements to reduce child mortality and to improve maternal health. In this paper, the significance, objective and principle of building the Health Information Management System (HIMS) for Women and Children were established. The authors will introduce the system framework, architecture and technology, the design of database and system security also included. The paper innovated and built HIMS for Women and Children to provide an intellectual support.

1   Introduction

"I want my leadership to be judged by the impact of our work on the health of two populations: women and the people of Africa."
--- Dr. Margaret Chan, Director-General, World Health Organization (WHO)

In the critical period of high-speed information technology promotion and reformation, it is necessary to build Health Information Management Systems (HIMS) to connect, coordinate, and manage the partnership for maternal, newborn and child health, thus ensure the government makes the scientific and effected decision and policy for women and children. Meanwhile, HIMS can improve the communication and cooperation between related organizations like multilateral organizations, Non-governmental organizations, professional health associations, bilateral donate and aid agencies, health foundations, academic and research institutions. The cooperation between multi-organizations helps the resources sharing and

Hangkun Ling
University of Arkansas for Medical Science, Little Rock, USA
and

University of Arkansas at Little Rock, Little Rock, USA
e-mail: lxhangkun@ualr.edu

Ning Ling
Jiangsu Administration Institute, Nanjing, China
e-mail: lingning6199@sina.com

F.L. Gaol et al. (Eds.): Proc. of the 2011 2nd International Congress on CACS, AISC 144, pp. 269–276.
springerlink.com   © Springer-Verlag Berlin Heidelberg 2012

rational allocations and making technical guidelines and best practices, in compliance with requirements of national development. It turns to be more important that government improves the cooperation and communication between partnerships for women and children.

Governments should enhance the constructions of HIMS for women and children; improve health to operate as a poverty-reduction strategy [9]. For this improvement, it is being addressed include the provision of adequate numbers of appropriately trained staff, sufficient financing, suitable systems for collecting vital statistics, and access to appropriate technology including essential drugs [1]. Thus, there is theoretical and practical significance for build national health information management systems for women and children.

2   Background

Being a man or a woman has a significant impact on health, as a result of both biological and gender-related differences. The health of women and girls is of particular concern because, in many societies, they are disadvantaged by discrimination rooted in socio-cultural factors [2]. The data from World Health Organization (WHO) indicates that there are about 500,000 died since pregnancy or childbirth in the world every year, and more than 1,000,000 children's mortality before 5 years old, 40% even before 1 month old [3]. In those deaths, at least 2/3 can be avoided by approved, effective intervention. By extending those intervention, and bringing it into the work of maternal, newborn and child health, around 7 million deaths of women and children can be avoid in the world every year.

It is no doubt that building the national and origins HIMS for Women and Children is a necessary technical support of the implementation of women and children's development. WHO/MPS the largest-ever gathering of Heads of State and Government ushered in the new millennium by adopting the United Nations Millennium Declaration. The Declaration was endorsed by 189 countries and was translated into eight Millennium Development Goals (MDGs) to be achieved by 2015 [1] and two of the eight are to reduce child mortality and to improve maternal health.

3   Objectives and Principle

Our goal is to build an advanced, stable and secure health information management system for women and children according to UN's general arrangements and requirements of exercising powers of project management to enhance the communication and cooperation between related organizations like multilateral organizations, Non-governmental organizations, professional health associations, bilateral donate and aid agencies, health foundations, academic and research institutions.

As the basis and regulatory, our project system adheres to the following principles: data and function standardization ease of use, humanist, security and reliability, information highly sharing.

4 Methodology

The principles will go throughout the process of HIMS system constriction and methodology. Firstly, at the beginning of designing system functions and database structures, we need consider information sharing. We had the index with women and children related keywords since the service data object (SDO) is used as foundation, thus, the SDO can be laterally shared. The information sharing is across different work groups by lateral communication method. Secondly, we had customized database in the information flows to help longitudinal sharing and future cooperation and improve efficiency.

4.1 Framework and Technical Route

The Health information Management System for Women and Children relies on National Health Administration Organization Network and Internet, and fully integrated the networks and resource database among WHO, multilateral organizations, Non-governmental organizations, professional health associations, bilateral donate and aid agencies, health foundations, academic and research institutions.

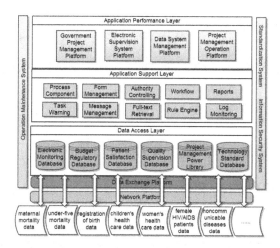

Fig. 1 System framework of HIMS for Women and Children

The framework of this system shows above in Figure1. The data collection and management are including maternal mortality, under-five mortality, registration of birth, children's health care, women's health care, female HIV/AIDS patients, non-communicable diseases. The databases in our system are all from practical projects with cure records held by the health organizations. Collected data records are extracted to upper system platform, system would save them into subject database.

As showed in figure 1, on data, the system create a subject database which is "classification on physical, unity on logical". On exchange, the system builds a standard Data Exchange Platform (DEP), and achieves a unite interface management in order to get data exchange on running system. On security, user will have unique and lock-only account.

4.2   System Architecture

In our system, concentration distribution deployment is being used. The deployment architecture is showed below in Figure 2.

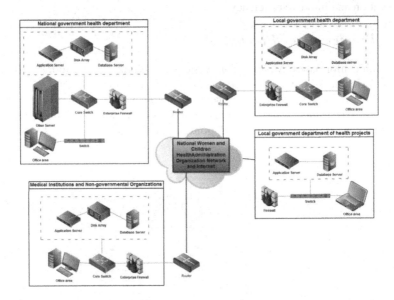

Fig. 2 System architecture of HIMS for Women and Children

4.3   Technical Key Tools

This system fully applies the data collecting and J2EE application skill. It based on the XML data exchange skill and current correspondence skill. The architecture method is based on SOA (Service Oriented Architecture), a flexible set of design principles used during the phases of system development [6]. Our system will package functionality as a suite of interoperable services that can be used within multiple separate systems. XML is commonly used for interfacing with SOA services.

In this Health information Management system, we used several technical tools as following.

- **Database Layered Technology.** The design of system database has layered by 3 levels: government coordination, medical institutions and non-governmental organizations, health projects. Those level are physical classification that can serve individually. The database can be submitted by lower one with unity standard.

- **WebService Technology.** Web services in our system are used to implement the architecture according to service-oriented architecture (SOA) concepts, where the basic unit of communication is a message, rather than an operation.
- **Rule Engine Technology.** There are some rules in the operation flow. It will be limitation for the system flexibility if the rules are written as a hard-code. Our system platform is provided a unity rule engine to define the rules. The rule engine will be used in operation sub-system and flows.
- **Form Technology.** The form technology in this system is used for solving unstable operations. Simple operation does not need coding as long as do some configuration that can be achieved on the input, editing, delete and query. The figure 3 below shows the form-tech.

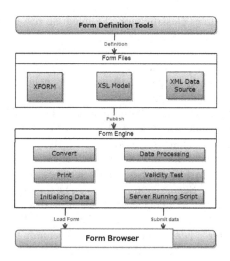

Fig. 3 Flow chart of Form Technology

By using this self-service form technology, the experts can participate in during the system development, compiling data form electronically instead of paper, reducing the risk of communication and requirement understanding.

- **System Monitoring Technology.** System monitor is hardware or software based system used to monitor resources and performance in our women and children management system. The monitoring system can help us to keep track of system resources.
- **Object/Relational Mapping Technology.** Using ORM can decoupling application system and database, thus make our system can support database like SQL Server, Oracle, DB2.

5   Database and System Security

The databases of the system are all from practical projects with cure records. This health project management system refers to personal privacy, government intervention and sensitive data, so system security is especially important. Not only we need consider all weak links of the network security protection, and also attend to the realization of the security policy.

5.1   *Information Resources Database Design*

According to the WHO's targets, we integrated management departments of countries' internal information resources, and provide a data support for the unity of the health information management system for women and children to provider data support.

1)   Data resources classification: To integrate resources of health related organizations mentioned in this paper.

2)   Fundamental Data: This includes basic information of the relevant department and organization of information and related agencies. All the data from running operation are engagement, preventive management, control and management, measurement and payment.

3)   Subject data set: The subject data set of this HIMS includes the following databases. a) Electronic Monitoring Database: To realize the covering of entire project, flows and domain of our system; b) Budget Regulatory Database: To more clearly understand the flow of funds or budget; c) Patients Satisfaction Database: To assess the performance of public services, and check the completion of objectives; d) Quality Supervision Database: The quality monitoring goes through the life-time of the project management, related to disease prevention and controlling for women and children; e) Project Management Power Library: To regulate women and children's health management; f) Technology Standard Database: For permanent protection of data for future retrieval of resource sharing service. In the process of building a database, it is according to international norms and standards for effective operation.

4)   Information collection standard: As the women and children health is complex data and structure, the standard is not uniformed yet and characteristics of the data itself needs to in-depth analysis. This system is required data collection standards and unified data exchange mechanism. The standardized collection of information resources is including the scope of the collection, collection methods, collection content, collection frequency and collection maintenance responsibility of the principal, etc.

5)   Database exchange standards: HIMS's sharing and exchanging data is defined according to data encoding rules, classification, metadata, data formats, etc, to follow the development standards. The main criteria include the following aspects: a) Sharing norms and interface standards; b) sharing information updates standards; c) sharing information updated interface implementation specification; d) the standard of subscription message; e) sharing information subscribe to bridge the interface specification; f) sharing information message publishing standard.

5.2   System Security Design

In order to ensuring system's security, we formulated a security management. It is mainly in the network security, application security management, security policy management, system strategy management.

The security of network is the foundation to ensure the safe operation of the system. It is primarily through the core switch, routers, firewall, and anti-virus products to provide security features [10]. In the operation system, configuration of the server administration, management of passwords that meets the requirements, timely repairing of the operation patches, stop and uninstall unnecessary services and procedures are being used.

Database data protection is used in this system. Database data protection includes the backup and recovery, and database log files. The backup database wizard and security management policy are also required for the system security.

6   Conclusions and Discussion

Women and Children's health are facing a challenge nowadays. WHO expressed the requisitions to reduce child mortality and improve maternal health on MDGs. It is necessary that governments and health organizations have an advanced, stable, secure health information management system for women and children.

In this paper, we discussed a HIMS for Women and Children. Our system should be based on the National Women and Children Health Administration Organization Network and Internet. It may integrate the resources from WHO, multilateral organizations, Non-governmental organizations, professional health associations, bilateral donate and aid agencies, health foundations, academic and research institutions. The system based on the J2EE three levels of structure and SOA/WebService technology, construction project management system. The databases in our system are all from practical projects with cure records held by the organizations motioned in this paper. Collected data records are extracted to upper system platform, system would save them into a "classification on physical, unity on logical" subject database which is including Electronic Monitoring Database, Budget regulatory database, Patients satisfaction database, Quality supervision database, Project management power Library, Technology standard database.

We also offered a security management for this HIMS. It is mainly in the network security, application security management, security policy management, system strategy management.

Reference

[1] World Health Organization, The WHO agenda, http://www.who.int/about/agenda/en/index.html (retrieved on May 2011)

[2] World Health Organization, Strategy for Integrating Gender Analysis and actions into the work of WHO. Sixth World Health Assembly, A/MTSP/2008-2013/PB/2008-2009 and Corr.1, May 23 (2007)

[3] World Health Organization, The Maternal and Children's health, the newborn Partnership, http://www.who.int/pmnch/en/index.html (retrieved May 2011)

[4] W3C Reference: XML 1.0 Origin and Goals, 5th edn., November 26 (2008) (retrieved on July 2011)

[5] Wohed, P., van der Aalst, W.M.P., Dumas, M., ter Hofstede, A.H.M.: Analysis of Web Services Composition Languages: The Case of BPEL4WS. In: Song, I.-Y., Liddle, S.W., Ling, T.-W., Scheuermann, P. (eds.) ER 2003. LNCS, vol. 2813, pp. 200–215. Springer, Heidelberg (2003)

[6] Seo, C., Han, Y., Lee, H., Jung, J.J., Lee, C.: Implementation of Cloud Computing Environment for Discrete Event System Simulation using Service Oriented Architecture. In: IEEE/IFIP International Conference on EMbeded and Ubiquitous Computing. IEEE Computer Society, HongKong (2010)

[7] Terwilliger, J.F., Bernstein, P.A., Unnithan, A.: Worry-free database upgrades: automated model-driven evolution of schemas and complex mappings. In: SIGMOD 2010 International Conference on Management of data. ACM, NY (2010) ISBN: 978-1-4503-0032-2

[8] Hersh, W.: The Health Information Technology Workforce. Applied Clinical Informatics 294, 203–204 (2010), doi:10.4338/ACI-2009-11-R-0011

[9] World Health Organization Regional Office for the Western Pacific. A selection of important health indicators, Manila, Philippines. World Health Organization (2000)

[10] Weitzman, E., Kaci, L., Mandi, D.: Sharing Medical Data for Health Research: The early Personal Health Record Experience. Journal of Medical Internet Research 12(2), e14 (2010)

[11] Baird, W., Jackson, R., Ford, H., Evangelou, N., Busby, M., Bull, P., Zajicek, J.: Holding personal information in a disease-specific register: the perspectives of people with multiple sclerosis and professionals on consent and access. J. Med. Ethics 35(2), 92–96 (2009)

How Hybrid Learning Works in the Workplace

Li Zhang

Abstract. This paper focuses mainly on with respective to hybrid learning in the workplace. Hybrid learning in the workplace means the adults to blend different learning ways for solving their work-related problems or adapting to new working environments. The ways involve relating work activities, mixing learning activities, blending learning technologies, and integrating learning environments. These four key components in the workplace are discussed. The factors affecting hybrid learning are also respectively analyzed from learning and context aspects.

1   Background

Adult learning is a fascinating subject recently. According to a study done in 2002 by KPMG (now BearingPoint), nearly 60 percent of corporate knowledge goes out of date within three years [1]. Lots of companies have spent great investments on training. They have developed e-learning program to train employees. The benefits of e-learning are obvious, such as efficiency, cost reduction, individualized instruction, and self-paced and so on. But e-learning is not the be all and end all to every training need. It does have limitations, among them: Up-front investment, technology issues, inappropriate content, and reduced social and cultural interaction [2].

Fortunately hybrid learning solves these problems. It provides the best combination of traditional learning with e-learning. In recent years, more and more managers and trainers of workplace accept the concept of "hybrid learning". Compared with single e-earning, hybrid learning can improve training efficiency, reduce the cost of workers, and reduce training time [3]. There are a few researches about hybrid learning in the workplace in china. This paper will focus on hybrid learning in the workplace in china. It will answer these questions: What is hybrid learning in the workplace? How hybrid learning works? And which factors affect hybrid learning in the workplace?

2   Definition of Hybrid Learning in the Workplace

Hybrid learning is a new term built on the profound consideration of e-Learning, which appeared in the field of education technology for a short time. However it

Li Zhang
Shanghai Institute for Youth Management, Shanghai, China
e-mail: lucky_zhang1@hotmail.com

F.L. Gaol et al. (Eds.): Proc. of the 2011 2nd International Congress on CACS, AISC 144, pp. 277–282.
springerlink.com   © Springer-Verlag Berlin Heidelberg 2012

has existed in the workplace for a long time of its essential meaning, for the nature of learning on work is "blended". For example, employees learn new skill, technology or knowledge by observing peer techniques, by being trained to operate new procedures, or by communicating with others to get experience. Technological innovations are also expanding the range of possible solutions that can be brought to bear on teaching and learning. So hybrid learning does not combine e-learning with face to face learning simply. It represents a return to the idea of learning, which reflects not only the information technology to support adult learning but also the tradition learning theory of adult.

There are many definitions about hybrid learning (blended learning). In the article of Blended Learning Environments: Definitions and Directions [4], the author concluded the three most commonly mentioned definitions. There are

1) BL=combining instructional modalities (or delivery media).
2) BL=combining instructional methods
3) BL=combining online and face-to-face instruction

But these definitions cannot claim clearly the similarities and differences between blended learning in traditional education and hybrid learning in corporate training. Hybrid learning can take place in its origin--the workplace, and can occur in the formal traditional education, such as higher education, primary and secondary education. The needs and purpose of hybrid learning in both areas are different. The purpose of hybrid learning in the workplace has involved improving the training efficiency, reducing training costs, and increasing corporation efficiency. The function of hybrid learning in traditional education includes improving pedagogy, increasing access/flexibility, and increasing cost effectiveness [5].

Besides, the term of "instruction" in these definitions limits hybrid learning in training program on work. In today's learning society, training has been unable to meet staff to keep on learning new skills and self-development needs. The traditional formal learning for adults, such as workplace learning, vocational training and participation in special task only consists of 20% new knowledge source for adults; while the informal learning such as experiencing on the job, network learning, mentoring & coaching, manuals & instructions accounts for 80% learning knowledge [6]. It is essential to return to the origin of hybrid learning, not just training.

Owing to the difference between hybrid learning of adult in the workplace and blended learning of students in traditional schools, and the difference between adult learning and staff training, it is necessary to give a new definition of hybrid learning in the workplace. Hybrid learning in the workplace is the combination of different learning of adults in the workplace to solve work-related problems or to adapt to new working environment, which consists of combination work activities, mixing learning activities, blending learning technologies, and integration learning environments.

3   Components of Hybrid Learning in the Workplace

How hybrid learning works in the workplace? There are many answers. Bersin proposed four hybrid learning processes; Rossett suggested "hybrid" took place in an organic combination of a number of ways [7]. Graham suggests blending should exist at four different levels: activity level, course level, program level and institutional level.

From author's opinion, the learning in the workplace can be blended in four ways: the combination of work activities, the mixing of learning activities, the blending of learning technologies, and the integration of learning environments. Figure 1 shows the four components of hybrid learning in the workplace.

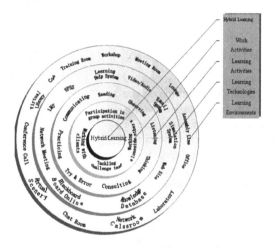

Fig. 1 Four Conponents of hybriding learning in the workplace

• Work activities in the workplace

The process of working is a process of practice in the workplace. And the work activities give the adult rise to learning. There are four main types of work activities in the workplace [8].

Participation in group activities included team working towards a common outcome, and groups set up for a special purpose such as audit, development or review of policy and/or practice, and responding to external changes. Working alongside others allows people to observe and listen to others at work and to participate in activities, and then to learn new practices, to get different knowledge, and to gain others' expertise. Tackling challenging tasks requires on-the-job learning and leads to increased motivation and confidence. Working with clients also entails learning about the client, or from any novel aspects of each client's problem or request and from any new ideas that got their joint consultation. These four work activities are not one from another. There are sandwiched on work.

- Learning activities in the workplace

Most learning activities were embedded within the working processes described above to a greater or lesser degree. The learning activities in the workplace included reading, listening, observing, thinking, training, consulting, practicing, trial and error, and communicating (see Figure 2).

These learning activities would merge individual learning with group learning, merge formal learning with informal learning, and merge F-to-F learning with e-learning, merge skill learning with culture learning.

Fig. 2 Mixing of learing activies

- Learning technologies in the workplace

The Hybrid learning activities are supported by information technologies. And the choice of technology is an important issue. The criterions for technologies selection are live or self-study, network requirements, pc configuration requirements, training type, development costs, and deployment costs and so on. Figure 3 lists the technologies applied in hybrid learning in the workplace.

Each technology has its own strengths and weaknesses. To make blended learning more powerful, we should start looking

Fig. 3 Blending of learning technologies

at all the technologies as options: learning help system, leaning management system, practice community, simulation lab, and so on. We can blend a more complex technology with one or more of the simpler technology. The most effective technology in hybrid learning system is an obvious mix.

- Learning environments of workplace

Learning environments of workplace consist of physical environments and virtual environments. Although work activities happen in the real environments, their learning activities can fulfill either in reality or in virtual environments.

Physical environments include training room, workshop, meeting room, lounge, assembly line, laboratory, office and so on. Virtual environments consist of network classroom, virtual scenery, chat room, video conference, virtual laboratory, practice community, etc. The mixing of learning environments is the base of hybrid learning. Employees can communicate with each other and learn face to face in reality, which gives them the feelings of socialization, belonging and reality. At the same time they can learn freely, instantaneously and effectively in virtual environments where they have more opportunities to learn from expert, from outside.

Nowadays hybrid learning means more than merely having the four components mixing in its own level. It means taking blended one level to the next [9],
including better integration and organization of work activates, learning activities,
learning technologies and learning environments.

4   Factors Affecting Hybrid Learning

Many factors affect hybrid learning directly or indirectly. We would divide them
into two sorts: learning factors and context factors, depicting the work context for
learning and the main factors influence learning within context, as Figure 4 lists.

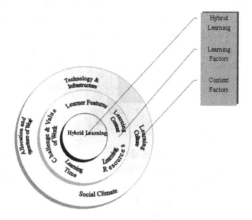

Fig. 4 Factors affecting hybrid learning

Learning factors consist of learner features, learning time, learning contents,
learning resources, challenge and value of the work. Both novices and experienced
employees should learn in the workplace, while their background of academic
knowledge, learning experience, motivations and the ability of IT are different.
Some people are willing to use more technology to learn while the others are
accustomed to learn face to face.

Learning time is also an important factor. When the learner was given a short
time to study, he usually learns by insulting expert, searching information from internet to solve the problem as soon as possible. When novices were given enough
time to learn corporation rules or office system, they would participate in classroom training, online learning, and network discussing in a longer duration.

Learning content on work is rich. Some knowledge may be well-structured, such
as corporate culture, operating specifications, or FAQ. Learners can access knowledge database, visit company's documents, or participate in training to get it. But
more knowledge is implicit and ill-structured, such as troubleshooting, operation
skills, tacit of communication, etc. So Employees would learn by talking to expert,
searching online, practicing in simulation system, trial and error, and so on.

The influences of learning resources include the diversity of resource, the frequency of update, and the adequacy of media developers. The challenge and value of the work is a motivating factor of learning. If the work is valuable and challenged, the learners are proactive in taking advantage of the learning opportunities.

Learning culture, social climate of the workplace, technology & infrastructure, allocation and structuring of work are main factors about context in hybrid learning. Firstly, learning culture is the most important element. Much hybrid learning on work occurs through doing things and being proactive in seeking learning opportunities, and this requires support and feedback, confidence and commitment. Secondly, good relationships and social climate of the workplace can promote learning culture and learning communication. In a harmonized workplace, learners are willing to interactive and learn from each other. Thirdly, hybrid learning is also facilitated by the allocation or structuring of work. It affects the challenge of the work, the extent of collaborating, and the opportunities of observing and working alongside others who had more or different expertise. Last but not least, a sound technology plan will make hybrid system work effective, while a poor technology plan will make program impossible. The principle of infrastructure is mixing and meeting different needs.

5   Conclusion

Hybrid learning in the workplace is the combination of different learning of adults to solve work-related problems or to adapt to new working environment. The learning on work is blended in four ways and these four components also take blended from one level to the next. The factors involved in hybrid learning should be analyzed by context factors and learning factors.

References

1. Bersin, J.: The Blended Learning Book: Best Practices, Proven Methodologies, and Lessons Learned. Pfeiffer, San Francisco (2004)
2. Kruse, K.: The Benefit and Drawbacks of e-Learning, http://www.e-learningguru.com/articles/art1_3.htm
3. Rossett, A., Frazee, R.: Blended learning opportunities, http://www.amanet.org/blended/pdf/WhitePaper_BlendLearn.pdf
4. Osguthorpe, R.T., Graham, C.R.: Blended Learning Environments: Definitions and Directions. The Quarterly Review of Distance Education 4(3), 227–234 (2003)
5. Bonk, C.J., Graham, C.R., Cross, J., Moore, M.G.: The Handbook of Blended Learning: Global Perspectives, Local Designs. Pfeiffer & Company, San Francisco (2005)
6. Harrison, M.: 13 Ways of Managing Informal Learning, http://www.kineo.com/shop/reports/kineo_informal_learning.pdf
7. Rossett, A., Douglis, F., Frazee, R.: Strategies for building blended learning, http://www.learningcircuits.org/2003/jul2003/Rossett.htm
8. Eraut, M.: Informal learning in the workplace, http://www.lerenvandocenten.nl/files/Eraut_2004.pdf
9. Weinstein, M.: A Better Blend. Training 45(9), 30–34 (2008)

Research on the Application of Twitter-Like Platform Based on Web 2.0 Technology in College English Extracurricular Learning: From the Perspective of Interactive Teaching

Xiao-Ling Zou and Ting Luo

Abstract. This study takes the non-English major students as the experimental subjects, conducting four-week extracurricular English interactive activities based on Twitter-like platform, whose feasibility has been positively proved by the data obtained from the pre-/post-questionnaire, the experimental study, the quantitative and qualitative analysis. To some extent, the interaction based on Twitter-like platform not only meets students' interest in and needs for extracurricular English learning, and promotes their control over and motivation for learning, but also increases their involvement in extracurricular English learning activities.

1   Introduction

Twitter, the social networking of Web 2.0, becomes increasingly appealing due to its powerful resource sharing function and speedy information transmission. Since Twitter can provide students with a P2P (peer-to-peer) communication, realize the true communication among learners, and to some extent meet students' emotional need, intrigue their learning interest and even encourage them to overcome difficulties in English learning (Liu & Meng, 2009), more attention has been given to its application in education. In recent years, some meaningful projects based on Twitter have been done in the field of education (Charlene, 2009; Dai & Yao, 2010; Ebner & Schiefner, 2008; Hao & Sun 2010). One program based on Twitter was carried out in the first-year seminar course for pre-health professional majors in South Dakota State University (Junco, Heiberger & Loken, 2010). The researchers found out using Twitter as a tool in university courses could increase student participation and increase grades. Another study is from Silver Spring International school in the US (Ash, 2008), where the English teacher George Mayo

Xiao-Ling Zou
Research Academia for Linguistics Cognition & Application, Chongqing University,
Chongqing China
e-mail: xiaolingzou@cqu.edu.cn

Ting Luo
College of Foreign Languages, Chongqing University, Chongqing China

F.L. Gaol et al. (Eds.): Proc. of the 2011 2nd International Congress on CACS, AISC 144, pp. 283–289.
springerlink.com   © Springer-Verlag Berlin Heidelberg 2012

used Many Voices as an online collaborative platform and invited students to participate in writing a story. In the period of six weeks, more than one hundred students finished the whole story with collaborative effort. Apart from the studies on Twitter use, a series of tentative applications in the field of education were conducted, such as classroom community, collaborative writing, cooperation between different areas and schools, links inside and outside the classroom, project management, interaction on a given topic, reflection, assessment, academic and scientific research, etc. (Grosseck & Holotescu, 2008; Kaplan & Haenlein, 2011; Liu & Meng, 2009). Although the majority of the researches are at the initial stage, Twitter, as an instructional tool, exerts its influence on students and offer students good chance to interact and communicate. Compared with blog or other Web 2.0 platforms such as QQ, Wiki, Twitter has incomparable virtues and powerful exploiting value. For this reason, the experiment has been conducted through answering the research questions below to explore its application effect in students' extracurricular English learning from the view of interactive teaching:

1) What is the feasibility for college students to carry out extracurricular English interactive activities based on Twitter-like platform?

2) Whether or not the extracurricular interactive activities based on Twitter-like platform are in line with students' interest in and needs of communication?

3) Whether or not the extracurricular interactive activities based on Twitter-like platform promotes students' control over and motivation for learning?

4) Whether or not the interactive learning based on Twitter-like platform increases students' involvement in extracurricular learning?

2   Research Methodology

48 non-English major students, taught by the same English teacher with 23 females and 25 males, took part in the experiment.

The pre-questionnaire involved four parts: 1) background information about the students' demographics, computer skills, English proficiency, and online frequency; 2) their experiences in and expectations for English learning via the Net after school; 3) their familiarity, attitudes and anticipation for micro-blogging application; and 4) the possibility of mobile learning based on micro-blogging.

Accordingly, four parts were designed in the post-questionnaire: 1) the subjects' background; 2) their remark on the interaction on micro-blogging platform from learner control, feedback and communication; 3) their comment on the interactive outcomes based on the experiment from satisfaction level and motivation; and 4) three open-ended questions aimed at getting further exploration into the interaction on micro-blogging platform. The second and third parts were developed on the basis of five-point Likert scales, in which response options ranged from strongly disagree to strongly agree.

The experiment on the most influential Twitter-like platform in China, Sina micro-blog, was carried out after the pre-questionnaire, which composed of one-day technical training and four-week extracurricular English interactive activities. Integrating the results of the pre-questionnaire into the experimental arrangement, the authors designed four theme activities: 1) discussion on such topics as campus life and community activities; 2) knowledge quiz of English culture between groups; 3) learning resource uploading and 4) learning experience exchanges. Each student was required to send at least 4 posts to Sina micro-blog in each theme activity. It was expected to explore their interactive status and effects, as well as record their thoughts, activities and lives (Goldstein, 2009). The feedback gathered from the experiment would offer suggestions on necessary revisions of the questions in the post-questionnaire for the sake of lucidity of the study.

In the quantitative analysis, Excel 2003 for Windows was used to illustrate the descriptive statistics. The qualitative analysis further elaborated the findings of the experimental study. Besides, structural and content analysis were conducted to gain insight into the structure of group interaction as well as the levels of involvement between teachers and students, and interaction differences (Wang, 2008). The message was the key part of structural analysis, while the messages themselves were the most important part of content analysis.

3   Results and Discussion

The descriptive statistics show of the 48 subjects, 39 (81.25%) come from urban areas, 45 (93.75%)/46 (95.84) are at the medium or above level in computer skills/English proficiency, 40 (83.34%) go surfing online 3 or over 3 times a week, and only one student surfs the web one-time-a-week to accomplish the listening task of the course. So the current situation of the students' using the internet provides the possibility for the authors to conduct the experiment.

Table 1 indicates that the students spend some time on the network course after class, but they are not very satisfied with their performance. Besides, there is not enough communication in English among them after class, and the micro-blog appears fairly new to them. However, the findings indicate that only a relatively small proportion of them (10.4%) show negative attitudes, while the other 89.58% show positive attitude toward the activities on the micro-blog.

Table 1 Students' extracurricular online learning

Levels of satisfaction	n=48	%	Using self-built platform	n=48	%
A) Very satisfied	2	4.17	A) QQ	15	31.25
B) Satisfied	4	8.33	B) MSN	5	10.42
C) Neither	24	50	C) Blog	5	10.42
D) Dissatisfied	12	25	D) others	7	14.58
E) Very dissatisfied	6	12.50	E) Never use	16	33.34
Attitudes toward the microblogging			Positive 43/89.58% Negative 5/10.42%		

It can be seen from Table 2 that the students rank their expectations for teachers, schools, classes or groups in order of importance as communicating with teachers, school providing resources, class developing own interactive platform, tutors' monitoring and groups making use of external website. In the aspect of students' anticipation for oneself, their choice order by percentage from high to low is getting fun from English learning, interacting with counterparts, getting more knowledge and teacher's tutorial, and improving learning ability. Their preferred activities are interesting topics discussion, followed by learning resource uploading, learning experience exchange and cultural knowledge quiz. The least selected is story solitaire, though it is a good form of collaborative writing. In short, the students are more inclined to communicate and share with each other.

Table 2 Students' expectation for extracurricular online English learning

Students' expectation for teachers, schools, classes or groups	frequency	%
A) instructor monitoring the platform	25	18.38
B) school website providing more English learning resources	31	22.79
C) class developing our own interactive platform	27	19.85
D) groups making use of external website for English learning	15	11.04
E) communicating with the instructor on the platform	38	27.94
Students' expectation for oneself		
A) interacting among peers	46	23.59
B) getting teacher's tutorial	32	16.41
C) getting fun from English learning	47	24.10
D) enlarging knowledge in English culture	41	21.03
E) improving learning ability	29	14.87
Students' expectation for extracurricular activities		
A) interesting topic discussion	47	23.73
B) quiz of cultural knowledge	34	17.18
C) story solitaire	29	14.65
D) learning experience exchange	43	21.72
E) learning resource uploading	45	22.72

The survey finds that all of the students' mobile phones have Wireless Application Protocol (WAP) and General Packet Radio Service (GPRS), which means that the students are able to get mobile networks to browse web pages as they like. 40 students can use IM tools such as QQ and MSN on their phones; 37 students' phones have functions of Multimedia Messaging Service (MMS) or ring tones. It is possible for students to log on the micro-blog with their mobile phones. On the other hand, 58.33% students would like to send 3-6 messages a week, 83.33% more than 6 times and 33.3% 1-2 messages.

After 4-week activities on Sina micro-blog, a post-questionnaire for the same subjects was presented and distributed for understanding whether the interactive activities are in line with students' interest in learning and needs for communication and promote their control over and motivation for learning.

It can be summarized from Table 3 that, in terms of satisfying students' interest in learning their comments are mainly on the medium/high level. As regards meeting their needs for communication, most students believed their interaction with learning resources did happen, and their connection with peers and instructors strengthened. Statistics in terms of promoting learner control show that the subjects had a feeling of control in using Sina micro-blog and in respect of stimulating motivation, over 80% students hold the favorable view. Thus the tentative experiment result seems relatively optimistic and promising.

Table 3 Effect of the interactive activities

Satisfying students' interest in learning	SA/%	A/%	N/%	D/%	SD/%
I prefer to learning English in such a flexible way	5/11.1	11/24.4	17/37.8	7/15.6	5/11.1
I like the online learning resources	6/13.3	7/15.6	15/33.3	9/20	8/17.8
The online learning materials are interest	11/24.4	9/20	15/33.3	6/13.3	4/8.9
I can get timely and useful feedback from peers.	3/6.7	7/15.6	18/40	12/26.7	5/11
I can get timely and useful feedback from teacher.	15/33.3	18/40	8/17.8	3/6.7	1/2.2
I can post my views and get timely feedback.	5/11	9/20	17/37.8	10/22.2	4/8.9
Meeting students' needs for communication	SA/%	A/%	N/%	D/%	SD/%
Microblog helps interact with learning resources.	9/20	11/24.4	14/31.1	8/17.8	3/6.7
Microblog enlarges communication with peers.	7/15.6	10/22.2	9/20	11/24.4	8/17.8
Microblog enlarges communication with instructors.	11/24.4	13/28.9	9/20	7/15.6	5/11.1
Promoting learner control	SA/%	A/%	N/%	D/%	SD/%
I can decide my login time and way.	8/17.8	11/24.4	20/44.4	4/8.9	2/4.5
I can check, collect, comment what I am interested in.	10/22.2	9/20	11/24.4	8/17.8	7/15.6
I can upload favorite videos, audios, texts or photos.	8/17.8	17/37.8	12/26.7	5/11	3/6.7
I feel free to share learning experience with others.	11/24.5	13/28.9	6/13.3	9/20	6/13.3
Stimulating motivation	SA/%	A/%	N/%	D/%	SD/%
Promoting positive attitude to English learning	9/20	8/17.8	15/33.3	8/17.8	5/11.1
Inspiring interest in English culture	6/13.3	10/22.2	17/37.8	8/17.8	4/8.9
Enhancing interest in English learning	8/17.8	11/24.4	14/31.1	7/15.6	5/11.1
Cultivating autonomous and cooperative ability.	5/11.1	7/15.6	18/40	9/20	6/13.3
Making up for the lack of classroom teaching	7/15.6	8/17.8	16/35.5	8/17.8	6/13.3
Encouraging participation in extracurricular learning	5/11.1	10/22.2	16/35.5	7/15.6	7/15.6

Notes: SA =strongly agree, A =agree, N=neither, D=disagree, SD=strongly disagree

Through structural and content analysis on Sina micro-blogging, the authors endeavor to ascertain whether interactive learning based on Twitter-like platform increases students' involvement in extracurricular learning. As posts sent by the participants to the micro-blogging platform are essential parts, the analysis centers on the number of them during the 4-week period. Totally, 752 posts were published from the students, ranging from original texts to video, audio and photo uploads, and 165 posts of online guidance and instruction from the instructor.

Apart from the structural analysis focusing on the number of posts on the micro-blogging, insight into the post content is another indispensable part of data analysis. According to Liebenau and Backhouse (1990), in the rule-based system, three sets of rules are applied: substantive, message, and control. The substantive rule refers to the message focusing on discussing the subject itself. The message rule represents the message providing information to advance discussion. The control rule identifies the message centering on norm or organization of discussion. In the period of the four weeks, it was found that control rule mainly exists at the early stage of the activity. Particularly, many a message is posted by the instructor with guidance and encouragement for the students. The substantive rule contributes to the posts most throughout the whole process as a result of discussion on the given tasks initiated by the instructor and some students. The message rule takes the second place of the three rules, owing to the uploader providing a certain portion of information by URLs to share with others.

Three open-ended questions about students' interactive activities based on the micro-blogging platform are put forward at the end of the post-questionnaire in order to provide some constructive suggestions for future study: 1) What do you like and dislike about the extracurricular English interactive activities on the micro-blogging? 2) What activities stimulated your participation in discussion on the micro-blogging? 3) What role do you think the instructor should play?

It is concluded from students' answers that most of them like the freedom without time and space limit, the friend atmosphere, the easy communicative style and the interesting topics, and the main negative comments are no oral English communication and taking too much time to write the posts. The discussing topics, the learning experience exchanges and the resource uploads spur their interest in interaction, and the comments, the video, the instant and timely reply from the instructor and other guys stimulate them to post. However, they wish the teacher would be their guider directing the activities and encouraging the less outspoken students, and play the roles of organizer, guider, supervisor and friend.

4   Findings and Conclusion

The pre-questionnaire findings show that students' major or gender accounts little for their willingness to take part in the extracurricular English interactive activities on the micro-blog. Although most of the students never experienced the micro-blogging platform before, they showed curious and active attitudes toward it and were ready to join in. So, it is feasible for college students to carry out extracurricular English interactive activities based on Twitter-like platform. The post-questionnaire findings suggest the interactive activities on the Twitter-like platform met up with students' interest in learning and needs for communication. Besides, the students demonstrated higher learner control over and motivation for the login way, the learning materials and the interactive activities. The structural analysis results indicate interactive activities on Twitter-like platform increased students' involvement in extracurricular learning. Students' posts made up the major proportion and most posts are substantive, which mean the students had a discussion around the topics or content indeed. The qualitative analysis findings reveal students' positive attitudes toward the activities on the platform.

Posted discussions and information exchange make Twitter-like platform possible for students to communicate with their instructors and other students, and everyone on the platform can make their individual contributions to his or her group. However, some limitations do exist in this research. A larger sample and longer period may offer more insights into the interactive learning style.

This research is merely designed to investigate the interaction and interactive effects between student and content, student and instructor, student and student. In future study, other aspects can also be taken into consideration, such as the interaction between platform and student, platform and instructor, instructor and content, instructor and instructor. Besides, other factors involved in implementing the extracurricular English interactive activities can also be considered, such as the instructor's perspective, to offer an integrate view of the interaction on the platform. Moreover, a comparison among web interface, mobile phone, as well as the IM tool may also be conducted in future study. Twitter-like platform applied in education is just in its infancy, whose use is interfered with other individual purposes. Therefore, it is necessary to establish a specialized Twitter-like platform on campus. We do believe Twitter-like platform will play its powerful role in the field of education.

References

Ash, K.: Educators Test the Limits of Twitter Microblogging Tool (2008), http://www.edweek.org/dd/articles/2008/06/24/01twitter_web.h02.html

Charlene, K.: Twitter for beginners (2009), http://www.CrowInfoDesign.com/downloads/twitter.pdf

Dai, X.Q., Yao, L.J.: The research on microblogging application at the web-based education. In: International Conference Networking and Digital Society, ICNDS 2010. IEEE (2010)

Ebner, M., Schiefner, M.: Microblogging — more than fun? In: Proceedings of IADIS Mobile Learning Conference, Portugal (2008)

Goldstein, A.M.: Blogging evolution. Evo. Edu. Outreach 2, 548–559 (2009)

Grosseck, G., Holotescu, C.: Can we use Twitter for educational activities? (2008), http://adlunap.ro/else/papers/015.-697.1.Grosseck%20Gabriela-Can%20we%20use.pdf

Hao, Z.J., Sun, Z.N.: Wear a "scarf" for College English teaching. Modern Educational Technology 6, 63–65 (2010)

Junco, R., Heiberger, G., Loken, E.: The effect of Twitter on college student engagement and grades. Journal of Computer Assisted Learning 2, 119–132 (2010)

Kaplan, A.M., Haenlein, M.: The early bird catches the news: Nine things you should know about micro-blogging. Business Horizons 2, 105–113 (2011)

Liebenau, J., Backhouse, J.: Understanding information: An introduction. Macmillan, London (1990)

Liu, Q.M., Meng, Q.: Twitter, what do you bring to education? Modern Educational Technology 10, 107–110 (2009)

Wang, X.W.: Interaction in a web-based English course. Shanghai Jiaotong University Press, Shanghai (2008)

A New Type of Voltage Transducer (EFVVT)

Qi Zengying and Xu Qifeng

Abstract. A new type of voltage transducer- Electrostatic Film Vibration Voltage Transducer (EFVVT) is presented in this paper in which three modes include single-end polarization one, dual-end polarization one and no polarization one are discussed. The fundamental principle is that an electric field force generated by a high voltage to be measured drives an electrostatic film to vibrate and the vibration is sensed by a reflecting fiber-optical displacement sensor. It can be used for both AC and DC voltage measurement including its true harmonics and transient non-periodic components. The mathematics models and the prototypes are discussed as well in the paper.

1   Introduction

There are three main type of voltage transformers include potential transformer (PT), capacitive voltage transformer (CVT) and optical voltage transformer (OVT). PT and CVT are the main tools for voltage measurement in current electric power systems. Along with much more electricity demanded and voltage level increased, PT and CVT are exposing some inherent disadvantages, such as bulky, very high insulation cost, low reliability and poor performance etc., PT and CVT are not qualified for modern smart grid[3]. Compared with PT and CVT, OVT has many significant advantages such as simple insulation, compact, light weight etc[7,8]. But only few OVT go to a trail run and no more applications so far caused by some big problems being very difficult to overcome[1,2]: (1) poor reliability and stability caused by a linear birefringence of optical crystal, (2) because of no way to sense the phase difference of polarized light directly it goes to measure the light intensity instead and results in nonlinear sensing and its measurement range should be less than 5% of the half wave voltage of crystal, usually the OVT based on current methodology is capable of sensing less than 3000 volts only, far below 110kV, 220kV or higher voltage level, (3) it is also effected by

Qi Zengying
Department of Power System and Automation, Fuzhou University, Fuzhou, China
e-mail: qizengying1125@163.com

Xu Qifeng
Department of Power System and Automation, Fuzhou University, Fuzhou, China
e-mail: fx9687@126.com

F.L. Gaol et al. (Eds.): Proc. of the 2011 2nd International Congress on CACS, AISC 144, pp. 291–296.
springerlink.com   © Springer-Verlag Berlin Heidelberg 2012

fluctuation of light source, power loss in light transmission and in photoelectric conversion, etc.

This paper puts forward an electrostatic film vibration voltage transducer (EFVVT). It is based on an electric field force generated by a high voltage to be measured to drive an electrostatic film to vibrate and then to sense the voltage with film vibration[5].

2   Physical and Mathematical Models

2.1   Single-Ended Polarization Electrostatic Film Vibration Voltage Transducer

Fig.1 shows the physical model of a single-end polarization EFVVT in which a fixed metal plate A is connected to an unknown voltage Vsig to be measured, the film is polarized by a DC power supply Vpol with an electrostatic charges. To define d as the distance from the plate to the film, x as the amplitude of film vibration, S as the effective area of the film, then the electrostatic force F between the plate and film is [4]

$$F = \frac{\varepsilon_0 S(V_{sig} - V_{pol})^2}{2d^2} \tag{1}$$

in which ε_0 is the vacuum permittivity ($8.845 * 10^{-12}$ F/m).

Except for the electrostatic force, the film also suffers the mechanical resistance and the elastic recovery force. Due to $x << d$, the film performs a full membrane vibration, so the mathematical model can be expressed as

$$m\frac{d^2x}{dt^2} + r\frac{dx}{dt} + kx = \frac{\varepsilon_0 S(V_{sig} - V_{pol})^2}{2d^2} \tag{2}$$

where m is the film mass, r is a mechanical resistance coefficient, k is an elastic recovery coefficient.

Fig. 1 The physical model of single-end system

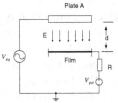

A stable vibration system should satisfy m<<k and r<<k, so the influence of m, r are negligible so (2) can be simplified as

$$kx = \frac{\varepsilon_0 S(V_{sig} - V_{pol})^2}{2d^2} \tag{3}$$

The film vibration amplitude x can be measured by a reflective optical fiber micro- displacement sensor. Fig.2(a) shows $Vsig$ to be measured. Fig.2(b) shows a signal is measured by the sensor and it is composed of two half sine wave and their difference between the half wave in red color and the half wave in green color is $2*Vpol$. Fig.2(c) shows $Vsig$ is recovered from Fig.2(b).

Fig. 2 The single mode sensing result

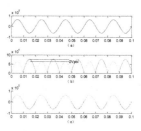

2.2   Dual-End Polarization Mode

In order to make a linear measurement with EFVVT a dual-end polarization mode based on the single-end one is presented in this paper. Fig.3 shows its physical model in which the plate A is polarized by a positive DC voltage and the plate B by a negative one. This means that phase difference between two plates is 180°.When the film vibration x and the distance d between the plates meet: $d>>x$ (Usually, x is one millionth of d) a sensing result is a linear function of x. The electrostatic force F applied to the film can be expressed as

$$F = \frac{2\varepsilon_0 S V_{pol} V_{sig}}{d^2} \tag{4}$$

Fig. 3 The physical model of single-end system

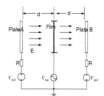

The mathematical model is

$$kx = \frac{2\varepsilon_0 S V_{pol} V_{sig}}{d^2} \tag{5}$$

The film vibration x is sensed by a reflective optical fiber micro-displacement transducer and the $Vsig$ can be recovered easily for it is linear with x, as equal (5) shows.

2.3   No Polarization Mode

This mode as shown in Fig.4 has a very simple construction due to the film being grounded directly. Its mathematical model is

Fig. 4 The physical model of no-polarization mode

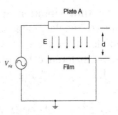

$$kx = \frac{\varepsilon_0 S (Vsig)^2}{2d^2} \tag{6}$$

It is known that the film vibration x can be measured by reflective optical fiber micro-displacement sensor. By (6) we can get the signal which contains the measured voltage shown as Fig.5 (b). This signal is composed of the same amplitude sine half waves and the *Vsig* can't be recovered directly. It should make a MCU to send out a reference pulse when line current is at zero point from positive half wave to negative half wave. In inductive grid the current is always lags behind the voltage with a displacement angle φ, as the voltage wave is always in the negative half wave at this moment so this pulse is used to trigger the DSP to inverse the sine half wave shown as Fig.5 (b). Then *Vsig* can be recovered as shown in Fig.5 (c).

Fig. 5 Voltage to sense and to be recovered

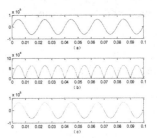

3   Prototype of EFVVT

Since the constructions of dual-end EFVVT is much more complicated then the other twos, the prototype of dual-end one is presented here. A dual-end EFVVT is shown in Fig.6 in which the film is connected to *Vsig*, the plate A to positive DC voltage and B to negative DC voltage. This prototype uses a reflective optical fiber micro-displacement transducer made by MTI Company. As shown in Fig.7 the probe tip is located in the Range 2 standoff point consistent with the direction of film vibration. This probe is immune to electromagnetic interference and it contains a set of light transmitting and receiving fibers which are arranged in a random configuration. Light signal is transmitted from a low potential to a high potential by transmitting optical fiber, then hits the film and is reflected from the film, captured by receiving fibers and transmitted to the sensor in a low potential.

The operating principle of the probe is shown in Fig.7. At contact, no light is exiting or received by the fibers, giving an output signal of zero. As the probe-to-film distance increases, increasing amounts of light are proportionally captured by the receive fibers. The result is a very sensitive, linear output response (Range 1 or 2). As the distance is further increased, the amount of light received approaches the maximum or "optical peak". After the optical peak is reached, a continued increase in probe gap will proportionally reduce the amount of light received. This results in a sensitive, linear output response (Range 1 or 2) with a large measurement range and standoff distance [6]. For a wider range of high-voltage measurement the probe tip is set in the Range 2 standoff point.

Fig. 6 The prototypes of dual-ended polarization mode

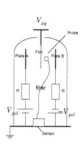

Fig. 7 The operating principle of probe

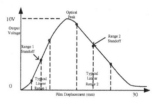

4   Conclusions

The EFVVT (Electrostatic Film Vibration Voltage Transducer) is presented, and its physical and mathematical models and three prototypes are discussed in this paper. The dual-end EFVVT has advantages of linear and accurate measurement and it is suitable for distribution networks where harmonics is required to sense. Single-end one has a simple structure and can be employed for a power system in which harmonics can be ignored. The no-polarization one has a very simple structure with the film is grounded directly and can be used in extra high voltage power systems.

References

1. Li, W.-K., Zheng, S.-X.: Study and Design of High Voltage Potential Transformer Using Precise Capacitive Voltage Divider. Chinese Journal of Sensors and Actuators 18, 634–637 (2005)

2. Li, W.-K., Zheng, S.-X.: Study on a New Type Optic-electric Transformer. Journal of Electronic Measurement 20, 28–32 (2003)
3. Shi, D.-G., Liu, Y., Zhang, L.-P.: Overview of High Voltage Potential Transformer. Transformer 40, 11–14 (2003)
4. Wang, Y.-Z.: Theory and Practice of Electrostatic Loudspeaker. Audio Engineering 2, 47–51 (2002)
5. Xu, Q.-F.: A Film Vibration Current Transducer. Automation of Electric Power System 33, 81–83 (2009)
6. Xu, Q.-F.: A Voltage Transducer Based on Electrostatic Film Vibration. In: International Symposium on Test Automation and Instrumentation, pp. 1–4 (2010)
7. Zou, J.-L., Liu, Y., Wang, C.-T.: Development of study of Optical Voltage Transducer Used in Power System. Power System Automation 10, 64–67 (2001)
8. Zhu, Y., Ye, M.-Y., Liu, J.: Design of 220kV Combined Optical Voltage and Current Transformer. High Voltage Engineer 6, 34–36 (2000)

Modelisation of Atmospheric Pollutant Emissions over the French Northern Region Using Database Management System

P. Lebègue, V. Fèvre Nollet, M. Mendez, and L. Declerck

Abstract. The development of atmospheric pollutants emission inventory is a critical step with regard to management of air quality at the regional scale. It involves the use of a Geographic Information System (GIS) software with the development of specific methodology and customized tools. The quality of input databases and the time and space scales specify the inventory effectiveness. But quality is often related to a large number of data that classical methodologies can hardly resolve. A specific methodology with integration step of different kinds of emission sources has been developed. The tools used are PostgreSQL[1] as DataBase Management System (DBMS) whitch is the kernel of the software, the Geographic Information System used is GRASS[2], finally the Graphical User Interface (GUI) is a web site using PHP within the PRADO[3] framework. The inventory concerns a large range of pollutants and Total Suspended Particles.

1 Background and Objectives

Since the eighties, the spatialized emission inventories of atmospheric pollutants have appeared. First, it was used for large spatial scale (global to continental), e.g., EMEP inventory for Europe [1]. Later, the research on this subject was focused on regional scale to provide for air quality modeling on big cities, e.g., the study in the Maurienne valley [2], REKLIP program on the East of France [3] and ESCOMPTE in the South [4].

    The present work took place in the framework of the CERPA[4] program and concerns the Nord-Pas-de-Calais region in the North of France.

P. Lebègue · V. Fèvre Nollet · M. Mendez · L. Declerck
Physico-Chimie des Processus de Combustion et de l'Atmosphère - PC2A, Université Lille1, UMR CNRS 8522, Bat C11, 59655 Villeneuve d'Ascq Cedex, France
e-mail: {patrick.lebegue,valerie.nollet}@univ-lille1.fr, {maxence.mendez,louise.declerck}@ed.univ-lille1.fr

[1] PostgreSQL : http://www.postgresql.org/

[2] Grass : http://grass.osgeo.org/

[3] Prado : http://www.pradosoft.com/

[4] CERPA : http://www.hygeos.com/en/exper-pf_cerpa.php

F.L. Gaol et al. (Eds.): Proc. of the 2011 2nd International Congress on CACS, AISC 144, pp. 297–302.
springerlink.com   © Springer-Verlag Berlin Heidelberg 2012

Fig. 1 Diagram of the types of sources: punctual, linear and surface-area sources

Regional air quality models need high resolution in time and space emission inventory. Our original inventory was designed by the PC2A and the Ecole des Mines de Douai in 2004 [6, 7, 5] and updated by the regional air quality network ATMO[5] of the Nord-Pas-de-Calais. In this inventory, the emission database has been constructed with spreadsheets and low level database (MS-ACCESS) and spatialized with the G.I.S. software ESRI ArcView8[6]. In order to refine time and space resolution, it was necessary to use new software with customized tools.

2 Approach and Methods

The general methodology used for the Nord-Pas-de-Calais spatialized emission inventory is divided into two steps. The first step is emission data production, It consists in specification, collection and organisation of the data in a DataBase Management system (DBM). The second step deals with spatialization with G.I.S. software and matrixes production.

In order to integrate pollutant emissions data in G.I.S., three types of sources are considered (Figure 1). Indeed, even if each source has its own emission point, it seems impossible to process all those data individually due to the lack of data.

Punctual sources are used for big industries or power plants due to their high emissions and we can add a third dimension in order to describe the emission height.

Linear sources are mostly applied to traffic. Instead of admitting a mobile point like a car as a pollutant emitter, it is easier to consider the traffic line as a linear source.

Surface sources are applied to almost all human activities. The most precise information on the localization of the pollution source is the administrative district area in which the source is situated.

The spatialized emission inventory is built from four main databases. These databases give information on population size, industrial, agricultural and transportation fields.

---

[5] ATMO : http://www.atmo-npdc.fr/home.htm

[6] Arcview : http://www.esri.com/software/arcview/

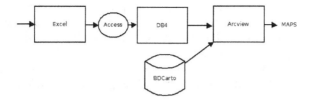

Fig. 2 First implementation of Nord-Pas de Calais spatialized emission inventory.

Some INSEE[7] databases gather the population, housing and heating characteristics (age and fuel type) for each administrative district.

The INSEE-SIRENE database contains economic information on companies and industries, especially the number of employees, the activities and the NAF (French Activity Nomenclatures) code of each company.

The AGRESTE[8] database reports surface-areas, type of agricultural areas and the type and number of livestock.

The transport database informs about the localization of the roads and the traffic of each type of vehicle. Moreover, information about rail, sluice and air traffic is available. This dataset is a vectorial information.

The databases give us information about all the human activities. Each activity is referenced to one or more SNAP (Selected Nomenclature for Atmospheric Pollution) code. The SNAP code is a 3-level code that permits to have accurate classification of each source of pollutants.

The SIRENE database has a specific nomenclature, each French company has a NAF code which is characteristic of its economic activities. Then, each NAF code is associated with one or more SNAP codes. For example, drug manufacturers are registered with the 244C NAF code. This NAF code is associated to the 03.01.00 (fuel), 04.05.22 (main activity) and 06.03.06 (solvent) SNAP codes. Each SNAP code is related to a specific emission factor for all chemical species considered : SO_2, NO_x, total non methane volatile organic compounds (**NMVOC**), CO_2, **CO**, CH_4, NH_3, N_2O, **HCl**, **HF**, **Pb**, **Zn**, **Cd**, **Hg**, dioxins (**PCDDs**) and Total Suspended Particles (**TSP**). Knowing the emission factor and the annual production of human activities, we can calculate the amount of annual emission by species and SNAP. Finally, all the calculated data can be integrated in a G.I.S. Software thanks to the BDCarto and GEORoute[9] databases that give us information on localization of traffic lines and the administrative district in the Northern French Area.

Further details about our procedure can be found in the Martinet et al. publications [7, 5, 6].

[7] INSEE : http://www.insee.fr

[8] AGRESTE : http://agreste.agriculture.gouv.fr

[9] BDCarto-GEORoute : http://www.ign.fr

The first approach used to construct the our spatialized emission inventory is illustrated figure 2. The most important problem is the great deal of data, spreadsheets are a very bad solution to manage such quantity of information. A lot of errors occur due to the copy/paste function and the multiple worksheets because of the limitations in size of each sheet (256 columns, 65536 rows). Each calculation is made by Excel, Access is only used to convert the results into DB4 format in order to use ArcView.

The use of a professional database management system is compulsory. We chose PostgreSQL because it is an open-source, free and very strong and efficient system. For the calculation we have tested three technologies: Python, Java and SQL. Python and Java have similar efficiency. But the shortest calculation time is obtained using pure SQL (Table 1).

Table 1 Computation time for SNAP 02.01 emissions in Nord-Pas de Calais region for different programming techniques (Pentium IV 1.3ghz machine).

Python with no optimization	2800 s
Python / Java with access optimization	50 s
Pure SQL (PostgreSQL views)	5 s

Figure 3 describes the actual architecture of our software. The POQAIR database contains all the inventory data, but also all the procedures which calculate the pollutant emissions. These emissions data can be exported directly in CSV format files. BDCARTO, GEORoute and the GRASS G.I.S permit us to transform administrative data into geographical data in order to export maps on one side and to export models formatted files on the other side.The administrative district code referenced

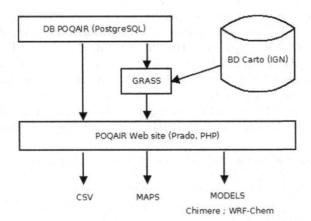

Fig. 3 Schematic diagram of the method used in the Nord-Pas de Calais spatialized emission inventory.

emissions are transformed into meshed data by a C++ program. This program uses the GRASS application to spatialize the data. The C++ program allows the choice of the spatialized domain and the mesh size depending on the needs of the air quality models.

The POQAIR web site were developed using a simple and very efficient RAD PHP-framework (Rapid Application Development) named PRADO.

3 Results

Air quality models are high-performance tools to predict the aging of gases and aerosols upon their release in the atmosphere. They need input to describe all air pollution sources in the study area. Air emission inventories provide this type of information. The spatial and temporal resolution of the inventories frequently depends on the spatial and temporal resolution of the models. For example, for spatial resolution a $100 \times 100\ km^2$ scale is used for cross-border air matter and a $5 \times 5\ km^2$ scale for regional air pollution study. For time scale, one hour temporal resolution is usually used.

Within this work, a code was developed to provide annual spatialized emission database with the appropriate input format for air quality models. Air quality models used in the laboratory are Chimere[10] developed by the IPSL and WRF Chem[11] by UCAR. It allows the user to specify his modelling domain, by giving the origin (X_0, Y_0), the maxima (X_{max}, Y_{max}) and the spatial resolution of the domain. So-defined, the domain-grid is framed on the GIS database and the code calculates for each mesh the surface percentages of towns or villages which are attributed to the mesh. The emissions are attributed according to these percentages and normalized from the GIS-base to a grid-base system (Figure 4). This process is applied to regional emission inventory. When the selected domain is wider than our regional inventory, the grid is completed by European emission inventories data EMEP.

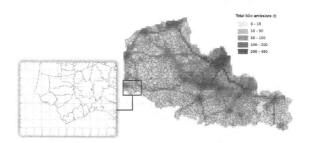

Fig. 4 Total annual emission of SO_X over the North French Region processed with $2 \times 2\ km^2$ griding

---

[10] Chimere : http://www.lmd.polytechnique.fr/chimere/

[11] WRF : http://www.acd.ucar.edu/wrf-chem/

4 Conclusion and Future Plans

Coupled with air quality modelling tools, emission inventory allows spatializing atmospheric pollution sources and answers to air quality settlement. It can be used for air quality forecasts, to contribute to develop air waste management systems and also for sanitary impact evaluation.

Its production involves association of various thematic and spatial database and geospatial tools. For regional to local studies, databases are much elaborated and heavy (in term of computer space) so its necessary to develop progressive software with a satisfactory response time for user.

In the framework of the CERPA project, we developed a methodology using PostgreSQL, G.I.S. Grass and C++ code to produce spatialized emission inventories. A more accurate study has been done on the emission data production with various tools (Excel, Access, PostgreSQL). Compared to Excel-Access solution, PostgreSQL efficiency permits a dynamic computational system. This method makes the data available through a website, POQAIR, which provides GIS emission maps and calculates matrixes with requested spatial and temporal scales.

In the framework of the CERPA project, we focused on PM (Particulate Matter) but the POQAIR database concerns a large range of pollutants. These pollutants are very toxic for human health. So, this database will be an important tool for the evaluation of risk impact and epidemiological studies. The computer methodology can be applied to other French or foreign areas provided that the emission database is available.

References

1. GENEMIS, Emission inventory database (EMEP-CORINAIR) for Europe for the year 1999, EUROTRAC-2 (1999), http://genemis.ier.uni-stuttgart.de/GENEMIS1999.pdf
2. Brulfert, G., Chollet, J.P., Jouve, B., Villard, H.: Atmospheric emission inventory of the Maurienne valley for an atmospheric numerical model. Science of the Total Environment 349, 232–248 (2005)
3. Ponche, J.L., Schneider, C., Mirabel, P.: Methodology and results of the REKLIP atmospheric emission inventory of the upper Rhine valley transborder region. Water, Soil and Air Pollution 124, 61–93 (2000)
4. François, S., Grondin, E., Fayet, S., Ponche, J.L.: The Establishment of the atmospheric emission inventories of the ESCOMPTE program. Atmospheric Research 74, 5–35 (2005)
5. Martinet, Y.: Conception, validation et exploitation d'un cadastre d'émission des polluants atmosphériques sur la région Nord Pas de Calais. PhD thesis Lille1 University, France (2004)
6. Martinet, Y., Wroblewski, A., Kergomard, C., Ponche, J.L., Nollet, V.: Inventaires spatialisés de polluants atmosphériques: rôle et utilisation des systèmes d'information géographique. Revue Internationale de Géomatique 15(1), 63–80 (2005)
7. Martinet, Y., Wroblewski, A., Nollet, V., Caplain, I., Kergomard, C., Déchaux, J.C.: Sensitivity of a spatialized emission inventory of atmospheric pollutants. In: Reconciling Environmental Concerns with Economic Aspects Air pollution. Witpress (May 2003)

Preliminary Study on Reliability Analysis of Safety I&C System in NPP

Chao Guo, Duo Li, and Huasheng Xiong

Abstract. Digital instrumentation and control (I&C) systems, such as digital Reactor Protection System (RPS), are being employed gradually at newly-built and upgraded Nuclear Power Plants (NPPs). The reliability analysis of digital I&C system turns to be a research focus. This paper proposes a preliminary study on reliability analysis for the digital RPS of High Temperature Gas-Cooled Reactor-Pebble bed Module (HTR-PM). We first introduce the function and structure of RPS in HTR-PM; then the processes of failure mode and effect analysis (FMEA) at system level are given; finally we present the FMEA of a computer device as an example and introduce the future work. This study analyses all possible safety-relevant implications arising from failures in RPS and promotes the reliability analysis of the digital RPS in HTR-PM.

Keywords: nuclear power plant, digital reactor protection system, reliability, failure mode and effect analysis.

1 Introduction

Digital instrumentation and control (I&C) systems are gradually being used in newly-built and upgraded nuclear power plants (NPPs). Different from conventional analog systems, digital systems are based on computers and are compounds of software and hardware. The Reactor Protection System (RPS) is one of the most important parts of I&C systems. The RPS monitors the condition of the plant during normal operation, and generates an emergency trip signal or an engineered-safety-function-actuation signal if a need arises. The hardware of digital RPS bases on discrete digital logic, while the safety functions are mainly realized by software coding.

Comparing with analog RPS, the advantages of a digital RPS are overwhelming, such as smaller size, decrease in components, stability and high precision for the setpoints, high tolerance for disturbances, low distortion in the process of information transferring and treatment, flexibility and ability to implement complicated functions, and so on [1].

Chao Guo · Duo Li · Huasheng Xiong
Institute of Nuclear and New Energy Technology
Tsinghua University, Beijing, China
e-mail: {guochao,liduo}@tsinghua.edu.cn

F.L. Gaol et al. (Eds.): Proc. of the 2011 2nd International Congress on CACS, AISC 144, pp. 303–310.
springerlink.com © Springer-Verlag Berlin Heidelberg 2012

The RPS is a safety-related system, which means its reliability must be investigated carefully. Digital RPS is very different with analog systems in application technique, developing tool, and failure modes, so the reliability analysis of digital RPS is of great difficulty. It has been a research focus that how to assess the reliability and failure modes of digital RPS [2, 3, 4].

Failure mode and effect analysis (FMEA) is a well-known qualitative method of reliability analysis. FMEA of RPS helps checking the designing process and finding the weaknesses of the design. The reliability data also helps to determine the maintenance period. Therefore the system reliability is effectively improved.

In this paper, we take the digital safety RPS of High Temperature Gas-Cooled Reactor-Pebble bed Module (HTR-PM) as the investigation object. After introducing the functions and structure of RPS, we will propose the meanings and processes of FMEA. Preliminary FMEA for RPS of HTR-PM will be presented in the end.

2   Functions and Structure of Safety RPS

Functions of RPS for HTR-PM include safety functions for the reactor trip system (RTS) and for the engineered safety features actuation system (ESFAS). Taking RTS for example, when designed basis accidents happen, this system generates emergency trip signal to guarantee the safety of reactor.

Structure features of RPS for HTR-PM include four-redundant protection-logic channels and two-level 2-out-of-4 voting logic structures. To decrease the risk of common-cause failures, every channel is divided into sub-system x and sub-system y. Both sub-systems have identical structures and components, while their protection variables and protection logics are partially different. Each sub-system outputs emergency trip signals and ESFAS signals independently. So sub-systems x and y act as the diversity of the safety function.

Devices of every channel in RPS include: Sensor/Transmitter, Signal Isolating Device, Signal Processing Device x, Signal Processing Device y, Logic Gating Device x, and Logic Gating Device y. Among them Signal Processing Devices and Logic Gating Devices are computer devices, while others are hardware devices. Structure of RPS is shown in Fig. 1.

Signal Isolating Device divides an input signal into four isolated output signals, and sends them to Signal Processing Devices x and y by hard-wired signals.

Signal Processing Device x serves the purpose of sampling, pre-conditioning, conversion of engineering unit, calculation of protection variable, and comparison with corresponding setpoint. Signal Processing Device y serves similar purpose with Signal Processing Device x, but it does not compare the protection variables with setpoints.

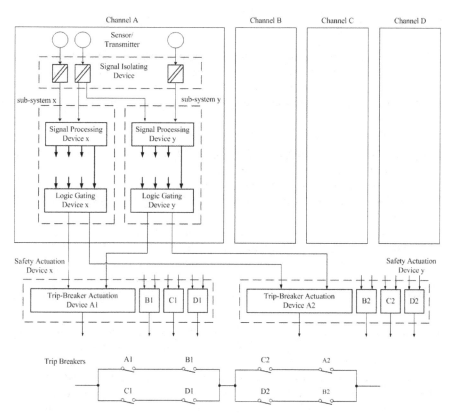

Fig. 1 Structure of Digital RPS.

Logic Gating Device x samples outputs of Signal Processing Devices x in four redundant channels. According to the setpoint-comparison results, 2-out-of-4 voting for each protections variable and 1-out-of-N voting for all protection variables are performed to generate the actuation signal. The output signals include reactor-trip-actuation signal and ESFAS signals.

Logic Gating Device y samples outputs of Signal Processing Devices y in four redundant channels. Different from Logic Gating Device x, 2-out-of-4 voting is realized by second maximum and second minimum comparisons. Then 1-out-of-N voting for all protection variables is performed and the actuation signals are calculated.

Logic Gating Devices x and y of four redundant channels output hard-wired signals simultaneously to Safety-Actuation Devices A and B. Every Safety Actuation Device receives both automatic triggering signals from Logic Gating Devices x and y and manual triggering signals from control console, performs logic "OR" calculation and outputs actuation signals for final actuators.

Each channel controls two trip breakers. All eight trip breakers construct the second-level 2-out-of-4 voting mechanism, which means if the trip breakers controlled by two or more channels break, emergency trip is triggered.

3   FMEA Basis

FMEA is a well-known qualitative reliability-analysis method. The function of an FMEA is to consider the potential failure mode and the corresponding effect on the system. The reliability of system can be improved in this process. For a detailed analysis result, the analysts should develop the widest possible list of potential failure modes.

According to [5], purposes of an FMEA are as follows:

1. To ensure that all conceivable failure modes and their effects on the operational success of the system have been considered;
2. To assist in selecting design alternatives with high reliability and high safety
3. To develop early criteria for test planning and the design of test and checkout systems
4. To provide a basis for quantitative reliability and availability analyzes
5. To provide historical documentation for future references to aid in the analysis of field failures and consideration of design changes
6. To assist in the objective evaluation of design requirements related to redundancy, failure detection systems, fail-safe characteristics, and automatic and manual override

To achieve the purposes above, an FMEA has to answer some basic questions like: a) How can each part conceivably fail? b) What mechanisms might produce these modes of failure? c) What could the effects be if the failures did occur? d) Is the failure in the safe or unsafe direction? e) How is the failure detected? f) What inherent provisions are provided in the design to compensate for the failure? [5]

Preparations for the FMEA include definition of the system mission, identification of the analysis boundaries and analysis depth, and identification of the components and their functions. A template of an FMEA is shown in Table 1. This table is the very heart of the procedure for documenting the FMEA.

Table 1 Template of Failure Modes and Effects Analysis documentation.

No.	Component Identification	Failure Mode	Failure Mechanism	Local Effect	Method of Failure Detection	Provision	Effect on System

"Component Identification" in Table 1, column 2 is the system component or part being analyzed. This could also be the whole system or device appropriate to the purposes of the FMEA. "Failure Mode" shows all the failure modes that the analyst can conceive for the component. All mechanisms of failure that could result in the failure mode are noted in "Failure Mechanism". "Local Effect" means the effect of the failure on the adjacent parts, while "Effect on System" describes the effects of the failure on the overall system functions. "Method of Failure Detection" and "Provision" describes the ways that failures can be detected and overcome.

4   FMEA of RPS for HTR-PM

In this section, we will describe the FMEA integrated with the RPS of HTR-PM. Detailed analysis process includes: identification of analysis scope, definition of failure mode, and failure analysis of each device.

4.1   Analysis Scope

Preliminary FMEA for RPS of HTR-PM is mainly focused on the FMEA analysis of reactor-trip function of RPS. The RPS of HTR-PM is composed of four redundant channels. Without loss of generality, the FMEA scope is limited to the devices of Channel A, together with Trip-Breaker Actuation Devices A1 and A2, and Trip Breakers A1 and A2. Devices of Channel A include: Sensor/Transmitter, Signal Isolating Device, Signal Processing Devices x and y, and Logic Gating Device x and y. Functions of each device are shown in Table 2.

Table 2 Functions of devices to be analyzed.

No.	Device Identi- fication	Type of Output Signal	Function
1	Sensor/ Transmitter	Analog signal	Convert the parameters like nuclear power, tempera- ture, pressure, and flow measured in situ to current signals
2	Signal Isolat- ing Device	Analog signal	Divide an input signal into four isolated output sig- nals
3	Signal Proc- essing Devices x and y	Network communi- cation signal	Sample and process the protection variables, calculate the indirectly measured protection variables. Signal Processing Device x also compares variables with setpoints and generates accident alarm signals.
4	Logic Gating Devices x and y	Switching signal	Perform 2-out-of-4 voting for each protection vari- able and the voting results are then performed 1-out- of-N voting to generate the trip triggering signal.
5	Trip-Breaker Actuation De- vices A1 and A2	Switching signal	Perform logic "OR" calculation for both manual and automatic triggering signals and output emergency- trip-actuation signal for corresponding trip breaker
6	Trip Breakers A1 and A2	Switching signal	Construct the 2-out-of-4 voting mechanism together with other 6 trip breakers. Cut off the power of all control rods and perform emergency trip

4.2   Failure Mode

There is no consensus about the failure-mode-definition method of digital system. An alternative is to define the failure modes of output signal as the failure mode of referred device, as the output signals of each device directly reflect the device's status [4]. In this paper we will define the failure modes of devices in this way.

Types of output signals have been listed in Table 2. Categorized by the signal type, output signals can be divided into three kinds: a) 4~20 mA analog signal; b) switching signal; c) network communication signal. Corresponding failure modes of these types of signals are listed in Table 3.

Table 3 Failure modes of three types of signals.

No.	Signal type	Failure mode
1	Analog signal	1) Lower accuracy
		2) Signal over range
2	Switching signal	1) Frozen signal output
		2) Faulted signal output
3	Network communication signal	1) Communication interruption or timeout
		2) Interfered communication data
		3) Internal error of packet

It should be emphasized that every device in Table 2 may output more than one signal, and each signal may have two or more failure modes. For example, Logic Gating Device x outputs two switching signals and each of them corresponds with two kinds of failure modes: Frozen signal output and Faulted signal output. Therefore the Logic Gating Device x has four kinds of failure modes in total.

4.3 FMEA Process

Analysis processes of FMEA for HTR-PM is as follows:

1. Analyze each device in Table 2 according to the template in Table 1. The FMEA result of each device should be expressed by an independent form.
2. Failure modes of each device are identified in accordance with Part B, Section 4.
3. List all possible failure mechanisms, local effects, and designed provisions.

Limited by the length of this paper, we only provide the FMEA of Logic Gating Device x, as shown in Table 4.

4.4 Significance of FMEA

Through the FMEA shown above, we can systematically list all failure modes of emergency-trip function of the RPS in HTR-PM, and analyze possible failure mechanism and effect, together with whether the failure mode has particular detection method and provision. Therefore, the FMEA results can assess or guide the design of RPS directly and is significant to achieve high-level reliability.

Table 4 FMEA Results of Logic Gating Device x

No.	Output Signal	Failure Mode	Failure Mechanism	Local Effect	Method of Failure Detection	Provision	Effect on System
1	Trip triggering signal	Frozen signal output	Failure of input module / Failure of relay contact / Failure of data-processing software	Sub-system x cannot trigger Trip Breaker A1.	Periodic test / Software check by system surveillance station	Triggering signals of both sub-system x and sub-system y in one channel are sent to Trip-Breaker Actuation Device after "OR" calculation	No effect on safety trip
		Faulted signal output	Failure of input module / Failure of relay contact / Failure of data-processing software	Trip Breaker A1 is incorrectly triggered.	Software check by system surveillance station	2-out-of-4 redundant structure of trip breakers	No effect on safety trip

For example, Logic Gating Device x, as shown in Table 4, has two kinds of failure modes, i.e. Frozen signal output and Faulted signal output. Against the failure mode "Frozen signal output", the input of Trip-Breaker Actuation Device A1 (or A2) is designed to be the "OR" result of the output of Logic Gating Device x and the Logic Gating Device y. In this way, the failure mode "Frozen signal output" of Logic Gating Device x or y can be avoided effectively.

In addition, failure modes obtained by FMEA provide a basis for future quantitative reliability and availability analysis, e.g. Fault Tree Analysis.

5   Future Work

The FMEA introduced in this paper belongs to the system level, at which the analysis objects are devices, and the research focuses on the failure of each device and the corresponding effect on the system. On this basis, we will go deep into the interior of each device and continue the FMEA of a lower level, at which we will analyze the failure mode and effect of each plug-in module, and the analysis might involve the software failure inside the digital processing module. As the complexity and potentiality of software failure modes, FMEA of next step will be difficult and important to the reliability analysis of the RPS in HTR-PM.

6   Conclusions

In this paper, we took the RPS of HTR-PM as the research object, and preliminarily analyzed the failure mode and effect of RPS at system level. The failure modes of a computer device were defined as an example according to the output signals of this device. The results of FMEA help to assess or guide the design of digital RPS directly and are significant to improve the reliability of the system.

Acknowledgments. This work was supported by the National Science and Technology Major Project of China under project ZX06901.

References

1. Li, F., Yang, Z., An, Z., Zhang, L.: The first digital reactor protection system in China. Nuclear Engineering and Design 218, 215–225 (2002)
2. Zhou, H.: Structure and Reliability Research of Digital Reactor Protection System in Nuclear Power Plant. PhD Dissertation, Harbin Engineering University (2007)
3. U.S.NRC, NRC Research Plan for Digital Instrumentation and Control, SECY-01-0155, August 15 (2001)
4. Chu, T.L., Martinez-Guridi, G., Yue, M., Lehner, J., Samanta, P.: Traditional Probabilistic Risk Assessment Methods for Digital Systems (NUREG/CR-6962), U.S.NRC (2008)
5. IEEE Std 352-1987. IEEE Guide for General Principles of Reliability Analysis of Nuclear Power Generating Station Safety Systems (1987)

A Traceability Model Based on Lot and Process State Description

Xuewen Huang, Xueli Ma, and Xiaobing Liu

Abstract. To solve the traceability problem in manufacturing process, especially for complex product of dynamic process route, on the basis of batch and process state description, a traceability model is proposed in this paper after the requirements for traceability are analyzed. In the model, each material can be viewed as a point in a three-dimensional space, and each point in the space can be taken as a trace unit. By registering all relations between input and output material lots, all the necessary information to carry out forward and backward traceability can be obtained. Some coding rules based on the theorem of the set named S(I) are stipulated to make the traceability easy and accurate.

1   Introduction

Over the last few decades, food deficiencies such as BSE have caused alert for the negative impact on consumers and businesses. And also in other industries, we can always find recall announcements to express some batches of product deficiency. Repeatedly occurring food safety and product recall affairs have aroused highly attention about traceability in manufacturing process. So, traceability is a needed strategic service to improve security and control quality or combat fraud [1].

Traceability includes two aspects. One is the ability to follow the path from a particular raw material to all the end products having consumed it by the where-used relations, which is called forward traceability. The other is the ability to follow the path from a particular product to all the raw materials lots consumed for it in manufacturing process by the records of where-from relations between objects, which is called backward traceability. This concept was first recognized by Petroff and Hill(1991) [2]. Forward traceability is the key approach for recall of end product, and backward traceability can be used to identify the origin of deficiency.

In most traceability models of materials in manufacturing process, lot is viewed as a traceability unit to make traceability possible. For example, M.H.Jansen-Vullers has proposed a reference data model, in which both forward and backward traceability can be carried out by the records of relations between lots and the

Xuewen Huang · Xueli Ma · Xiaobing Liu
Dalian University of Technology
e-mail: huangxuewen@tsinghua.org.cn, maxueli1020@163.com,
xbliu@dlut.edu.cn

F.L. Gaol et al. (Eds.): Proc. of the 2011 2nd International Congress on CACS, AISC 144, pp. 311–319.
springerlink.com   © Springer-Verlag Berlin Heidelberg 2012

related information of manufacturing operations [3]. Alessio Bechini introduced a traceability data model to record lot behavior and responsible actors [4]. Fucheng Pan proposed an event-based production process traceability model[5]. Massimo Bertolini adopted failure model and effects criticality analysis methods to analysis the failure, outrage, and equipment accident during the manufacturing process, so traceability depend on those key frames can be implemented [6]. C.Dupuy developed a batch dispersion model to minimize the quantity of recall [7].

Currently available models are always oriented to certain industries, such as the food industries [10-13]. And always traceability of end product is emphasized in the model, whereas the traceability of work-in-process product in the workshop is ignored. For the complicated process and uncertainty of complex product, traceability of end and work-in-process products cannot be satisfied by the current models.

A traceability model based on lot and process state is proposed in this paper. By this model, traceability during manufacturing process of both end and work-in-process product can be implemented, and the traceability information can be acquired by the records of relations between lots and change of process state.

2   Requirements of Traceability

Requirements of traceability are similar in different industries. An effective traceability model must include lot and activity as key entities to be able to trace both lots and activities and allow both forward and backward traceability [8]. So the following information must be achieved by traceability:

- *Activity and the related information:* Such as resources, the properties affected by the activity and the in-out relation brought about by the activity;
- *Change of basic properties of materials:* If some properties of a material are not in accord with its basic properties, the material should be given a new identifier;
- *Change of traceability lot:* A lot is a product unit processed or packaged under the same conditions, or a batch of products that share such characteristics such as type, category, size, package and place of origin [4]. Generally, changes of lot are as follows:

  - Lot division : A lot is split into a number of lots;
  - Lot integration: A number of lots are integrated into a unique lot;
  - Lot movement: Different places may have different lot coding scheme. For example, for the same lot, supplier and manufacturer have different code rules, so the lot identifier should be changed after bought from the supplier;
  - Lot alteration: A new lot is generated if new material is produced;

- *Change of process state:* Process state is used to describe the position a material is in its process route. Once an activity takes place, the process state of the materials involved in the activity will change.

3   Traceability Model Based on Lot and Process State

3.1   Definitions

Definition 1. *Material* involved in this paper refers to all the raw materials, intermediates, parts, subassemblies, which are transformed into an end product. *Material* is defined as *Material=< MaterialID,MaterialName, StaticAttributeCollection >*

MaterialID is the unique identifier of a certain material. *MaterialName* is the name of it. *StaticAttributeCollection* is the collection of static attributes that do not have any changes in the whole production process, such as the place of origin, date of production etc.

Definition 2. *ProcessRouting* is the process route of a material within the producing enterprise, which is comprised of a list of activities. Each activity can be expressed as: *Activity =< ActivityID , ActivityName , ActivityType,ResourceCollection>*

ActivityID is the unique identifier of an activity; *ActivityName* is the name of it; *ActivityType* means the type of the activity. *ResourceCollection* is the collection of the resources involved in the activity, such as human resources and equipments.

Then *ProcessRouting* of a material can be defined as follows:

$$ProcessRouting =< MaterialID , List < Activity >>$$

List < Activity > is a list of activities that compose the *ProcessRouting* . The sequence of activities is the same as the sequence in the list.

For some product, dynamic process design is always employed, so the actual process route is always inconsistent with the route pre-defined, which will make the traceability in manufacturing process more difficult [9].

3.2   Feature Modeling for Traceability

Definition 3. The attributes that change with the happen of activities and have effects on traceability objectives are called *TraceAttribute*. It can be expressed as *TraceAttribute =< AttributeID, AttributeName, AttributeType,DesignValue,UpperDeviation, LowerDeviation,Description >*

AttributeID is the unique identifier of a *TraceAttribute* . *AttributeName* is its name. *AttributeType* is the type of the attribute. If *AttributeType* is numeric, *DesignValue* is the designed value of the attribute, and *UpperDeviation* and *LowerDeviation* are the maximum upper and lower deviation of the *DesignValue* respectively. If *AttributeType* is string, the attribute can be described by *Description* .

Definition 4. *MaterialTraceFeature* is the collection of *TraceAttrbute* of a material. It can be expressed as

MaterialTraceFeature =< MaterialID,TraceAttributeCollection >

Definition 5. Changes of some *TraceAttribute* of a material are caused by some activity. The relation between activity and the collection of *TraceAttribute* affected by it is named *ActivityTraceFeature* .

ActivityTraceFeature =< MaterialID, ActivityID, TraceAttrbuteSubColection>

TraceaAttributeSubCollection is the collection of *TraceAttrbute* of *MaterialID* that are affected by *ActivityID* .

The relations of the definitions can be expressed by Fig. 1.

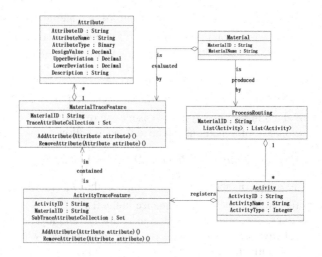

Fig. 1 Model of Material trace feature

MaterialTraceFeature is a sub collection of *TraceAttribute* that is relevant to the traceability objective. Traceability of a material is performed along its actual process route forward or backward. The activities within the route lead to some changes of *TraceAttribute* within the collection *MaterialTraceFeature*, and then the relation between activity and sub collection of *MaterialTraceFeature* it affected is recorded by *ActivityTraceFeature* . By the collection of *ActivityTraceFeature* of all the activities involved in the actual process route, all the factors that have effect on the traceability objective can be obtained.

3.3   Key Techniques for Material Traceability

3.3.1   Three-Dimensional Space

The analysis of traceability requirements shows that there are three types of changes of a material in the manufacturing process:

- Change of basic attributes
- Change of lot
- Change of process state

The three types of changes can be identified by *MaterialID* , *LotID* and *ProcessState* respectively. One activity may bring about one or more types of changes. Each state of a material during the manufacturing process can be viewed as a point in the three-dimensional space, as Fig. 2 illustrated. Activity leads to the change of location in the space. Then, material traceability can be converted to location traceability. For example, the movement in Fig. 2 from $P(M,L,P)$ to $P(M',L',P')$ is caused by an activity. Forward traceability is in the direction from P to P', and backward traceability is reversed.

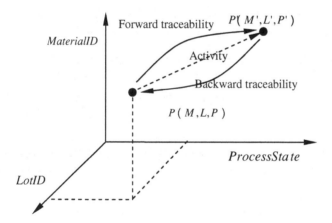

Fig. 2 Definition of three-dimentional space

There are some encoding rules for the three dimensions:

MaterialID **coding rule**: *MaterialID* should adhere to the principles of uniqueness, generality, scalability and efficiency, so it should be simple and clear to be quickly processed by computer.

LotID **coding rule**: Each *LotID* is divided into two parts. The first part is called the basic lot code, which doesn't have any changes. Our rules are mainly for the second part, which always changes in the manufacturing process.

On the basic idea of $S(I)$, any natural number β can be uniquely expressed as $\beta = (x_0 \, x_1 \, ... x_n)(2^0 \, 2^1 \, .. 2^n)^T$ [7], $x_1 = 0 \, \text{or} \, 1$. $(x_0 \, x_1 \, ... x_n)$ is called the coefficient string of natural number β. The *LotID* coding rules are as follow:

- Lot division rule: Suppose a lot XXXX should be divided into n lots, $n = 1, 2...$, the *LotID* after divided is expressed as $XXXX + '-' + 2^{n-1}$.

- Lot integration rule: The period of the code numbers before the last '-' of *LotID* is called main code, and the period of the code numbers after the last '-' of *LotID* is called auxiliary code. If the main codes of the lots to be integrated are not identical, a new *LotID* will be created for the new lot. Else, the lot after integrated can be expressed as main code +'-'+the coefficient string of the sum of all the natural numbers of the auxiliary codes that to be integrated. Suppose the *LotID* of the lots to be integrated are $XXXX - 001$ and $XXXX - 011$ respectively, the natural numbers of the auxiliary codes are 4 and 6 respectively. The new *LotID* should be XXXX–0101.

- Lot movement and alteration rule: New *LotID* should be created.

ProcessState **coding rule**: Process state can be described by process state vector M_s and process sate value V_s. M_s can be expressed as $M_s = (X_0 \quad X_1 \quad \cdots \quad X_{n-1}) . n$ is the number of activity in the process route. $X_i = 0$ means that activity i has not been processed. $X_i = 1$ means that activity i has been processed, $i \in (0, n-1)$. V_s can be expressed as $V_s = M_s \times (2^0 2^1 ... 2^{n-1})^T$. After activity i is finished, X_i in M_s will be updated as 1. So the process state value V_s will be updated as $V_s = V_s + 2^i$.

Process state value could reduce the space resources needed to store. And process state can be inferred easily by simple calculation. Both forward and backward traceability can be realized by this method. So the WIP can be controlled well-rounded.

3.3.2 *InOutRelation*

Any activity will bring about a relation between the input materials and output materials, which is named as *InOutRelation*.

$InOutRelation = < InOutID, InPutCollection, OutPutCollection, ActivityID >$

InOutID is the unique identifier of the relation; *InPutCollection* is the collection of the input materials, and *OutPutCollection* is the collection of the output materials; *ActivityID* is the identifier of the activity that brings about the relation.

$\forall m \in InPutCollection$ (or $m \in OutPutCollection$) can be expressed as:

$m = < MaterialID, LotID, ProcessState, Number, Unit >$

LotID is the lot identifier of m, and *ProcessState* is the process state value of m. *Number* is the input or output quantity of m. *Unit* is the quantity unit.

3.4   Traceability Model Based on Lot and Process State

On the basis of traceability requirements in manufacturing process, with the coding rules of material, lot and process state, the model as Fig. 3 illustrated is established to make traceability in manufacturing capable.

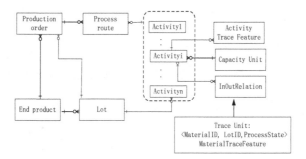

Fig. 3 Traceability model based on lot and process state

When there are requirements for an end product, production order will be generated according to the requirements. Each order has a correspondence lot for the end product of it. Each activity in the process route of the end product should take some resources and bring about *InOutRelation* and change of location and trace features of the materials in the three-dimensional space. So the *InOutRelation*, *ActivityTraceFeature* and activity resources should be recorded. For a certain trace unit, inquire about all the the *InOutRelation* that the *InPutCollection* of which includes the material. And then the activities and related information corresponding to the *InOutRelation* inquired about will be identified. Take each material in the *OutPutCollection* of the *InOutRelation* as the object material, search for the activity and *OutPutCollection* corresponding to the *InOutRelaton* that the *InPutCollection* of which includes the new trace object. Repeat the steps until none of the *InPutCollection* includes the last object material.

4   Examples

The high- carbon chromium bearing round steel with the steel number GCr15 has several types of order delivery status- hot rolling, hot-rolled spheroidizing, hot-rolled softening annealing, etc. Currently there are two orders with the order no G01051606002 and G01051606003. The order G01051606002 needs 10tons GCr15 high- carbon chromium bearing steel bar of specification $\phi 20$, with the delivery status of hot-rolling; and the order G01051606003 needs 15tons GCr15 high- carbon chromium bearing steel of specification $\phi 20$, with the delivery status of hot-rolled spheroid zing. The electric furnace and LF/VD can smelt 30 tons of molten steel one time; therefore, the two orders can be produced at the same time. For the production process of the two orders, the activities, input and output materials of each activitie and the changes of materials in three-dimensional space is showed in Fig. 4. Now it needs to conduct forward and backward traceability of the round steel blanks with batch no B10524006, that is, the attribute value of this material in the three-dimensional space is (M100415006,B10524006,1). Take the physical property as the traceability feature, the result of traceability is shown in Fig. 5.

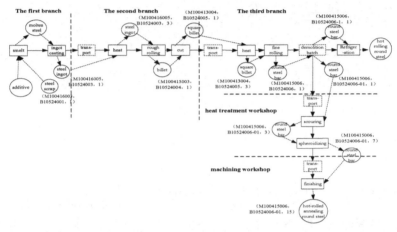

Fig. 4 Traceability model based on lot and process state

Activity	Furnace no.	Steel no.	Standard	Form	Batch	Quantity	starting time	ending time	equipment	team	working center
Fine Rolling	2010104 57	GCr15	GB/T1825 4-2002	Round steel	B1052466	25t	2011.5.26 9:15	2011.5.26 9:45	MB_101	T_021	WC_Z02

Sample no.	Item	Sub-item	Value
1	Physical	Tensile strength(MPa)	680
2	Physical	Tensile strength(MPa)	675
3	Physical	Tensile strength(MPa)	665
4	Physical	Tensile strength(MPa)	650
5	Physical	Tensile strength(MPa)	675

Fig. 5 Results of activity traceability attributes

5   Summary

The material traceability model based on lot and process state description regards the material as a point of the three-dimensional space composed of attribute, lot and process state, which can determine the material status in the production process uniquely. The material traceability is converted to the traceability of the material position in the space. In the model, each point in the space can be re-garded as a traceability unit. By the recording the material *InOutRelation* , *ActivityTraceFeature* and the resources and information related to the *MaterialTraceFeature* , the forward and backward traceability of materials in the production process can be realized more easily and accurately. And the WIP can be controlled well-rounded.

Acknowledgments. This research work has been supported by the National Natural Science Foundation of China (Grant No. 70772086), the National Science & Technology Pillar Program of China (Grant No. 2006BAF01A01) and the Doctoral Fund of Ministry of Education of China (Grant No. 200801411063).

References

1. Jansen-Vullers, M.H., van Dorp, C.A., Beulens, A.J.M.: Managing traceability information in manufacture. International Journal of Information Management 23, 395–413 (2003), doi:10.1016/S0268-4012(03)00066-5
2. Bechini, A.: Patterns and technologies for enabling supply chain traceability through collaborative e-business. Information and Software Technology 50, 342–359 (2008), doi:10.1016/j.infsof.2007.02.017
3. Sengupta, S.: Requirement Traceability in Software Development Process: An Empirical Approach. In: Proceedings of the 2008 The 19th IEEE/IFIP International Symposium on Rapid System Prototyping, pp. 105–111. IEEE Press (2008), doi:10.1109/RSP.2008.14
4. Pan, F., Peng, H., Shi, H.: Event-Based Production Process Traceability Model, Intelligent Control and Automation. In: The Sixth World Congress on WCICA 2006, October 2006, pp. 7211–7215. IEEE Press (2006), doi:10.1109/WCICA.2006.1714485
5. Bertolinia, M., Bevilacquab, M., Massinia, R.: FMECA approach to product traceability in the food industry. Food Control 17, 137–145 (2006)
6. Petroff, J.N., Hill, A.V.H.: A framework for the design of lot-tracing systems for the 1990s. Production and Inventory Management Journal 32, 55–61 (1991)
7. Liu, X., Huang, X., Ma, Y., et al.: Describing the material process state based on the theory about the set named S(I). Chinese Journal of Mechanical Engi-neering 39, 62–65 (2003)
8. Huang, X., Fan, Y.: Description method of material process state based on binary & hex system. Computer Integrated Manufacturing Systems 12, 280–284 (2006)
9. Zhang, N., Bo, H., Liu, A.X.: Batch-based description method of material process state for iron & steel industry. Computer Integrated Manufacturing Sys-tems 14, 785–792 (2008)
10. Moe, T.: Perspectives on traceability in food manufacture. Food Science & Technology 9, 211–214 (1998)
11. Alfaro, J.A.: Buyer–supplier relationship's influence on traceability implementation. Journal of Purchasing & Supply Management 12, 39–50 (2006)
12. Regattieri, A., Gamberi, M., Manzini, R.: Traceability of food products General framework and experimental evidence. Journal of Food Engineering 81, 347–356 (2007)
13. Bello, L.L., Mirabella, O., Torrisi, N.: Modeling and Evaluating tra-ceability systems in food manufacturing chains. In: IEEE Xplore. Computer Society, pp. 207–214 (2004)

Research on Three-Dimensional Sketch Modeling System for Conceptual Design

Jingqiu Wang and Xiaolei Wang

Abstract. In this paper, a three-dimensional sketch modeling system is presented. This system is composed of some modules, such as gesture recognition, sketch profile recognition, three-dimensional modeling, and free-form surface modeling, etc. In this study, pattern recognition is adopted to identify different gestures, and a tolerance ring algorithm is adopted to identify primitives, such as rectangles and circles.

Keywords: Sketch, Gesture, Three-Dimensional Modeling.

1 Introduction

Computer aided conceptual design (CACD) is an important branch of Computer-aided design(CAD), which involves the design methodology, ergonomics, artificial intelligence, cognition and thinking science and so on [1]. At present, manufacturing industry is facing tremendous challenges to improve product development capabilities, shorten development cycles, reduce cost, and introduces high-quality and high performance new products. CACD can be used to provide designers with ideas to solve problems and help designers to accelerate product design process, and support innovative product design. So, how to develop three-dimensional sketch-based modeling system is one of the main issues in CACD system. It can be expected, with the development and improvement of CACD system, it will be welcomed more and more by the concept designers.

With the development of computer hardware and software, computer becomes more and more important in people's daily life. At the same time, human-computer user interface is being developed towards more transparent, flexible,

Jingqiu Wang
Nanjing University of Aeronautics and Astronautics, 29 Yudao Street., Nanjing 210016, China
e-mail: meejqwang@nuaa.edu.cn

Xiaolei Wang
Nanjing University of Aeronautics and Astronautics, 29 Yudao Street., Nanjing 210016, China
e-mail: xlei_wang@nuaa.edu.cn

F.L. Gaol et al. (Eds.): Proc. of the 2011 2nd International Congress on CACS, AISC 144, pp. 321–327.
springerlink.com © Springer-Verlag Berlin Heidelberg 2012

efficient, natural direction. Users have become the subject in human-computer interaction, and the human-computer interaction patterns are being developed in the "users' freedom "direction. Computers are expected to become more simple and efficient tools for designers, instead of the traditional pen and paper, so as to improve design efficiency, and help designers carry out innovative design.

Natural interaction based on pen gesture is efficient for innovation in conceptual design [2]. In this paper, a computer aided concept design system is implemented, a sketch input interface is adopted to create 3D model.

2   User Interface

Users convey instructions to computer and get feedback through human-computer interface. Human-computer communication process is often referred to as human-computer interaction. The media, including software and hardware resources, on which the process depends, is called the Human-Computer Interface. The user interface of an application is the part that the person using the software sees and interacts with [3].

The traditional graphical user interface based on the WIMP (Windows, Icon, Menu, and Pointing Device) is not suitable for the conceptual design, mainly in the following areas:

- Users need to use the mouse to click constantly emerging windows, buttons and menus, so that the user's continuity innovative thinking and free expression are hindered.
- Too many input parameters which make the expression of fuzzy concepts have been affected
- To build a precise 3D objects, all operations are applied based on the specific steps and specific rules, so the modeling steps can't be fully consistent with the cognitive processes of human thinking.

With the development of pen-based interactive tools, pen-based user interface is constantly evolving. Advanced interactive tools and interactive pen-based system has emerged. Igarashi et al present a sketching interface for quickly and easily designing freeform models such as stuffed animals and other rotund objects [4]. SketchUP system can be used to quickly complete the 3D architectural model based on model segmentation, drag and drop, editing tools [5].

In this paper, a pen-based user interface was developed. It includes a 2D sketch user interface, and 3D modeling display and editing interface.

As shown in Fig.1, the 2D sketch input interface, which mainly includes main menu, shortcut menu, sketch gesture input area, feature profile input area and control information input area. The main menu contains some necessary configuration and parameter settings, and provides alternative options for the modeling commands, which allows users to correct errors in gesture recognition. Shortcut menu provides a screen toggle button, some modeling commands shortcut buttons, and a number of standard buttons, such as file operations, etc. Sketch gesture input area can be used to receive information of the user's gesture sketch, identify the gesture to obtain user's modeling intent. In the feature profile input area, the user is free to sketch the object contour, and then, the system will automatically combine the semantic understanding and gestures to generate a three-dimensional model. Control information input area can be used to receive control information and other information such as the scanning track.

Fig.2 shows the 3D display window, in which, the 3D model can be freely viewed, edited by users through a simple drag and drop way.

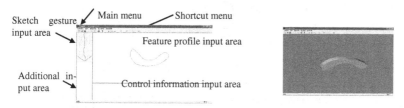

Fig. 1 2D sketch input interface

Fig. 2 3D modeling display and editing interface

3 Gesture Recognition and Primitive Contour Recognition

The key issues in computer-aided conceptual design system are how to build an effective interaction tool between user and computer, and how to make the system understanding user's design intents fast and accurately. The understanding process includes two aspects. The first one is the understanding of user's modeling intention that is gesture understanding. The second one is the understanding of modeling, that is the recognition or regularization of sketch primitives.

- Gesture Recognition

Sketch gesture is one kind of interactive command, which means gesture should be identifiable and unambiguous. Otherwise, it is likely to cause misuse or error. Therefore, during design process of gestures, following factors need to be considered:

- Gestures should be easily recognized by computer
- Gestures should be easily learned and understood by the user
- Gestures should be easily remembered by the user
- Gestures should be easy to be drawn by the user

Considering above factors, as well as the specific application of sketches in the 3D modeling system, some commonly used gesture commands for 3D modeling are designed in this system, such as Extrude, Revolve, Sweep and Variable Cross-Section Sweep, as shown in Table.1. These gesture commands can basically meet the requirements of 3D modeling application for some common objects.

Gesture recognition can be attributed to the application of pattern recognition technology; different methods are used by researchers for gesture recognition, which mainly includes the following methods: identification based on pattern matching [6], statistics analysis [7], heuristic rules [8], and geometric properties [9]. In this study, the gesture recognition method based on pattern matching is adopted, as shown in Fig.3, and have achieved good result [10].

Table 1 Basic Gestures

Extrude	↓	Sweep	∼
Revolve	⟳	Variable Cross-Section Sweep	⬠⬡

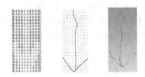

Fig. 3 Gesture recognition

- Primitive Contour Recognition

Common primitives, such as rectangle, circle, arc, etc., are often used in computer-aided design, especially in the design of mechanical parts. However, due to the ambiguity of the sketch input, the input primitive is usually not complete and accurate. In order to quickly identify and understand these irregular, fuzzy primitives, the identification of basic primitives is studied in this paper.

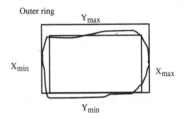

Fig. 4 Rectangle identification

When the primitive contour input by the user is close to common regular primitive, the system should automatically identify the irregular input as regular primitive, otherwise keeping it as irregular one. In this paper, a tolerance ring algorithm [10] is adopted to identify primitives, such as rectangles and circles. Fig.4 shows the process of identifying a rectangle: Firstly, preprocessing the user's input data, extract the maximum and minimum X, Y value. And then, construct the outer ring according to the maximum and minimum X, Y values. After construct the inner ring which scaled down from the outer ring based on user-set threshold, calculate the proportion of the sampling points to total pints between the inner and outer ring. Lastly, according to this ratio determine how close the input sketch primitive and the regular one as to identify it.

4   Three-Dimensional Modeling

Humans have the extraordinary ability to understand and cognize 3D object from 2D information, with which the computer can't compare. Complex calculations and conversion rules must be implemented by computer to create 3D model.

From 2D sketch to 3D modeling, the third dimensional information could be obtained under the relevant rules contained in 2D space. It is necessary for the computer to get efficient and rapid understanding of the information input by the user in order to build 3D model in accordance with relevant rules and semantics.

In order to create different surfaces, the following two different approaches are studied for 3D modeling.

- Three-Dimensional Modeling Based on Sketch Gesture

In our study, the input model feature profile is separated from gesture, in this way, gesture can be used to clearly express the user's design intent. User can input modeling intent in the separate gesture input area, such as extrude, revolve, sweep, etc. Feature profile is input in the profile input area; then, the modeling intent expressed by gestures was act on the feature profile to create 3D model. Fig.5 shows the extrude gesture and the corresponding 3D model.

- Free Surface Modeling Based on Sketch Profile

The Teddy system developed by the University of Tokyo is an epoch-making product, it mainly adopts free surface modeling method based on contour, which has opened up a new way for the design of three-dimensional free form. In this paper, Teddy thought is taken as reference; surface modeling based on the input 2D hand-drawn contours is studied. By using the CGAL computational geometry library, irregular contour of the skeleton is extracted from the contour profile, and then, the skeleton line is raised and sewn up to create surface, as shown in Fig.6.

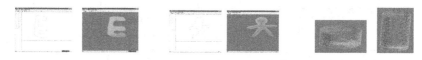

Fig. 5 Gesture and 3D model.   **Fig. 6** Free surface modeling   **Fig. 7** 3D model deformation

- 3D model deformation

An initial model can be constructed after the correct semantic understanding for 2D sketch; however, due to fuzzy description of the size of the modeling in the conceptual design, the initial model does not necessarily meet the requirements of user's modeling. So a simple adjustment in 3D space is needed to achieve rapid deformation, especially for adjusting the size of a particular direction to meet the design requirements. The surface of the object is expressed by using 3D point composed of triangular facets, during the interaction process, 3D coordinates of points of the surface can be changed by moving the mouse or light pen in order to achieve quick and easy 3D deformation, as shown in Fig.7.

5   Conclusions

In this paper, a 3D modeling prototype system was developed based on sketch. The main function of the system is the following:

- Gesture recognition: to identify gestures commonly used in the design process, such as extrude, revolve, etc.

- 2D sketch recognition and understanding: pre-processing strokes, identifying the basic primitives such as lines, rectangles and circles, etc.
- 3D modeling: based on sketch, to construct the corresponding 3D model based on the recognition results and semantic understanding of 2D sketch.
- Rapid deformation of 3D model.
- The results display.

At first, in the system, the basic gestures and primitive contours can be identified. Then according to the recognition result, the system can automatically create 3D model based on gesture, or free-form surface based on the profile. The 3D model can also be deformed as needed by the user. In this system, 3D modeling for conceptual design is implemented based on sketches, and it can better support for computer aided conceptual design. Fig.8 shows the framework for 3D modeling system.

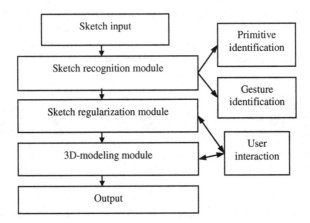

Fig. 8 Framework for 3D modeling system

Acknowledgments. The authors would like to thank NUAA Research Funding (No: NS2010136).

References

1. Hsu, W., Woon, L.M.Y.: Current Research in the Conceptual Design of Mechanical Products. Computer-Aided Design 30(5), 377–389 (1998)
2. Ma, C., Dai, G., Teng, D., Chen, Y.: Research of Interaction Computing Based on Pen Gesture in Conceptual Design. Jounal of Software 16(2), 303–308 (2005)
3. Myers, B.A.: User interface software technology. ACM Computer Survey 28(1), 189–191 (1996)
4. Igarashi, T., Matsuoka, S., Teddy, T.H.: A Sketching Interface for3DFreeform Design. In: ACM SIGGRAPH 1999 Conference Proceedings, pp. 409–416 (1999)
5. Williams, R.: Sketch-based 3D design (OL) (2008), http://www.sketchup.com

6. Wang, Y., Yuan, B.: Pen-Based Gesture Recognition in Multimodal Human-Computer Interaction. Journal of Northern Jiaotong University 25(2), 10–13 (2001)
7. Rubine, D.: Specifying Gestures by Example. In: ACM SIGGRAPH 1991 Conference Proceedings, Computer Graphics, vol. 25(3), pp. 329–337 (1991)
8. Fonseca, M.J., Pimentel, C., Jorge, J.A.: CALI:An Online Scribble Recognizer for Calligraphic Interfaces. In: AAAI Spring Symposium on Sketch Understanding, Palo Alto, pp. 51–58 (2002)
9. Hammond, T., Davis, R.: Tahuti:A Geometrical Sketch Recognition System for UML Class Diagrams. In: AAAI Spring Symposium on Sketeh Understanding, Palo Alto, vol. 68, pp. 59–68 (2002)
10. Han, H., Wang, J.: Research on a Three-Dimensional Modeling Method Based on Freehand Sketch. Computer Applications and Software 27(1), 215–217 (2010)

MC9S12XDP512-Based Hardware Circuit Design and Simulation for Commercial Vehicle's Electric Power Steering System

Liu Jingyu, Wang Tao, and Hu Zhong

Abstract. Taking into account the large steering shaft load of the commercial vehicle, this paper aims to design a commercial vehicle electric power steering system. Firstly, a framework of MC9S12XDP512-based hardware circuit design is presented, and then the high-power motor drive circuit and signal processing circuits are introduced in detail. In virtue of building the DC Motor reversible speed regulation system simulation models, the simulation results show that the assisted motor has been proven having a good torque response.

Keywords: Commercial Vehicle, EPS, H-bridge, PWM, Simulation.

1 Introduction

With the development of modern automotive electronics, the EPS technology has been maturing, and its application scope has grown from the original passenger vehicles to commercial vehicles direction. For the light medium-sized commercial vehicles, the steering shaft at full load is about 3 to 5 times than passenger cars, which requires much more motor power than passenger vehicles. Therefore, the key technology of the commercial EPS is the driving high-power and high power motor to provide torque assist. So developing electric power steering system for commercial vehicles is a high-new technology that follows modern vehicle development topic closely, which will be the inevitable trend for automobile power steering system.

2 The Hardware Circuit of Commercial Vehicle's EPS System

2.1 The Overall Hardware Logical Diagram of the EPS System

The main structures of the Commercial Vehicles Electric Power Steering system are as follows:

Liu Jingyu · Wang Tao · Hu Zhong
School of Automobile, Chang'an University, Xi'an, Shannxi, China
e-mail: 7883617a@chd.edu.cn, wangtaochd@yahoo.cn,
     13991918542@163.com

F.L. Gaol et al. (Eds.): Proc. of the 2011 2nd International Congress on CACS, AISC 144, pp. 329–334.
springerlink.com   © Springer-Verlag Berlin Heidelberg 2012

Fig. 1 The hardware circuit of the EPS system mainly includes microprocessor and its peripheral circuits, power conversion circuits, the H-bridge driving circuit module1 and logic control circuit, as well as signal processing circuits, such as torque signals, vehicle speed signals, the engine ignition signals processing circuits. In additional, there are over-current protection circuits, fault diagnosis circuits.

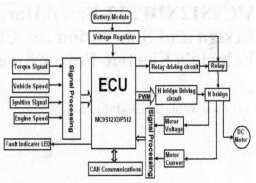

2.2 New Function Features of the EPS Microprocessor MC9S12XDP512

In this paper, the electronic control unit (ECU) of the EPS system is MC9S12XDP512, which is a 16-bit high performance micro controller that based on an enhanced S12 core, and the maximum operating frequency is 80 MHz, it also includes 512 K bytes of Flash EEPROM, 32 K bytes of RAM, 4 K bytes of EEPROM. The S12X ECU is designed to retain the low-cost, low power consumption, excellent EMC performance and code-size efficiency advantages over previous 16-bit MC9S12 ECU family. The S12X Family features the performance boosting XGATE co-processor. The XGATE module, which is programmable in "C" language, has an instruction set which is optimized for data movement, logic and bit manipulation instructions. It runs at twice the bus frequency of the S12X and off-loads the CPU by providing high speed data transfer between any peripheral module, RAM and I/O ports. This is particularly useful in applications such as automotive gateways where there are multiple busses carrying heavy data traffic which would otherwise exert a heavy interrupt/processing load on the CPU.

2.3 H-Bridge Driving Circuit of the Power Motor and Its Logic Control

The power electronic device of the H-bridge drive circuits are shown in figure 2:

Fig. 2 The H-bridge drive circuits mainly consist of four high-power MOSFETS IRF3710 and two half-bridge drive chips IR2104, by controlling the turn-on and cut-off of the four IRF3710, the motor can be achieved forward and reverse controlling.

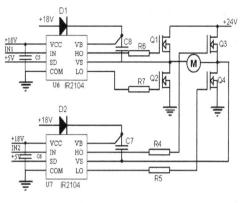

The logic control of the H-bridge driving circuit: as shown in figure 2, IN1, IN2 is the PWM control signals, C7, C8 are the bootstrap capacitors. U6 determines the power switch Q1and Q2's turn-on and cut-off according to the input control signal of IN1, U7 determines the power switch Q3 and Q4's turn-on and cut-off according to the input control signal of IN2. When IN1 input is high-level, the HO output of U6 is high-level, the switch Q1 turns on, meanwhile IN2 input is low-level, the LO output of U7 is high-level, the switch Q4 turns on, the motor armature current gets through from the power supply side via the power switch Q1 through the motor M and the power switch Q4 to the ground, in terms of the motor, the direction in the graph is from left to right, the motor forward controlling is accomplished. When IN1 input is low-level, IN2 input is high-level, the switch Q2 and Q3 become turn-on, the motor armature current gets through from the power supply side via the power switch Q3 through the motor M and the power switch Q2 to the ground, in terms of the motor, the direction in the graph is from right to left, the motor reverse controlling is accomplished. The motor reversing control logic is shown in table 1.

Table 1 The Motor Reversing Control Logic

α_1 、 α_2	Motor Action	Q1	Q2	Q3	Q4
$\alpha_1 > 0\ \alpha_2 = 0$	Stop	OFF	ON	OFF	ON
	F- rotation	ON	OFF	OFF	ON
$\alpha_1 = 0\ \alpha_2 > 0$	R-rotation	OFF	ON	ON	OFF
	Stop	OFF	ON	OFF	ON
$\alpha_1 = 0\ \alpha_2 = 0$	Stop	OFF	ON	ON	OFF

Note: α_1 stands for the duty cycle of IN1; α_2 stands for the duty cycle of IN2.

2.4 Signal Processing Circuit Unit

The steering wheel torque and angle value are the main factors in the target current output of the controller, the torque-angle sensor output signals often contain noises and other interference glitch, which need to be adjusted and filtered before entering the ECU A/D conversion port, meanwhile, the speed signal must be converted from 24V to 5V:

Fig. 3 Circuit diagram of torque signal's processing As shown in the figure 3, the signal output is added by two diodes D4 and D5 to form clamp circuit, clamping the torque signal level error in no more than two diode voltage drop voltage level, resistor-condenser circuit aims at low pass filtering, and removing the high frequency signal interference.

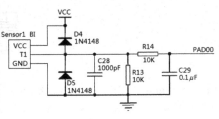

Fig. 4 The vehicle speed signal is a series of 24V single-polarity pulses from the speedometer. ECT port can handle the signal's high voltage is about 5V, here we use photoelectric coupled methods convert the 24V signals to 5V digital signals, the processing circuit is shown in figure 4.

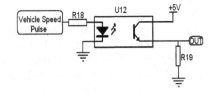

3 PWM-M Control Reversible Speed Regulation System Simulation

Full-controlled power electronic devices are used in the DC PWM-M speed control system with the features of high frequency modulation, fast response, which is widely applied in the servo control system. Bidirectional unipolar control DC PWM-M reversible system increased speed regulator ASR and current regulator ACR, the ASR and ACR are used with output limiting PI regulator.

The DC PWM-M reversible system simulation model is shown in figure 5.

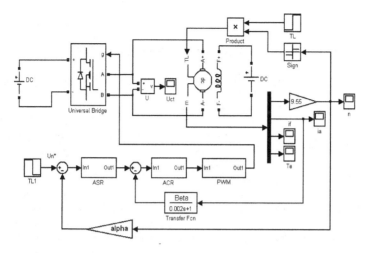

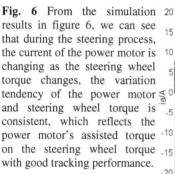

Fig. 5 The PWM-M reversible system simulation model

Fig. 6 From the simulation results in figure 6, we can see that during the steering process, the current of the power motor is changing as the steering wheel torque changes, the variation tendency of the power motor and steering wheel torque is consistent, which reflects the power motor's assisted torque on the steering wheel torque with good tracking performance.

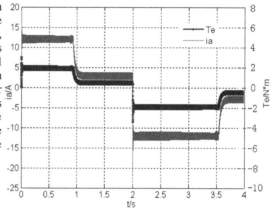

4   Conclusion

Aiming at the commercial vehicle steering shaft load larger characteristics, in this paper, we mainly design the electronic control unit hardware circuit based on the MC9S12XDP512 microcontroller, through the establishment of the PWM-M reversible speed control system simulation model, the simulation results show that the bidirectional unipolar H-bridge PWM driving circuit is not only meet the requirement of high-power motor driving but also the motor has a smooth torque pulsation, the power torque and power current have a good following performance.

References

[1] Wu, J., Li, Q., Song, D.-Y., Yang, L.-K., Cheng, F., Liang, X.-J.: School of mechanical Automotive Engineering. In: Energy Saving Analysis Using Pulse Width Modulation Techniques Controlling Electromagnetic Clutch in EPS System. Proceedings of 2009 Asia-Pacific Conference on Information Processing, Zhejiang University of Science and Technology Hangzhou, China (2009)

[2] Qun, Z.: The Development of Control Unit in Automotive Electric Power Steering System. In: Proceedings of 2009 Asia-Pacific Conference on Information Processing, Department of Information Engineering University of Nanchang Nanchang, China Huang Juhua Department of Mechanical & Electrical Engineering University of Nanchang Nanchang, China (2009)

[3] Chen, K., Ma, X., Ji, X.: Research on control method of electric power steering system. Journal of Jiangsu University(Natural Science Edition) 25(1), 21–24 (2004)

[4] Li, W.-G., Lin, Y., Wang, Y.-C.: Electric Power Steering system hardware design. Journal of South China University of Technology (Natural Science Edition) 34(2) (2006), 李伟光,林颖,王元聪. 汽车电动助力转向系统硬件设计 华南理工大学学报（自然科学版）

[5] Li, S.-L., Xu, C., Yang, Z.: Electric Power Steering system hardware design. Mechanical & Electrical Engineering Magazine 21(1) (2004), 李叔龙,许超,杨智. 汽车电动助力转向系统硬件设计 机电工程

[6] http://www.freescale.com

A Research on the Urban Landscape Value Evaluation Based on AHP Algorithm

Jun Shao, Junqing Zhou, Junlei Yang, and Hanxi Liu

Abstract. Based on the analytic hierarchy process (AHP), this paper constructs the multi-target and multilevel comprehensive evaluation model of urban landscape value evaluation from three perspectives of landscape, ecology and social effects. By adopting landscape ecological unit grid-type analytical method and by virtue of GIS technique, the paper performs lattice-based treatment, establishes landscape value evaluation system database, performs comprehensive grading of and assignment calculation of the landscape value evaluation factor by integrating weighted index model and determines the high and low each unit landscape value and its distribution. The transformation of the traditional qualitative analysis method into the combination of qualitative analysis with quantitative analysis increases the objectivity and scientificity of the urban landscape value evaluation. And the evaluation system and evaluation model is applied into the urban landscape value evaluation of Huashan, Wuhan.

1 Research Background

As the important support for the normal operation of the urban socioeconomic system, the urban landscape has obvious visual features and functional relationship. It carries the deepest urban cultural connotation, revealing the urban vitality and intelligence. With the rapid advance in China' s urbanization process , the construction of urban landscape is not only closely associated with the people's city life but also associated with the city's core competitiveness and future development space. The inter-urban competition is the competition of human resource and capital input in the final analysis while the human resource and capital input usually depends on whether a city has attraction, cultural charm and residential suitability. The beautiful urban landscape is an important evaluation criterion. It is thus clear that the urban landscape has multi-value of the economy, ecology and culture.

Jun Shao · Junqing Zhou · Junlei Yang · Hanxi Liu
Huazhong University of Science and Technology
e-mail: sj1984hust@gmail.com, 250542685@qq.com,
304403332@qq.com, 906228289@qq.com

F.L. Gaol et al. (Eds.): Proc. of the 2011 2nd International Congress on CACS, AISC 144, pp. 335–340.
springerlink.com   © Springer-Verlag Berlin Heidelberg 2012

The evaluation of the multi-value is the basis and premise of the development and protection of all landscapes, thus it should be highly valued.

2 Research Feasibility

Currently, the universally applicable measurement scale still hasn't been found for the evaluation of landscape value. The intercomparsion, therefore, among many factors appears exceptionally difficult. The AHP establishes the scale by adopting the measurement method of pairwise comparison, so that the corresponding mathematical models could be introduced and it is possible to transform the qualitative analysis into the quantitative analysis. In the meantime, when introducing the mathematical model, avoid introducing too complicated and seeking after perfect parametric mathematical model so as not to be bogged down in huge and complicated mathematical difficulty when seeking after solution. Consequently, it is technically feasible to evaluate the urban landscape value by adopting AHP technique.

3 Research Method

3.1 AHP Analytical Method

3.1.1 Technical Route

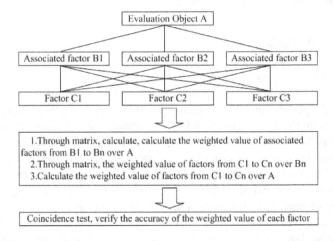

Fig. 1 AHP Technical Route

3.1.2 Constitution of Landscape Value Evaluation System and Establishment of Hierarchical Structure Model

Table 1 Hierarchical Chart of the Factors of the Landscape Value Evaluation System

Destination layer	Criterion layer	Factor layer
Landscape value (A)	Landscape (B1)	Landscape level(C1)
		Unique feature(C2)
		Landscape matching degree(C3)
		Overall coordination(C4)
	Ecology (B2)	Ecological influence(C5)
		Biological diversity (C6)
		Ecological integrity(C7)
		Water conservation capacity(C8)
		Ecological sensitivity(C9)
	Social effect (B3)	Popular science education value(C10)
		Historical scientific research value(C11)
		Economic value(C12)

3.2 Landscape Ecology Unit Grid-Type Analytical Method

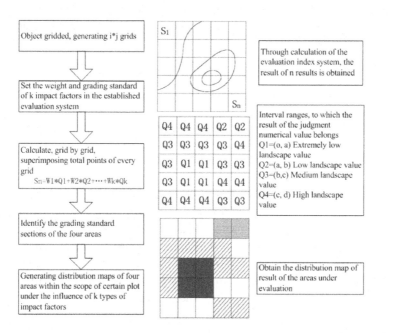

Fig. 2 Technology roadmap

3.2.1 Mesh Generation

The size of each grid could directly affect the accuracy of the evaluation result. The information will be easily lost if the grid is too large and if the grid is too small, a lot of redundant data will be generated. Thus the process of the spatial analysis is affected. According to the actual situation of the study area, to ensure the accuracy of the final evaluation result, take 250m*250m for the landscape evaluation under discussion this time as the landscape evaluation criterion/unit.

3.2.2 Outcome Processing

Divide it into four levels from the great to the small according to the total value of each grid and connect it with the ID code of the map layer of the original grid and display it at sub-layers for check-up & calibration to draw the hierarchical diagram of the landscape value.

4 Empirical Analysis

The study area is located in Huashan of Wuhan. It has a total area of 65.95 km2. This area pertains to the area of Dadong Lake, a part of green wedges in the east of Wuhan. There are hills and lakes closely distributed in the area, which posses Yanxi Lake and Yandong Lake. The town area is naturally divided into three parts by two mountain ranges (Chang Mountain and Dawu Mountain ranges), significantly meaningful to construct the city's ecological safety and landscape pattern.

Applying the AHP evaluation method and model set up previously to calculate the index weights at all levels and the exponential value of the urban landscape evaluation of Flower Mountain.

Table 2 Form of Index Weights at All Levels of the Evaluation System

Criterion layer	Weights	Factor layer	Weights
Landscape (B1)	0.24	C1 landscape level	0.06
		C2 unique feature	0.13
		C3 landscape matching degree	0.02
		C4 overall coordination	0.03
		C5 ecological influence force	0.05
Ecology (B2)	0.64	C6 biological diversity	0.16
		C7 ecological integrity	0.31
		C8 water conservation capacity	0.10
		C9 biological sensitivity	0.03
Social Effects (B3)	0.12	C10 popular science education value	0.01
		C11 historical scientific research value	0.08
		C12 economic value	0.03

$CR_{total} = 0.073 < 0.1$.

By using GIS technology to complete grid-based treatment, establishing the database (type data) of landscape value assessment system and applying weighted index model to comprehensive grade and assignment calculation of factors of the landscape value evaluation. Make grading of the evaluation result by combining the experts' opinions according to the division principle of normal distribution, and divide the Flower Mountain urban landscape value into four partitions from the top down levels: high urban landscape value, medium urban landscape value, low urban landscape value, and extremely low urban landscape value. Finally, the evaluation chart (Fig. 3) of the Flower Mountain urban landscape value will come into being. For the levels and area of each type of areas, refer to the table 3.

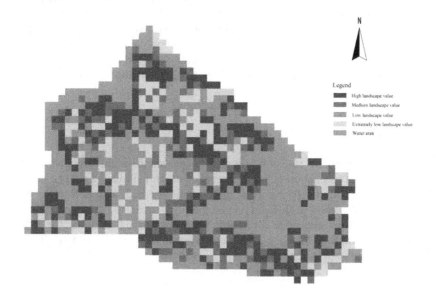

Fig. 3 The Zone Chart of Huashan Urban Landscape Value Evaluation

Table 3 Hierarchical Form of Huashan Landscape Value Evaluation

Scores	Area (km^2)	Ratio (%)
High landscape value (including water area)	32.26	49.06
Medium landscape value	7.44	11.29
Low landscape value	16.49	25.00
Extremely low landscape value	9.66	14.65
Total	65.95	100.00

From the table above, it can be seen that the whole region under evaluation shows up forms zonal landscape pattern, in which, the area with high landscape quality accounts for 49.06% of the whole region under evaluation, they mainly refer to east-to-west mountain ranges of Chang Mountain range and Jiufeng Mountain as well as the area of Yanxi Lake and Yandong Lake. Such types of the areas should keep the integrity of the original ecological system and away from the activities of urban development and construction; The area with medium landscape value accounts for 11.29% of the whole region, mainly distributed in transition zones of the mountain ranges of Chang Mountain and Jiufeng Mountain as well as the green line control zone of the water area of Yanxi Lake and Yandong Lake. Such types of areas could combine the construction of landscapes such as: urban greening and wetland park; The area with low landscape value takes up 25.00% of the whole region, mainly distributed in transition zones between the water area of lakes and land for construction as well as the canyonland among the mountain ranges, these areas should make full use of terrain, water body and primary vegetation to strengthen the greening, and urban exploitation could be performed to certain extent; The area with extremely low landscape value accounts for 14.65% of the whole region, they mainly refer to the urban development area. Great efforts should be paid to improve the ecological situation to create amicable living environment for people.

5 Conclusion

From the aforementioned research, it is easy to see that the multi-target and multi-level analysis and evaluation on the elements of urban landscape could be carried out vividly and intuitively by combining AHP method and GIS technique. Through optimal control of different landscape value areas, the damage and influences of urban development & construction activities to and on urban landscape are effectively controlled and the purpose of rationally developing and making use of the urban landscape is realized.

References

1. Fu, B., Chen, L., Ma, K.: Principle and Application of Landscape Ecology. Science Press, Beijing (2001)
2. Zhao, H., Xu, S., Jin, S.: AHP—A New Proposed Method for Making Decision. Science Press, Beijing (1986)
3. Ye, J., Niu, X., Li, X.: Geographic Information and Planning Support System. Science Press, Beijing (2006)
4. Yu, K.: Evaluation and Analysis of Landscape Sensitivity and Its Threshold. Geographical Research 10(2), 38–51 (1991)
5. Xiao, D.: Assessment on Landscape Value and Protection. Geographical Science 26(3), 505–507 (2006)

Regional Degraded Trend through Assessing of Steppe NPP on Remotely-Sensed Images in China

Suying Li[*], Wenquan Zhu, XiaoBing Li, and Junjiang Bao

Abstract. The ecological crisis derived from NPP decline in grassland region not only endangers the regional stockbreeding development, but also aggravates the contradiction between ecological environment and social economy development. This paper developed a methodology to estimate the degradation process of grasslands at large scale based on the assessment of regional NPP on TM/ETM images. The improved CASA Model was used to simulate regional NPP in the short grassland (typical steppe) region on account of NPP frequently considered to indicate the degrees of grassland degradation in China. The results displayed that the average regional NPP gradually decreased from 1991 to 2000, 2005 due to impacts of climatic change and the grazing manners. Image analysis showed that the ecosystem had been in an incompetent' environment over the past decades. Continued degradation will be intense unless grazing activities are changed.

1 Introduction

In recent years, NPP (Net primary production) of grassland ecosystem has become a hot-point research as a principal indicator for steppe degradation in North China [2,4,6]. NPP is a key component of the terrestrial carbon cycle, and it is defined as the rate at which an ecosystem accumulates energy or biomass, excluding the energy it uses for the process of respiration [5]. Most research on grassland NPP was carried out at the level of communities or at the scale of large regions [1]. Hence, there is an imperative need to accurately monitor the spatio-temporal patterns of grassland NPP at local scales. The CASA (Carnegie-Ames-Stanford Approach) model is a typical and extensive model of ecosystem NPP based on

Suying Li · Junjiang Bao
Department of Environmental Science, College of Energy and Power Engineering, Inner Mongolia University of Technology, Huhhot 010051, China
e-mail: syli2010@hotmail.com, lisuying70@yahoo.com.cn

Wenquan Zhu · XiaoBing Li
College of Resources Sciences and Technology, Beijing Normal University, Beijing 100875, China

[*] Corresponding author.

F.L. Gaol et al. (Eds.): Proc. of the 2011 2nd International Congress on CACS, AISC 144, pp. 341–347.
springerlink.com   © Springer-Verlag Berlin Heidelberg 2012

light use efficiency [3], but the classic CASA model has some weaknesses, such as the estimation of maximum light use efficiency (ε_{max}), impact to NPP from vegetation classification accuracy, and parameter calculation of the soil water model. Thereupon, this study used the improved CASA model [5] to analyze the spatial and temporal dynamics of ecosystem NPP in the typical steppe region of Inner Mongolia, one of world largest grassland regions, in order to illuminate the ebb and flow of grassland NPP under global climatic change and different local policy during the past twenty years.

2   Study Area and Method

2.1   Selecting Study Area

Xilinhot near the geometric center of Inner Mongolia Autonomous region (IMAR) was selected as the study region because it represents the most typical temperate steppe in north China [4].

2.2   Pretreating the Images Required

1991, 2000, 2005 TM/ETM and the other relative images for the research region were attained by false color composition, mosaic, geometric rectification, and re-sample of gray values, followed by transformation into UTM in ERDAS software. The spatial resolution of those images was 30m × 30m.

2.3   Obtaining the Regional NPP

Following image processing, NPP (gC/m^2) was computed as the amount of photosynthetically active radiation absorbed by green vegetation (APAR) (MJ/m^2) multiplied by the actual light use efficiency (ε) (gC/MJ) by which the radiation is converted to plant biomass increment.

The formula of Net primary production:

$$NPP(x, t) = APAR(x, t) \times \varepsilon(x, t) \tag{1}$$

Where x is a pixel in a remote sensing image, and t is the period that NPP is cumulated, such as a month in this paper. The technology flow chart for APAR and ε are provided by Zhu et al. [5] (Fig.1).

In addition, thirty-two samples in the study area were selected to test the result of CASA. The error was evaluated by linear regression model in SPSS, and was found to be acceptable ($R^2=0.375$, $p<0.05$).

2.4   Reclassing the Regional NPP

Some reclassification of data was made to facilitate the analysis. The regional NPP was classified into six classes: $0{\sim}25gC/m^2$/month, $25{\sim}50gC/m^2$/month, $50{\sim}75gC/m^2$/month, $75{\sim}100gC/m^2$/month, $100{\sim}125gC/m^2$/month and $125{\sim}150$ gC/m^2/month because the regional NPP in the study area varies between 0 and 150

$gC/m^2/month$ in three years. As NPP was a good indicator of vegetation situation, the developing trend of grassland degradations was exhibited by analyzing on NPP conversion between the different NPP classes of grasslands in the time serial years.

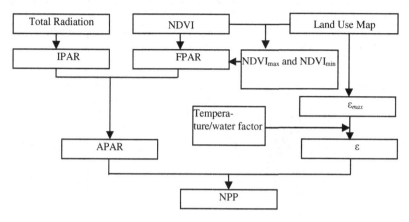

Fig. 1 Technology flow to calculate the regional *NPP*

3 Results and Analysis

Through integrating remote sensing and geographic analysis methods, the study evaluated the dynamic transform of different NPP classes between years. This paper made an in-depth analysis of the temporal and spatial trends of NPP in order to reflect the degradation situation in typical steppe region over the last twenty years.

3.1 Regional Distribution of NPP

The spatial distributions of NPP of three years in Xilinhot are shown in fig. 2. The higher NPP of 1991 were mainly located in the southeastern part of Xilinhot, Baiyinxile Village, and the southern borders of study area, Baiyinkulun Village. At the same period, the other grasslands in Xilinhot also had a > 80 $(gC/m^2/month)$ output in production of forage grass.

In 2000, it was similar that the higher NPP were mainly located in the south of research region, but most of northern region of Xilinhot had a less than 50 $(gC/m^2/month)$ production. Compared with 1991, it was reflected that 2000 was really a droughty period in which there was an approximate 200mm precipitation. At the same time, it was also seen how serious the degradation of grasslands in Xilinhot were already droved.

For 2005, the higher yields were mostly distributed in the south and southeast of Xilinhot region. However, these high values of regional NPP were dispersed into a discontinuous, fragmented pattern. Most of the northern section had a <50 $(gC/m^2/month)$ production. Therefore, to some extent, the vegetation of study area in 2005 was turned for the better developing trend for the regional ecological environment. Comparatively speaking, the regional NPP of 1991 showed the best situation while a kind of unfavorable state of NPP was revealed in 2000. Up to 2005, the regional NPP started to turn for the better status in some areas. However, there still existed deterioration of the grasslands in some areas where the pattern of NPP were fragmental and low-yield. Also, it was clearly proved that the ecological situation of the typical steppe was still not optimistic as far as the overall pattern of vegetation was concerned.

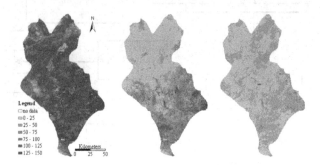

Fig. 2 Distribution of NPP classes in Xilinhot

3.2   Transformation between NPP Types

The transformation between NPP types from 1991 to 2000 was represented in Table 1. During the first ten years, from 1991 to 2000, grassland NPP sharply decreased, and the main transforms were taken place from higher grassland yield in 1991 to lower grassland yield in 2000, for example, the largest transferring was from NPP classes (25-50, 50-75, 75-100, 100-125 $gC/m^2/month$) in 1991 to NPP class (0-25 $gC/m^2/month$) in 2000. The second transferring was brought from the higher NPP types (50-75, 75-100, 100-125, 125-150 $gC/m^2/month$) to the second lowest NPP type (25-50 $gC/m^2/month$). On the contrary, the conversion from the lower yield to the higher yield was rarely happened during 1991-2000.

Table 1 Transformation between NPP classes during 1991- 2000

Unit: %

2000 \ 1991	0-25	25-50	50-75	75-100	100-125	125-150
0-25	79.03	12.36	4.78	1.74	2.07	0.03
25-50	66.46	20.42	8.11	2.71	2.27	0.02
50-75	54.94	24.97	12.62	4.53	2.91	0.02
75-100	52.62	23.93	14.70	6.32	2.39	0.04
100-125	52.62	23.37	14.44	6.50	2.97	0.10
125-150	25.93	26.76	21.80	12.31	11.81	1.38

For instance, the value of the 100-125 or 125-150 NPP types switched from every NPP types in 1991 were almost closed to 20% or much lower proportion. During the ten years, grassland NPP in Xilinhot region was cosmically deteriorated due to warm-arid climate and intensive grassland use.

However, after 2000 (Table 2), the degradation trend of NPP has been partly mitigated. The 25-50 gC/m^2/month class in 2005 became the main transformed part instead of the 0-25 gC/m^2/month class in 2000. And the second biggest switch between 2000 and 2005 was still 0-25 gC/m^2/month NPP type. This trend of regional NPP possibly implicates that the measures taken by the government, such as the limit of livestock density on the grasslands, played an important role.

Table 2 Transformation between NPP classes during 2000- 2005

Unit: %

2000 \ 2005	0-25	25-50	50-75	75-100	100-125	125-150
0-25	67.20	30.53	2.02	0.24	0	0
25-50	55.54	41.63	2.42	0.41	0	0
50-75	41.65	54.12	3.69	0.54	0	0
75-100	29.88	62.68	6.57	0.88	0	0
100-125	17.70	49.76	27.39	5.14	0	0
125-150	1.73	38.49	54.76	5.02	0	0

In 1991, NPP of Xilinhot region in 0-25, 25-50 and 50-75 (gC/m^2/month) level took a very small proportion (14.37%) so as to make the map color green featuring the conversion of the different NPP during 1991-2000 (Figure 3). Transformation of NPP were mainly consisted in the high-value areas (75-100,100-125, 125-150 gC/m^2/month) in 1991 to the low value areas (0-25, 25-50, 50-75 gC/m^2/month) in 2000. And this proportion accounted for 85.63% of total NPP in 1991. These low values of the transformation zone were mainly distributed in the northern and central section of the study area, and just because of this distribution of NPP, the local NPP in the study area were ordered into the pattern of decreasing trend from southeast to northwest. This regional distribution of NPP was basically consistent with both topographic elevation and regional pattern of rainfall because the distribution of the regional landform and rainfall were shown a decreasing trend in the study area from southeast to northwest. It was basically said that in the ten years, from 1991 to 2000, the NPP in most parts of the region tended to yield attenuation. In other words, from the 90's years of the 20th century to the actual beginning of the 21st century, Xilinhot region in the typical steppe zone was experiencing the overall deterioration of grassland ecosystems characterized by steppe production.

As for the diversion of different NPP from 2000 to 2005, the entire study area were fundamentally constituted by low NPP making the conversion map gray and yellow (Figure 3). The low value areas (0-25, 25-50, 50-75 gC/m^2/month) occupied 87.43% of the total NPP in 2000, mainly distributing in the northwest and north, and through the middle of Xilinhot, and then the high value areas of NPP (75-100, 100-125, 125-150 gC/m^2/month) were displayed only by 12.57% of the total NPP in 2000. Similarly, NPP of 2005 were mostly made up of low-value areas (0-25, 25-50, 50-75 gC/m²/month), 99.37% of the total NPP of 2005, mainly because the extreme lack of rain caused to the lower NPP over the entire region. However, it was clear that there are dark gray patches massively existing in the gray area of the north.   It was the result of NPP conversion which was occurred from the low grass yield (0-25, 25-50 gC/m²/month) of 2000 to the little low value of grassland (25-50, 50-75, 75-100 gC/m²/month) in 2005, contiguously existing in the north of the study area.   This means that the policy to reduce livestock and to protect grasslands made the eco-environment of Xilinhot region to a better situation, and gradually restored a good grassland ecosystems characterized by the higher NPP throughout the region.

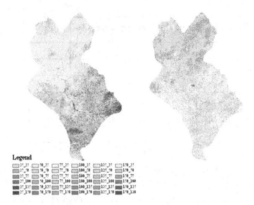

Fig. 3 Conversion between NPP types in different years (Left: 1991-2000; Right: 2000-2005)

4   Summary and Conclusion

This study used the improved CASA model to simulate regional NPP in the short grassland (typical steppe) region, a significant indicator for grassland degradation in China. The results indicated that the average regional NPP gradually decreased from 1991 to 2000, 2005 due to impacts of climatic change and the grazing manners. Image analysis showed that the ecosystem had been in poor condition over the past decades.   Continued degradation will be intense unless grazing activities are changed.   Remote sensing techniques are effective tools for detecting changes in the regional ecological conditions.   The case study demonstrates that it can synthesize the grassland information to commendably evaluate the grassland status at regional scale.

Acknowledgment. This study was supported by National Natural Science Foundation of China (Contract No. 31060078), by Natural Science Foundation of Inner Mongolia (Contract No. 200804040MS0514), and by the postdoctoral Research Fund of Inner Mongolia University (Contract No. Z20090141).

References

[1] Prieto-Blanco, A., North, P.R.J., Barnsley, M.J., Fox, N.: Satellite-driven modelling of Net Primary Productivity (NPP): Theoretical analysis. Remote Sensing of Environment 113(1), 137–147 (2009)

[2] Tong, C., Wu, J., Yong, S., Yang, J., Yong, W.: A landscape-scale assessment of steppe degradation in the Xilin River Basin, Inner Mongolia, China. Journal of Arid Environments 59(1), 133–149 (2004)

[3] Potter, C.S., Randerson, J.T., Field, C.B., Matson, P.A., Vitousek, P.M., Mooney, H.A., Klooster, S.A.: Terrestrial ecosystem production: a process model based on global satellite and surface data. Global Biochemical Cycle 7(4), 811–841 (1993)

[4] Li, S.Y., Li, X.B., Fu, N., Wang, D.D., Wang, H., Long, H.L.: Analysis on the Degradation Pattern of the Steppe Grassland on Different Slope in North China. In: IGARSS, pp. 433–3436 (2007)

[5] Zhu, W.Q., Pan, Y.Z., He, H., Yu, D.Y., Hu, H.B.: Simulation of maximum light use efficiency for some typical vegetation types in China. Chinese Science Bulletin 51(4), 457–463 (2006)

[6] Liu, Z.L., Wang, W., Hao, D.Y., Liang, C.Z.: Probes on the degeneration and recovery succession mechanisms of Inner Mongolia Steppe. Journal of Arid Land Resources and Environment 26(1), 84–94 (2002)

Examining the Effects of and Students' Perception toward the Simulation-Based Learning

Peng-chun Lin, Shu-ming Wang, and Hsin-ke Lu

Abstract. Teaching computer networking concepts is difficult for its abstract concepts and complex, dynamic process among various devices and protocols. To improve students' understanding, instructors seek the facilitation of technology. Among computer-assisted pedagogical methods, simulation-based learning (SBL) tool is regarded as a highly flexible and effective computer instruction application. While most previous studies mainly addressed the functionalities and design pattern of SBL tools, little attention has been devoted in examining what learner's perceptions of the SBL are and its relationships with learner's individual differences. However, teaching is an interactive process, how learners perceive the novel instruction tools be may further influence their usage of instruction tool and learning outcome subsequently. This study seeks to fill the literature gap by investigating students' perception toward the SBL and its relationships with students' learning outcome. Data were gathered through a survey conducted in two classes of a university in Northern Taiwan. Results suggest that simulation tool's appeals to student (positive perceptions) are highly associated with their learning outcome as well as sophisticated learning conception. Meanwhile, students with higher learning outcome tend to engage with simulation tool more and possess more cohesive learning conception than those with lower learning outcome do. Implications of results are also discussed in this study.

1 Introduction

As the world is getting more and more connected, computer networking concepts have become a fundamental part of modern information technology education. However, computer networking involves abstract, complex, and dynamic concepts

Peng-chun Lin
Graduate Institute of Information and Computer Education,
e-mail: pclin@sce.pccu.edu.tw

Shu-ming Wang · Hsin-ke Lu
Graduate Institute of Information Management, School of Continue Education, Chinese Culture University
e-mail: scottie.wang@gmail.com, hslu@sce.pccu.edu.tw

F.L. Gaol et al. (Eds.): Proc. of the 2011 2nd International Congress on CACS, AISC 144, pp. 349–354.
springerlink.com © Springer-Verlag Berlin Heidelberg 2012

and processes which, in turn, lead to the difficulty in learning effectively for students. In order to improve students' learning performance, many computer assisted pedagogical approaches have been introduced. Among them, simulation-based learning has been shown to be an effective approach in improving student's learning performance [1-2].

Packet Tracer (PT), introduced by CISCO Networking Academy, is known to be a powerful tool in assisting learning of computer networking concepts through simulation-based learning. Frezzo et al. [3] provided guidelines in applying PT in simulation-based learning and assessment. Previous study also demonstrated the effectiveness of using PT in achieving better learning outcomes [1]. However, while most previous studies mainly addressed the functionalities and design pattern of PT or simulation-based learning tools, little attention has been devoted in examining what learner's perceptions of the simulation-based learning are and its relationships with learner's individual differences.

Learning is an interactive process, whether interact with instructor, materials, or supporting tools. Thus, using innovative technology in teaching doesn't facilitate deep learning by itself. To achieve deep and effective learning, students have to form a positive perception toward the technology at first. Innovation adoption literature pointed out that the favorable beliefs toward the innovation are the influential factor of adoption intention [4-5]. In other words, the innovation must appeal to the potential adopters. Simulation learning tools, such as PT in this study, can be taken as an innovation for students. How students perceive PT is and its relationship with students' individual differences worth further investigation in order to achieve better learning outcome. Hence, to fill the literature gap, the purposes of this study are to investigate students' perception toward the simulation-based learning tool and its effect on students' learning.

Results from this study can serve as a guideline for instructor to design courses that incorporate PT as teaching support. Furthermore, this study also seeks to explore the effects of incorporating simulation-based learning tools in students learning outcome. Implication of this study could be a reference for teacher to decide how and to what extent to use simulation-based learning tool in their course based on their own specific needs and students' individual differences.

2 Literature Review

This study adopted Packet Tracer (PT), a software application of Cisco networking academy that support teaching computer networking skills and concepts, as a simulation-based learning tool. According to Frezzo et al. [6], incorporating PT into instruction provides structured, guided, and simulated real situations learning activities for students. Moreover, PT also visualizes the dynamic process of computer networking. Students are thus easier to gain a more comprehensive view of computer networks. In comparison with animated learning materials, PT provides flexibility for instructors to depict the complex and abstract concepts. Also, PT provides simulation of how actual networking devices work. For example, PT simulates the real time delay of a hub. To be specific, when you plug a wire to a hub, the time that one need to be waited for the network to be ready is exactly simulated by PT. Numerous studies have demonstrated the functionalities and effectiveness of PT [1, 3, 6].

To measure how student perceive simulation-based learning tool, this study adapted Wrzesien and Alcanniz Raya's instrument [7]. Six-dimensions, namely perceived usefulness, perceived educational value, enjoyment, intrinsic motivation, engagement, and intention to participate, were used to measure PT's appeal to student. PT's appeal to student can be defined as the extent of student's favorable beliefs toward the simulation learning tool which is Packet Tracer in this study.

Conceptions of learning (CoL) can be defined as "students conceive of what learning is and what its outcomes might be" [8, p. 268]. Tsai [9] investigated the possible differences of students' conceptions of learning between regular courses and web-based courses. His results suggested that sophisticated conceptions are more likely associated with web-based learning. This perspective suggested that varied learning environment may be associated with different conceptions of learning. Previous research has demonstrated different factors that comprise the CoL. Lin and Tsai (2009) depicted seven factors of conceptions of learning engineering (CoLE) and further divided these factors into two perspectives, namely the quantitative and qualitative view. In Ellis et al.[8]'s work, similar idea has been proposed. Ellis et al. suggest that the CoL can be divided into fragmented and cohesive view. The former involves the accumulation and use of pieces of knowledge while the later includes the deeper and more holistic conceptual change. In accordance with this view, this study thus categorized the conceptions of learning computer networking into fragmented and cohesive view.

3 Research Method

The survey was conducted in a university of northern Taiwan. Participants were students who attend the "Computer Network Engineering" course in two classes. There are forty-four students enrolled in the two classes collectively. Most of the students were major in Information Management. Both classes employ Packet Tracer as primary instruction supporting tool. Specifically, each computer networking concepts were firstly lectured by instructor. Next, a corresponding Packet Tracer simulation session or tasks were assigned to students as practical exercises.

The questionnaire used in this study is comprised of four major parts. Measurements were primarily adapted from previously validated instruments. The first part collects the demographic information of participants. The second part measures students' conceptions of learning computer networking (CoLCN) using 23 items adapted from Lin and Tsai 's work (2009).

The third part uses 18 items from Wrzesien and Alcanniz Raya (2010) to gather students' perceptions toward the simulation-based learning tool. The last part, six items were used for students to self-evaluate his/her learning performance. All the survey items were jointly checked by two experts with more than five years of computer and IT teaching experience to ensure the content validity. The questionnaire was distributed to students at the end of semester (18 weeks course).

Participants were told that the questionnaire is totally irrelevant to the overall evaluation and were asked to response freely based on their own opinions.

4 Data Analysis and Result

A total of thirty-seven valid responses were collected. Among them, 75% (or 28) were male. Most of the respondents were in age between 21-25 (43%), followed by those who are of age between 26-30 (32%).

Due to the limited number of sample and the pilot phase that this study is in, some advanced statistical analysis, such as factor analysis or regression, might not be applicable. This study thus used Pearson correlation and t-test to explore the associations and differences between research variables. SPSS for Windows 17.0 is used to perform the analysis for the data collected.

To understand the associations among PT's appeal to students, CoLCN, and learning outcomes, Pearson correlation analysis was conducted.

First of all, for PT's appeal to student and learning outcome, results indicated that student's learning outcome is highly associated with each of six dimensions of appeal to student. (correlation coefficient ranging from 0.426 to 0.570, all significant at level of p-value < 0.01)[1]. Secondly, for students' CoLCN and learning outcome, this study found that only cohesive conceptions is positively associate with learning outcome ($r=0.334$, $p < 0.05$). Lastly, results showed that cohesive conception is highly correlated with engagement of PT ($r=0.587$, $p < 0.01$) while fragmented conception is not.

Further analysis was conducted using t-test to explore the differences of PT's appeal to student and CoLCN between students of higher and lower learning outcome. Based on the average of six items of self-evaluation, students whose mean score is greater than 3 are labeled as "High" (n=25) in learning outcome while the rest are labeled as "Low" (n=12).

For PT's appeal to student, results showed that there are significant differences in all six sub-dimensions (all at level of $p < 0.05$). Students with higher learning outcome generally exhibit higher average score in the six dimensions of PT's appeal. The difference between students' different CoLCN is further explored. This study found no difference in fragmented conception between students of high and low learning outcome (mean difference = 0.11, $p > 0.05$). However, result of this study provided support for the difference of cohesive conceptions between students of different level of learning outcome (mean difference = 0.3833, $p < 0.1$). To be specific, students with higher learning outcome possessed stronger cohesive conceptions than those with lower learning outcome do. Interpretation of results is presented in discussion and conclusion section.

---

[1] Due the limited space of the manuscript, detailed correlation table and results of t-test are available upon request.

5   Discussion and Conclusion

In order to better understand how students perceive the simulation-based learning tool, this study adopted a well-known computer networking simulation learning tool, i.e. Packet Tracer, to explore students' perceptions.

Results showed that PT's appeal to student is highly correlated with students' learning outcome. This finding is in accordance with previous innovation adoption literature. The higher the appeal of innovation that one perceives, the more likely one will engage in using the innovation which may result in better learning outcome. For conception of learning computer networking (CoLCN), this study found that cohesive conception is associated with better learning outcome. This is also consistent with previous CoL study. Ellis et al. [8] found that students with cohesive conceptions tend to adopt deep learning approaches such as seeking deeper and holistic understanding of subject to learn. Hence, they may achieve better learning outcome. Result of this study also provided further support for this view. That is, student's extent of engagement with PT is positively associated with cohesive conception.

This study also revealed the differences in perception toward the simulation-based learning tool and CoLCN between students with higher and lower learning outcomes. Aligned with views of learning approach and adoption literature, students with better learning outcome generally showed higher perception of PT's appeal. This result suggested that the benefits of PT, such as usefulness, educational value and enjoyment, are more highly valued by students who learn better than those who don't. The differences in CoLCN between students of different level of learning outcome are further examined. For fragmented conception, this study found no difference between students of higher and lower learning outcome. However, there is difference in cohesive conception between students of different level of learning outcome. Specifically, students with higher learning outcome also score higher in cohesive conception than that of those with lower learning outcome. The deep learning approach that accompany with cohesive conception might be the reason for this result.

In conclusion, this study found that students with better learning outcome tend to be highly appealed by the simulation-based learning tool, i.e. Packet Tracer in this study. Meanwhile, in the context of simulation-based learning, students with higher learning outcome tend to possess cohesive conception of learning. Moreover, students with more cohesive conceptions are more likely to have deeper engagement with PT. As simulation-based learning is known to be effective in facilitating active learning, this study also found similar result with previous study[1]. Furthermore, previous research indicated that simulation-based learning provides support for student to learn highly dynamic and complex subjects such as computer networking concepts in this study[1-2]. Tsai (2009) also provided support for the relationship between sophisticated conception and computer-mediated learning environment, i.e. web-based learning. Results from this study suggested that simulation-based learning could be taken as a deep learning approach for students with cohesive concepts of learning. Since students with cohesive view might engage more deeply and interact more frequently with PT to seek deeper and better understanding. This approach could further in turn lead to better learning outcome.

The major limitation of this study apparently is the sample size. However, as exploratory in nature and the pilot phase this study is in, preliminary results of this study could serve as a guideline for following research. The second limitation is the development of instruments used in this research. Nonetheless, instruments in this study were primarily adapted from previously validated measurement, two experts who major in computer science and have plenty years of experience in teaching computer and information technology courses were joined in the adaption of instruments. Thus, the content validity can be basically ensured. Lastly, for better clarifying the causal effect among appeal to student, CoLCN, and learning outcome, a more elaborated experiment design is needed.

References

1. Goldstein, C., et al.: Using a network simulation tool to engage students in active learning enhances their understanding of complex data communications concepts. In: Proceedings of the 7th Australasian Conference on Computing education, vol. 42, pp. 223–228. Australian Computer Society, Inc., Newcastle (2005)
2. Yehezkel, C., Eliahu, M., Ronen, M.: Easy CPU: Simulation-based Learning of Computer Architecture at the Introductory Level. International Journal of Engineering Education 25(2), 228–238 (2009)
3. Frezzo, D., Behrens, J., Mislevy, R.: Design Patterns for Learning and Assessment: Facilitating the Introduction of a Complex Simulation-Based Learning Environment into a Community of Instructors. Journal of Science Education and Technology 19(2), 105–114 (2010)
4. Chin, W.W., Gopal, A.: Adoption intention in GSS: relative importance of beliefs. In: DATA BASE Advances, vol. 26(2-3), pp. 42–64 (1995)
5. Wang, W.-T., Wang, C.-C.: An empirical study of instructor adoption of web-based learning systems. Computers & Education 53(3), 761–774 (2009)
6. Frezzo, D.C., et al.: Psychometric and Evidentiary Approaches to Simulation Assessment in Packet Tracer Software. In: 2009 Fifth International Conference on Networking and Services, Valencia, Spain (2009)
7. Wrzesien, M., Alcañiz Raya, M.: Learning in serious virtual worlds: Evaluation of learning effectiveness and appeal to students in the E-Junior project. Computers & Education 55(1), 178–187 (2010)
8. Ellis, R.A., et al.: Engineering students' conceptions of and approaches to learning through discussions in face-to-face and online contexts. Learning and Instruction 18(3), 267–282 (2008)
9. Tsai, C.-C.: Conceptions of learning versus conceptions of web-based learning: The differences revealed by college students. Computers & Education 53(4), 1092–1103 (2009)

Automated User Analysis with User Input Log

Jae Min Kim and Sung Woo Chung

Abstract. Many studies are on progress in the field of digital forensics. However, most analysis methods lack from complexity as the size of data to be investigated enlarges. Thus, automated ways of analyzing the data is required to reduce the work done by the analysts. In our study, we propose an automated user analysis method that works based on the user input log. From the automated analysis, we provide priority on the further user classification, which helps reduce the total number of potential user to 21% of the total users, even in the worst case. In average cases, the exact matching user is found within the 10.5% highest priority users. By combining our proposed method with other existing methods, it would be possible to further reduce the complexity of jobs need to be done by the analysts.

1 Introduction

An ideal way of facing the digital crimes would be, of course, preventing the crimes in prior with security solutions. Many crimes, indeed, are prevented with the effort of many researchers. Unfortunately, some of the crimes are not found in prior, resulting in over millions of dollars of damage. Thus, digital forensics is being highlighted. There are many different types of footprints we can find from the computer usage logs. Some make attempts on analyzing the memory dump or digital files to figure out the type of job that has run on a certain time [2][3][4]. However, a binary file would usually contain multiple types of data, it cannot be one-piece solution. Thus, we need to extract more information to support the analysis. In [1], hard disks analysis methods are proposed to provide quick classification of the seized computer user.

Yet, not many studies focus on classifying the user by analyzing their input pattern. In our study, we provide a user analysis method that is based on user input speed. From our study, we find that there is high probability that a same user would make similar usage log in terms of input speed. Although analyzing the input speed cannot provide the direct information of the user, it may provide priority for the analysts. Furthermore, as prior works are not complete solutions

Jae Min Kim · Sung Woo Chung
Korea University, Anamdong, Sungbukku, Seoul, South Korea
e-mail: joist@korea.ac.kr

F.L. Gaol et al. (Eds.): Proc. of the 2011 2nd International Congress on CACS, AISC 144, pp. 355–360.
springerlink.com

themselves, combining different solutions together can also result in more accurate analysis.

2 User Input Speed Analysis

2.1 Input Per Minute

In order to analysis user input speed, the exact method should be defined in prior. In our study, we use the term IPM (input per minute) to analyze user input speed. IPM is the total number of user inputs (including keyboard inputs and mouse clicks) made in each minute. Fig. 1 shows the IPM log of a user during 76 minutes usage. Thus, there are 76 IPM values, each representing number of inputs per single minute of usage. The user made around 80 inputs per minutes on the time 10 to 30 (minute), and nearly 180 input at the peak usage time (in the beginning). The user did not make any input on time from 5~10 and 36~68 (minute), thus the IPM value around corresponding time is 0. Throughout our study, we use the IPM log of users, which contains 76 minutes usage each. In total, we analyze 20 IPM logs, which are collected from 10 different users (2 IPM logs per each person).

2.2 Analytic Techniques

To analysis the IPM logs, we considered many different statistical tests as a basis, but applied four for our initial study: Arithmetic Mean, Standard Deviation, Arithmetic Mean (valid), and Standard Deviation (valid). The prior two basis (arithmetic mean and standard deviation) are obtained from the entire usage time, including the period where the user did not make any input. On the other hand, the latter two bases are obtained from the usage time with user input; in other words, IPM is 0. For instance, in Fig. 1, as there is no user input between 5:00~10:00 and 36:00~68:00. Since the latter basis exclude these time duration, the arithmetic mean (valid) is the average user input made per minute during 0:00~5:00, 10:00~36:00, and 68:00~76:00. As the arithmetic mean (valid) does not add 0s for calculation, it is always higher or same compared to the arithmetic mean. On the other hand, the standard deviation (valid) is lower than the standard deviation in most cases, for the same reason. However, there are some exceptional cases where the standard deviation (valid) is higher than the standard deviation. It occurs when the arithmetic mean itself is very low, and taking the '0's into calculation can lower the standard deviation. The values of four bases for the user in Fig. 1 are;

Arithmetic Mean : 44.51 (0.29)
Standard Deviation : 49.76 (0.71)
Arithmetic Mean (valid) : 86.74 (0.52)
Standard Deviation (valid) : 40.61 (0.71)

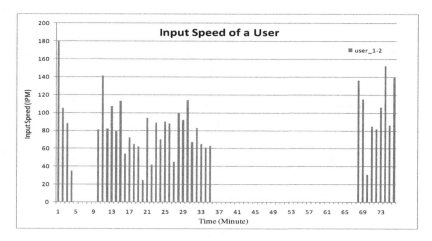

Fig. 1 An Example of User Input Speed (IPM)

To efficiently evaluate the four bases with the same weight, we normalized each range, 0~1. For each basis, the user with highest value is turned to value 1, and others are reduced with the same ratio. Above values in the parenthesis are the transformed value (normalized) used in our evaluation.

Though we use these four bases to analysis the user in our initial study, we do not guarantee that those are the ideal possible set. However, these four bases are relatively simple and showed fairly good results in our evaluation. Thus, we select these four bases in our study.

2.3 User Analysis

Table 1 shows the normalized statistic test values of each user IPM logs. Ideally, each of the users should have the same values from each of the statistic tests in the first usage and the second usage. However, the values are not exactly the same in most cases. For instance, we obtain (0.4096, 0.3773, 0.4319, 0.5121) in user_1's first usage, while the values are (0.2875, 0.5160, 0.7090, 0.7150) in his second usage. Thus, when we are given statistical values from two different usages, we need a method to determine how similar those two usages are. In our study,

Table 1 Static Test Values of User IPM Logs

	1st usage										2nd usage									
	user_1	user_2	user_3	user_4	user_5	user_6	user_7	user_8	user_9	user_10	user_1	user_2	user_3	user_4	user_5	user_6	user_7	user_8	user_9	user_10
Arithmetic Mean	0.4096	0.1241	0.3466	0.4229	0.0984	0.6639	0.9768	0.2395	1.0000	0.8340	0.2875	0.2143	0.3195	0.5739	0.0744	0.7836	0.9476	0.1436	0.9644	0.6412
Arithmetic Mean (Valid)	0.3773	0.1639	0.3978	0.6884	0.1325	0.6639	0.9154	0.3644	1.0000	0.7682	0.5160	0.2632	0.6391	0.7439	0.3474	0.7455	0.8844	0.4570	0.9064	0.7124
Standard Deviation	0.4319	0.5273	0.8019	0.9928	0.5043	0.8845	0.6067	0.6503	0.8874	0.8138	0.7090	0.7249	0.9585	1.0000	0.5124	0.8432	0.5539	0.6145	0.6064	0.9405
Standard Deviation (Valid)	0.5121	0.8755	0.8482	0.9180	0.7136	0.9763	0.7342	0.8302	0.8807	0.7265	0.7150	0.8287	0.8842	0.9264	1.0000	0.9127	0.6719	0.8224	0.8912	0.9330

we measure the Euclidean distance (from 4 dimensional space) in order to determine how similar each usage logs are. For instance, when we measure the distance between the first and second usage of user_1, we calculate the distance (D) as following:

D(first(user_1), second(user_2))
= sqrt((0.4096-0.2875)2 + (0.3773-0.5160)2 +
(0.4319-0.7090)2 + (0.5121-0.7150)2)
= 0.3909

In our paper, we call this calculated distance as the "usage similarity distance".

3 Evaluation

In this Section, we evaluate how accurately the usage similarity distance can classify the user. For the first step, we calculate usage similarity distance between every two IPM logs, which is shown in Table 2. We have 190 usage similarity distances, from every single usage logs to the other usage logs ($n2/2$, $n=20$). Each value in a cell (except the values on the "Rank" row) represents usage similarity distance. For instance, the value in the cell outlined with red color represents the usage similarity distance between the first usage of user_1 and the second usage of user_1, and the value is 0.3909 as we calculated in the previous Section.

To determine how meaningful this metric is, we assume the situation where we have 19 user logs which we already know whose usage it is. Then, we are given a new usage log and we need to figure out to which user the new usage log is collected from by analyzing the usage logs. In each column of Table 2, user_1 through user_10 are the "new usage log", respectively. Then, we calculate the

Table 2 Usage Similarity Distance

| | | 1st usage | | | | | | | | | | Average |
		user_1	user_2	user_3	user_4	user_5	user_6	user_7	user_8	user_9	user_10	
1st usage	user_1	0	0.5286	0.5057	0.7602	0.4511	0.7512	0.835	0.4244	1.0397	0.722	
	user_2	0.5286	0	0.5636	0.7643	0.1764	0.821	1.1602	0.267	1.2687	0.9885	
	user_3	0.5057	0.5636	0	0.2929	0.6237	0.2846	0.6708	0.313	0.7184	0.4483	
	user_4	0.7602	0.7643	0.2929	0	0.8368	0.2724	0.7396	0.5148	0.6664	0.493	
	user_5	0.4511	0.1764	0.6237	0.8368	0	0.901	1.1879	0.3308	1.3221	1.0193	
	user_6	0.7512	0.821	0.2846	0.2724	0.901	0	0.5526	0.5846	0.4912	0.3292	
	user_7	0.835	1.1602	0.6708	0.7396	1.1879	0.5526	0	0.9342	0.3282	0.2915	
	user_8	0.4244	0.267	0.313	0.5148	0.3308	0.5846	0.9342	0	1.0234	0.7428	
	user_9	1.0397	1.2687	0.7184	0.6664	1.3221	0.4912	0.3282	1.0234	0	0.3327	
	user_10	0.722	0.9885	0.4483	0.493	1.0193	0.3292	0.2915	0.7428	0.3327	0	
2nd usage	user_1	0.3904	0.4688	0.2435	0.4102	0.482	0.5082	0.8041	0.2147	0.8921	0.6037	
	user_2	0.4908	0.2381	0.3787	0.5568	0.303	0.6408	1.025	0.1257	1.0969	0.814	
	user_3	0.7025	0.6748	0.2775	0.1225	0.7409	0.3639	0.8118	0.4283	0.7726	0.5675	
	user_4	0.8102	0.8731	0.3403	0.1587	0.9441	0.1772	0.6266	0.6211	0.5168	0.3803	
	user_5	0.6025	0.2311	0.5358	0.6891	0.3654	0.7626	1.1117	0.2784	1.2026	0.9575	
	user_6	0.7758	0.9385	0.3891	0.3947	1.0012	0.1703	0.3927	0.6973	0.3381	0.1903	
	user_7	0.7626	1.1188	0.6567	0.757	1.1354	0.5809	0.0985	0.9001	0.4202	0.3126	
	user_8	0.4559	0.3169	0.3817	0.5345	0.367	0.6395	0.9619	0.1473	1.0551	0.7876	
	user_9	0.8719	1.129	0.6401	0.6966	1.18	0.4832	0.1616	0.9089	0.297	0.3195	
	user_10	0.7735	0.8623	0.3114	0.2267	0.9327	0.0949	0.5555	0.6137	0.4677	0.3108	
Rank		1	3	2	2	4	2	1	3	1	3	2.2

value, Rank of each user's second usage, which is shown in the bottom row. Rank represents the index of the distance when sorted in ascending order. Thus, the best Rank value for a user's second usage log is, one.

For instance, in Table 2, Rank of the user_1 column represents the usage similarity distance between user_1's first and second IPM logs. The value is 1 since 0.3904 is the smallest distance among the column (except for 0, which is the distance between the same IPM logs). In this case, we can exactly find out that the new usage log belongs to user_1 in the first attempt. Similarly, the Rank of user_2 column is 3 since 0.2381 is the third smallest among that column (after 0.1764 and 0.2311). Three out of ten users has Rank 1, and the average Rank is 2.2. Even in the worst case, the Rank is 4 (user_5).

We also achieve meaningful result by looking at the absolute distance (instead of rank). The average distance between the two logs of the same user is only 0.24, and still less than 0.4 even for the worst case (user_1). On the other hand, if we randomly choose two logs and calculate the distance, the average comes out to be 0.61.

Although the evaluation is done with limited number of usage logs, we were able to find the user after examining 21% (4 logs) of the potential IPM logs, even in the worst case. In average case, we only need to examine 11.5% (around 2 to 3 logs). The absolute distance also stays very low when the logs are from the same user. Thus, by applying our method, it can significantly reduce the time required by the analysts to determine the user of a certain usage.

4   Conclusions

In our paper, we proposed an automated user analysis method that works based on the user input speed (IPM). After applying four different statistical tests, we calculate the usage similarity distance between every two usages using the Euclidean distance. The results show that the usages from the same user are most likely to have the shortest usage similarity distance and at least the fourth shortest similar distance, even in the worst case (21%). By applying this analysis method in prior to a précised analysis method done by the analysts, it would help reduce the job and required time significantly. Our study is independent to the prior works, and can be used together to further improve the accuracy.

Acknowledgments. This research was supported by the project of Global Ph. D. Fellowship which National Research Foundation of Korea conducts from 2011 under contract NRF-2011-0007278, the Basic Science Research Program of the National Research Foundation of Korea, funded by the Ministry of Education, Science and Technology under contract NRF-2011-0004917, the Ministry of Knowledge Economy, Korea, Information Technology Research Center support program, supervised by the National IT Industry Promotion Agency under contract NIPA-2011-C1090-1121-0010, and the Brain Korea 21 Project in 2011.

References

1. Grillo, A., Lentini, A., Me, G., Ottoni, M.: Fast User Classifying to Establish Forensic Analysis Priorities. In: The Fifth International Conference on IT Security Incident Management and IT Forensics (2009)
2. Conti, G., Dean, E., Sinda, M., Sangster, B.: Visual Reverse Engineering of Binary and Data Files. In: Workshop on Visualization for Computer Security (2008)
3. Conti, G., Bratus, S., Shubina, A., Sangster, B., Ragsdale, R., Supan, M., Lichtenberg, A., Perez-Alemany, R.: Automated Mapping of Large Binary Objects Using Primitive Fragment Type Classification. In: The Proceeding of Tenth Annual DFRWS Conference on Digital Investigation, August 2010, vol. 7(suppl. 1), pp. S3–S12 (2010)
4. Calhoun, W.C., Coles, D.: Predicting the Types of File Fragments. In: The Proceeding of the Eigth Annual DFRWS Conference on Digital Investigation, September 2008, vol. 20(suppl. 1), pp. S14–S20 (2008)

Developing an Online Publication Collaborating among Students in Different Disciplines

Kirsi Silius, Anne-Maritta Tervakari, Meri Kailanto, Jukka Huhtamäki, Jarno Marttila, Teemo Tebest, and Thumas Miilumäki

Abstract. In order to offer opportunities to practice collaboration skills to students of hypermedia engineering and students of journalism and mass communication, a cross-university course "Developing an Online Publication for Journalism" based on simulations of real-world design projects was organised in spring 2011. According the results, the simulations motivated both hypermedia and journalism students to learn and to develop their professional competencies and collaboration skills. A web service design project with special working methods was unfamiliar to the journalism students, who were dissatisfied with the division of work tasks and asked for more instructions and guidance.

1 Introduction

Design and production of web-based services is increasingly carried out on a project basis in collaboration with various stakeholders. An ability to work together with people with different competencies is increasingly required. Therefore, students need opportunities to develop their interaction skills throughout their education. A richer variety of opportunities to develop such skills can be offered to the students by anchoring traditional lectures into authentic real-world situations. Activities and problems drawn from the real-world professional practices can promote e.g. active participation, problem solving and collaborative interaction [1].

In order to promote students' collaboration and communication skills by simulating real-world design situations, the Hypermedia Laboratory of Tampere University of Technology (TUT) organised a cross-university course called "Developing an Online Publication for Journalism" in spring 2011 in co-operation with the School of Communication, Media and Theatre of University of Tampere (UTA). Students from different disciplines can be said to belong to different

Kirsi Silius · Anne-Maritta Tervakari · Meri Kailanto · Jukka Huhtamäki · Jarno Marttila ·
Teemo Tebest · Thumas Miilumäki
Tampere University of Technology, Korkeakoulunkatu 10, 33720 Tampere, Finland
e-mail: {kirsi.silius,anne.tervakari,meri.kailanto,
      jukka.huhtamaki,jarno.marttila,teemo.tebest,
      thumas.miilumaki}@tut.fi

F.L. Gaol et al. (Eds.): Proc. of the 2011 2nd International Congress on CACS, AISC 144, pp. 361–367.
springerlink.com   © Springer-Verlag Berlin Heidelberg 2012

academic tribes according to what they study. This may cause challenges in communication and collaboration between students from different disciplines [2, 3].

In this research we aimed at finding out whether students from two different tribes (hypermedia engineering and journalism and mass communication) were able to collaborate and if so, at which level and how they experienced it. During the course students both met face-to-face and worked in a socially enhanced learning environment. We were also interested in how they experienced the use of social media in a studying context.

2 Academic Tribes

Different disciplines at universities can be categorised into different academic tribes [2]. Disciplines favour and require different learning styles which students adopt along with their studies. Some of the needed learning skills are assured through the selection of students in entrance exams, but teaching methods used during their studies are said to have the greatest effect. Kolb-Biglan's classification divides academic knowledge into a typology of disciplines: abstract - reflective (hard pure), abstract - active (hard applied), concrete - active (soft applied) and concrete - reflective (soft pure). Kolb also found that the main learning styles go hand-in-hand with this division. Respectively the learning styles are assimilators, convergers, accommodators and divergers [3].

The discipline and learning styles of hypermedia students in this research are closest to engineering students in the hard-applied, convergent quadrant and more specifically to computer science as they have a strong background in it. However, hypermedia as a major includes softer issues as well, so hypermedia students' learning styles and orientation could be a little softer than with engineers in general. Journalism students could be equated with the communication discipline in the soft-applied quadrant and the faculty from the journalism field in soft-pure quadrant. Common to hypermedia and journalism students is that at university level they mostly create applied solutions, but theory lies behind their decisions. The classification is done based on the division in [3].

3 Teaching Experiment

There were 11 students of journalism from UTA and 5 students of hypermedia from TUT, who worked together during the course with the aim of designing and implementing a journalistic online publication using Drupal, an open source content management framework. During the course, a completely new, modern online publication named "Intro" was created with the purpose of substituting for the website of the earlier "Utain", a weekly publication run by the students of journalism at UTA.

The instructional design of the course was based on the idea of simulating professional practices as far as possible. During the course, the students were given the possibility of experiencing a real-life development project, to learn how to communicate and create ideas in a multidisciplinary environment similar to real-world design situations. In order to promote collaboration among the student groups, eleven workshops with activating group exercises and stimulating lectures given by well-known experts were organised weekly. For supporting students' teamwork outside the workshops, TUT Circle, a social media web service created by the Hypermedia Laboratory at TUT, was introduced as the primary learning environment. TUT Circle is also built with Drupal and thus introduced a chance for the students to familiarise themselves with the possibilities of the technology with which "Intro" was going to be built.

TUT Circle allows students, among other things, to do private messaging, contribute content, exchange opinions and create communities for different needs. In community groups, students can, for example, collaboratively chat, write news, manage events, write blogs, posts and edit wiki pages and share resources (files, images, videos etc.) [4]. On the course, TUT Circle was used appropriately for needs that came about. Firstly, all the important documents were stored to TUT Circle as stories or wiki pages depending on how and by whom the information was maintained. For example, a vision document that defined the aim of the course was stored as a wiki page so that all the students were able to edit it.

4 Data Collection, Analysis and Results

Data for the study were collected in parts from two different sources. Firstly, the students who participated in the course responded to a survey. Secondly, log data collected from TUT Circle were analysed and visualised for verifying online interactions among students.

4.1 Survey of the Students' Experiences in Co-operation among Disciplines within TUT Circle

The students completed a web survey with open-ended questions on such themes as "TUT Circle as a learning environment" and "Collaboration among disciplines". Eight students out of 16 responded to the survey. Four respondents were from UTA and the other four from TUT. It can be assumed that the students who responded to the survey were those who had the most positive or most negative learning experiences. Although the total number of respondents was quite low, the study highlighted whether the students' experiences of the teaching experiment were mainly negative or positive. The study also revealed students' conceptions of collaboration in a multidisciplinary environment.

The qualitative data collected by the survey were analysed by using content analysis. An aim of the analysis was to examine meanings, themes and patterns that were manifest or latent in students' answers, and to identify similarities and differences among the answers. The researchers read the material through and coded the data separately (researcher triangulation), and afterwards the classifications were compared and the results formalised. According to the respondents, TUT Circle supported collaboration and interaction among students by offering useful tools for document delivery, resource sharing, collaborative writing and instant messaging. The respondents also made suggestions for the development of TUT Circle, mainly for improving usability.

Almost every respondent had a positive experience of collaboration. Only one student of journalism found the collaboration among disciplines ineffective and saw that the hypermedia students were working in a separate group from the journalism students, though (s)he considered that lunchtime discussions with hypermedia students were useful. However, all the respondents, whatever their academic background, agreed that collaboration enhanced their learning.

> It is essential for journalists to understand technological principles and learn to work as a team with people having different ways of thinking. (journalism student 5)

The journalism students seemed to have different experiences in collaboration from the hypermedia students. In contrast to the hypermedia students, all the journalism students thought that the division of work was uneven and partly unfair, causing a greater workload to the hypermedia students. Some journalism students seemed not to fully understand basic ideas behind the division of the work tasks during the design process.

> "Hour after hour hypermedia students had to listen to the journalism students' opinions about the Utain journal, electronic version of it and all the things which are wrong in it. ... It was also curious that when the layout of the electronic journal was designed, only students of journalism were asked to make suggestions for it. Isn't it the case that better suggestions might be made by students of hypermedia because they were those who implemented a site technically?... Now it felt like the students of journalism said what to do and the students of hypermedia just did it." (journalism student 6)

Almost every respondent thought that a lack of strict decisions and a clear project plan at the beginning of the design project caused problems, uncertainty and even frustration among the students, and that had a negative influence on collaboration. In particular the journalism students were critical. They would have needed more support, more information, more guidance and more controlled project management.

> I wanted more concrete guidance to this course and study results about which kind of electronic journal should be produced... Discussing about same issues time after time without making a decision frustrates. (journalism student 4)

Some of the respondents (in both disciplines) mentioned, that there could have been more collaboration among the students from different disciplines.

> The collaboration could have been worthwhile if everybody had taken part and if, for example, creation had been done in small mixed groups right from the beginning. (journalism student 6)

4.2 Visualisation of Students' Activity within TUT Circle

The log data collected from TUT Circle was analysed utilising social network analysis (SNA), which studies the social structures of actors. SNA and graph theory-based models and metrics can be used to measure and visualise network-like phenomena [5]. Visualisation of social configurations "allows investigators to gain new insights into the patterning of social connections, and it helps investigators to communicate their results to others" [6]. In this case, the modelled network is two-mode user-to-content type of directed networks. Connections between the nodes are established when a student or teacher creates or modifies content.

The evolution of the user-content network is presented in Fig. 1. Students and teachers from different universities are divided by different nodal colours. Light blue is used for people from TUT and an orange colour represents people from UTA. Gray nodes represent content which has been created either by students or teachers. Nodes have been enlarged by nodal degree. Edge size is determined by the amount of edits a user has made to a content node.

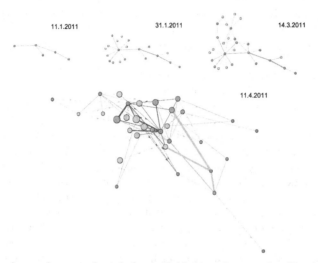

Fig. 1 In the picture, the network evolution is divided into four samples. The three samples at the top of the image represent the evolution of network topology and the collaboration of students by co-creation of content. The bigger image represents the final status of the course including co-creation of content.

Student collaboration starts slowly and most of the network activity - co-creation of content - happens towards to the end of the course. On January 11th only UTA and TUT teachers have contributed content. On January 31st it can be noticed that both journalism and hypermedia students start to collaborate on a certain single topic. In March, more topics are introduced and participants start to separate into groups consisting of students from both disciplines and to groups that only have contributions of students from one discipline.

In the bigger picture, co-creation of content and reading of content produced by others is visualised. Light green edges represent contributions to the content and dark green edges represent reading of the content. Width of the edge is proportional to the amount of contributions or readings. This visualisation shows, in comparison to network evaluation visualisation, that even though students seem to divide into their own groups they still do read content produced by other people and especially from other disciplines. It can be noticed that hypermedia students read more content produced or co-created by journalism students than vice versa.

5 Discussion and Conclusions

During the course the students were given the possibility of experiencing a real-life development project, to learn how to communicate and create ideas in a multidisciplinary environment similar to real-world design situations. The journalism students were exposed to platform and concept development in a web context whereas the hypermedia students got the chance to work together closely with the stakeholders. As in a real web design process, the design process during the course was not linear, brief or even very well-defined. The requirement phase alone generated numerous problems. Interviews of potential users, analysis of related web-services and the requirements represented by stakeholders produced plenty of information that needed to be taken into account in the design and development activities. The student groups needed to self-organise, communicate, collaborate and coordinate their purpose to solve unstructured problems and to integrate individual efforts for producing collective outcomes [1].

For supporting collaboration and interaction, TUT Circle, a social media service, was used as the learning environment. The students in both disciplines adopted TUT Circle easily and thought that TUT Circle supported the task by offering useful tools and facilities for collaboration. The students also succeeded in achieving the main target of the design course; a beta version of the Intro online publication using Drupal framework was implemented. Both hypermedia and journalism students mentioned that simulations of a real-world design practice motivated them to learn and to develop their professional competencies.

Both hypermedia and journalism students had positive experiences of collaboration, and they both agreed that collaboration enhanced their learning. The hypermedia students mentioned that the course offered them a good opportunity to practice real-world-like web design. For the journalism students it was important to develop a better understanding of technology and practice collaboration with people from a different educational background. However, there could have been more opportunities for collaboration and support for grouping. The cross-discipline collaboration started slowly (see Fig. 2) during the first weeks of the course both hypermedia and journalism students focused on collaborating with individuals with the same background. The cross-discipline collaboration happened only when all the students were asked to contribute to the vision document. However, the cross-discipline collaboration increased towards to the end of the course.

During the whole course the journalism students criticised the management of the project, the division of the work tasks and the working methods noticeably more than the hypermedia students. The journalism students seemed not to fully understand the purpose of collaboration and the division of the work tasks during the design process. However, that was understandable. Hypermedia students have studied the design of web services and the management of web service projects; therefore they know the principles and working methods of user-centred design. That is why the hypermedia students, for example, more often read the content produced by the journalist students than vice versa, see Fig.2. For developing requirement specifications they needed to know, what their customers (journalism students) required. Web service design process was unfamiliar to the journalism students. That might be the reason for their dissatisfaction with work tasks and the the workflow, and also for their demand for instructions and guidance. Further research is needed to better understand how to support cross-discipline collaboration within social media-enhanced learning environments in a university context.

During the analysis, other TUT Circle users (not attending the course) were also interested in the activities and the documents. In the future, we are interested in investigating the patterns related to the cross-course interest that the students and other actors show for course material and activity in a socially enhanced learning environment.

References

1. Carrol, J.M., Rosson, M.B.: Towards Even More Authentic Case-Based Learning. Education Technology 45, 5–11 (2005)
2. Becher, T., Trowler, P.R.: Academic Tribes and Territories: Intellectual enquiry and the culture of disciplines, 2nd edn. Open University Press (2001)
3. Kolb, D.A.: Learning Styles and Disciplinary Differences. In: Chickering, A. (ed.) The Modern American College. Jossey-Bass Publishers, San Francisco (1981)
4. Silius, K., Miilumäki, T., Huhtamäki, J., et al.: Students' Motivation for Social Media Enhanced Studying and Learning. Knowl. Management & E-Learning: An International J. The Special Issue on Technology Enhanced Learning 2(1), 51–67 (2010)
5. Wasserman, S., Faust, K.: Social Network Analysis Methods and Applications. Cambridge University Press, New York (1994)
6. Freeman, L.C.: Methods for Social Network Visualization. In: Meyers, R.A. (ed.) Encyclopedia of Complexity and Systems Science. Springer, Berlin (2009)

Programming of Hypermedia
Course Implementation in Social Media

Kirsi Silius, Anne-Maritta Tervakari, Jukka Huhtamäki, Teemo Tebest,
Jarno Marttila, Meri Kailanto, and Thumas Miilumäki

Abstract. In order to promote students' peer learning skills and to motivate them to learn, the Hypermedia Laboratory at the Tampere University of Technology (TUT) organised a course entitled Programming of Hypermedia, which attempts to solve authentic, real-world problems. TUT Circle, a social media web service, was used as the web-based learning environment for the course. It has most of the basic functions that are common in modern social media and Web 2.0 services. The research focused in finding out, first, if the teaching experiment with TUT Circle was pedagogically usable and, second, if students voluntarily utilised peer learning to learn with and from each other without immediate teacher intervention. Both a survey and network analysis of the log data support the fact that TUT Circle supported peer learning and does not have major defects from a pedagogical usability viewpoint.

1 Introduction

According to Carroll and Rosson [1] students can be motivated and supported to learn by offering them interesting and challenging real-world problems to solve. In order to make students' learning relevant to real-life experiences, learning environments must be authentic. Authentic learning is a process of interacting with the real world and re-analysing and reinterpreting new information and its relation to the real world problem [2, 3]. Authentic learning allows students to explore, discuss as peer learners and meaningfully construct concepts and relationships together in real-world contexts that are relevant to their learning processes [4].

Kirsi Silius · Anne-Maritta Tervakari · Meri Kailanto · Jukka Huhtamäki · Jarno Marttila ·
Teemo Tebest · Thumas Miilumäki
Tampere University of Technology, Korkeakoulunkatu 10, 33720 Tampere, Finland
e-mail: {kirsi.silius,anne.tervakari,meri.kailanto,
       jukka.huhtamaki,jarno.marttila,teemo.tebest,
       thumas.miilumaki}@tut.fi

F.L. Gaol et al. (Eds.): Proc. of the 2011 2nd International Congress on CACS, AISC 144, pp. 369–376.
springerlink.com

The aim of peer learning is to promote collaboration, teamwork, becoming a member of a learning community, critical enquiry and self-reflection, communication skills and skills for learning to learn. Peer learning drives students to work together and develop skills of collaboration while combining theory with students' own understandings. It opens up possibilities for students to engage in the reflection and exploration of ideas when a teacher is not present [5].

In order to promote students' peer learning skills and to motivate them to learn, the Hypermedia Laboratory organised a course in the spring of 2011 entitled Programming of Hypermedia, which uses authentic, real-word problems that must be solved. TUT Circle, a social media web service created in the Hypermedia Laboratory, was used as the web-based learning environment for the course. It has most of the basic functions that are common in modern social media and Web 2.0 services [6].

In this research, we aimed at finding out, firstly, if the teaching experiment with TUT Circle is pedagogically usable. The term "pedagogical usability" is denoted as meaning that the tools, content, interface and tasks of the web-based learning environments support the different learners to learn in various learning contexts according to selected pedagogical objectives [7]. Secondly, the purpose of the research is to study if students voluntarily utilised peer learning in order to learn with and from each other without immediate teacher intervention.

2 Teaching Programmatic Hypermedia

Programmatic hypermedia is a very broad subject. At the end-user level, the Web operates with a small set of technologies, such as HTML and JavaScript, but there are numerous technologies that can be used to build the back end of a Web application. More important than individual technologies are the features and functionalities that programmatic hypermedia enables, including syndication, mashups, information visualisation and collaborative information management, to name a few.

When learning programmatic hypermedia, one has to focus on the individual implementation of a technology, such as PHP or Django, to be able to acquire enough technical skills to start prototyping features. Moreover, it should be acknowledged that there is no silver bullet in Web development.

Thus, we see that when giving a course on programmatic hypermedia, it would be useful to a) let the students select a technology from a variety of alternative technologies and b) allow the students to exchange information regarding the utility of different technologies in the context of developing high-end Web applications with each other.

3 Teaching Experiment

Sixty eight students from TUT participated in the course during the Spring 2011 semester. As a requirement for passing the course, each student had to build his/her own Web-service prototype by using a free-choice technology. Before the students started building their prototypes, they familiarised themselves with the basics of Web programming in PHP and MySQL, after which they started implementing predefined features into their own Web services using the selected technology. After implementing each feature, the students were asked to report to the other students regarding the findings gained from their implementation either by giving an oral presentation or by writing a blog post using TUT Circle. The instructor motivated the students to share their information by offering that extra points be added to their score on the final exam. During the course, only a few oral presentations were given. Most of the students shared their information by writing blog posts using TUT Circle, with the posts totalling 138.

4 Data Collection, Analysis and Results

Data for the study were collected, in parts, from two different sources. Firstly, the students participating in the course responded to a survey. The aim of the survey was to gather information about the students' experience of utilising TUT Circle for peer learning. Secondly, log data collected from TUT Circle were analysed and visualised to verify interaction among students.

4.1 Survey of the Students' Experiences

The students completed a web survey with fifteen structured and five open-ended questions about the pedagogical usability of TUT Circle regarding the organisation of teaching and studying, learning processes, interaction, sense of community, tutoring and peer learning.

The response rate was 49% (19 students out of 39). The total number of respondents was quite low; thus, statistical generalisation is not possible. Nevertheless, the study revealed the students' conceptions of the teaching experiment. The data were collected using a five-level Likert scale. Results for the Likert-scale questions can be seen in table 1. The qualitative data collected by using open-ended questions were analysed by using content analysis. One aim of the content analysis was to examine the meanings, themes and patterns that were manifest or latent in the student's answers and to identify similarities and differences among answers. The researchers read the material through and coded the data separately (researcher triangulation); afterwards, the classifications were compared and the results formalised.

According to the results of the analysis, TUT Circle seemed to support teaching and studying organisation. A major proportion of the respondents (68%) agreed or strongly agreed that instructions were easy to find in the web-based learning environment. The learning materials were also easily found, according to 78% of the respondents. The respondents had different thoughts on the findability of the schedules. A quarter of the respondents (26%) disagreed or strongly disagreed that the schedules were easy to find, while one third of them (32%) agreed or strongly agreed with the claim.

The respondents also had different thoughts regarding the teacher's guidance within TUT Circle. Half of the respondents (53 %) agreed or strongly agreed that their teacher's guidance encouraged them to ask for advice concerning their exercises. Half of them (52%) neither agreed nor disagreed that their teacher's guidance encouraged them in problem solving.

The respondents had different opinions concerning whether TUT Circle promoted interaction among the students. Forty-two per cent of respondents agreed or strongly agreed that TUT Circle offered them facilities to support and help other students. Thirty-six per cent of respondents agreed or strongly agreed, and one quarter of them (26%) disagreed or strongly disagreed, that TUT Circle offered functionalities that support interaction among students. Forty-two per cent of respondents agreed or strongly agreed that TUT Circle offered functionalities that support interaction between students and teachers.

As a learning environment, TUT Circle seemed to support students' learning processes. A major portion of the respondents (61%) strongly agreed or agreed that TUT Circle allowed them to present to other students the problems they encountered with exercises. Almost half of the respondents (47%) agreed or strongly agreed that TUT Circle encourages students to receive feedback and comments from other students in order to solve problems. A quarter of the respondents (26%) agreed or strongly agreed that TUT Circle offered them functionalities to receive support and guidance for solving problems. In addition to interaction required by teacher, there was no voluntary interaction between the students in TUT Circle. In the open-ended questions, one respondent mentioned that an evaluation done by peer learners would increase the interaction between students:

> If commenting and evaluating other students work had been required, the social properties of the TUT Circle would have played a greater role (Respondent 7).

One of the main research aims was to discover if the students found that TUT Circle supports peer learning. Almost all of the respondents (90%) agreed or strongly agreed that TUT Circle offered functionalities for presenting their final works to the other students. A majority of the respondents (84%) agreed or strongly agreed that TUT Circle offered good possibilities to learn from other students' works. However, only half of the respondents (53%) reported that they took advantage of being able to learn by studying other students' works. During the open-ended questions, 14 out of 19 mentioned that they had written blog messages for others to read or had read other students' blogs. Some students sought more knowledge about the same technology they had used themselves, and some were interested in the implications of different technologies in order to widen their perspective.

> I also read a few blog messages that other students had written on exercise work features; it was interesting to see how others had implemented the same feature (Respondent 15),

Thirteen out of nineteen respondents stated that they had found reports written by other students useful to their learning.

Table 1 The division among the students according to their level of agreement to statements. (N=19)

Statement	A^a	B^a	C^a	Av.	Stdv.
Instructions were easy to find in TUT Circle	16	16	68	3.95	1.13
Learning materials were easy to find in TUT Circle	11	11	78	4.05	0.97
Schedules were easy to find in TUT Circle	26	42	32	3.11	1.05
Guidance of teachers in TUT Circle encouraged me to ask for advice	26	21	53	3.32	1.11
Guidance of teachers assisted in problem solving	21	52	26	3.00	0.82
Guidance of teachers encouraged interaction between students	26	42	32	3.00	1.20
TUT Circle offers functionalities that support interaction among students	26	37	37	3.11	0.81
TUT Circle offers functionalities that support interaction between students and teachers	21	37	42	3.21	0.98
TUT Circle allows students to present problems to other students	6	33	61	3.67	0.77
TUT Circle allows for receiving feedback from and giving comments to other students	16	37	47	3.32	0.75
TUT Circle offers functionalities for finding support and guidance	16	58	26	3.11	0.66
TUT Circle offers functionalities for students to support and help other students	16	42	42	3.26	0.73
TUT Circle offers functionalities for presenting final works to other students	0	11	89	4.16	0.60
TUT Circle offers good possibilities for learning from other students' work	0	16	84	4.11	0.66
I took advantage of the chance to learn by studying other students' work	47	0	53	3.00	1.37

A=disagreed or strongly disagreed, B= neutral, C= agreed or strongly agreed

Blog messages and exercise work instructions that were clearly at hand and contents of the lectures helped completing the course (Respondent 12).

4.2 Visualisations of Students' Activity

The log data collected from TUT Circle were also used as a source of research data. Log data consist of user interactions carried out in TUT Circle. Everything that users do in TUT Circle (e.g., write and post messages or read each others' writings) are recorded. This data can be gathered using tailored data collection methods and modelled into a network that represents the interactions between students. In particular, we were interested in the patterns that emerge regarding students' readings of the feature reports written by their peers. Social network analysis (SNA) methods were used to model and analyse the gathered log data. SNA studies the social structures of actors. Visualisation of social configurations "allows investigators to gain new insights into the patterning of social connections" [8].

During the spring of 2011, 30 students reported a total of 138 different feature implementations in TUT Circle. In addition, eight students preferred to present their

Fig. 1 Network representation of students and their reading habits; two nodes, each representing a student, are connected if a student has read the reports written by another student. Connection weight, visualised as edge width, represents the number of individual reading events. The colours of the nodes distinguish the different technologies used by the students (CodeIgniter=cyan, Django=green, PHP=blue, Ruby on Rails=red, Vaadin=yellow, users not attending the course=grey). Visualisation was produced with the network visualisation tool Gephi and the Fruchterman-Reingold force-based layout algorithm.

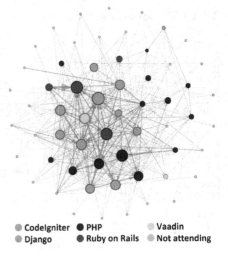

exercise work by means of a physical presentation. The reports in TUT Circle were read 1189 times. Students' readings of their own reports are excluded from the data.

The network in Fig. 1 represents the students who took part in the Programming of Hypermedia course in the spring of 2011. The connections between the nodes, which represent the students, are created on the basis of students' readings of the other students' reports. The weight of the connection, visualised as edge width, equals the number of times a student has read another student's report.

This kind of modelling leads to a directed graph, or a digraph. Thus, when the arrows point away from the readers towards the authors of the reports, we can use node indegree, i.e., the number of arrows pointing towards a node, to see which students receive the most attention from other students. When investigating the reports of the top indegree students, we can see whether the attention focuses on reports that are of high quality and promote the course's learning objectives.

From the network we can conclude that the students did, indeed, read the reports of other students quite intensively. The network is relatively dense, with a tight core of students sharing information. The density value for the directed network is 0.140. This means that more than 10% of all the possible connections appear on the graph. When compared to our previous findings on friendship networks [6], we can see that the network is now much denser. The measured phenomena—friendships between students and the reading of exercise work reports—are, of course, somewhat different by nature.

The students with the the the largest indegree values—most readers—reported about different technologies. The technologies taught during the course were PHP and CodeIgniter. The students were also using alternative technologies that were not covered in the course content. This indicates that using TUT Circle offered the students an environment in which to share information among each other for the purpose of peer learning.

The connection with the most weight is between two students who were using the same alternative technology (Ruby on Rails). Both of the students mentioned that they are already experienced in PHP and wanted to learn something new. In this case, one student clearly followed the work of the other—one published all of his/her reports before the other, and they both produced reports of high quality.

5 Discussion and Conclusions

In order to develop teaching methods of programmatic hypermedia to support authentic and meaningful learning the Hypermedia Laboratory organised a teaching experiment during which the students had chances to engage in peer learning within a social-media-enhanced web-based learning environment called TUT Circle.

As this was a qualitative research study, the results cannot be generalised, but they give valuable information about the use of TUT Circle as a tool for supporting peer-learning. According to the results, it can be asserted that TUT Circle is pedagogically usable. Its functionalities, content and tasks allow learners to study, interact and learn by means of peer learning. TUT Circle also offers functionalities that permit students to present to other students the problems that they encounter with exercises as well as their final works. It also offers students a chance to learn from other students' work.

Peer learning is difficult to monitor and investigate without visualising the network of student-written reports and their readers. According to the analysis and visualisations of network activity, half of the students seemed to read others' reports quite intensively. It was especially those students who used technologies not covered in the course content who benefited from peer learning by obtaining information about new technology from other students.

Students' descriptions of implementing different hypermedia features with alternative technologies were a valuable resource for other students; thus, only a few students read reports that discussed technologies other than those the reader was dealing with. One reason could be that all three deadlines for reporting the features were scheduled for the last week of the course, which was not a decision that promoted peer learning.

We agree with Boud et al. (1999) that interaction between students could be increased if assessment indicates that peer learning activities are valued. An appropriate assessment practices address important educational outcomes, add the value of peer learning and increases students' commitment in active and productive learning [5]. A major future improvement for supporting the sharing of information is to refine the course timeline and more equally distribute the deadlines. This would give students more opportunities and more time to learn from each other.

The positive results from this one course encourages us to do further research in other hypermedia courses regarding how a social-media-enhanced web-based learning environment like TUT Circle can better support student peer learning. It would be interesting to see for example how smaller study groups work in TUT Circle while some of the group's content can be made visible to others. And as the TUT Circle becomes familiar to the students it is interesting to see if the students begin to use it voluntarily to support their studies over course boundaries.

References

1. Carroll, J.M., Rosson, M.B.: Toward even more authentic case-based learning. Educational Technology 45, 5–11 (2005)
2. Brown, J.S., Collins, A., Dugui, P.: Situated cognition and the culture of learning. Educational Researcher 18(1), 32–42 (1989)
3. Lave, J., Wenger, E.: Situated Learning: Legitimate Peripheral Participation. Cambridge University Press, Cambridge (1991)
4. Donovan, M.S., Bransford, J.D., Pellegrino, J.W.: How People Learn: Bridging Research and Practice. National Academy Press, Washington, DC (1999)
5. Boud, D., Cohen, R., Sampson, J.: Peer learning and assessment. Assessment and Evaluation in Higher Education 24(4), 413–426 (1999)
6. Silius, K., Miilumäki, T., Huhtamäki, J., Tebest, T., Meriläinen, J., Pohjolainen, S.: Students' motivations for social media enhanced studying and learning. Knowledge Management & E-Learning: An International Journal, the special issue on "Technology Enhanced Learning" 2(1) (2010)
7. Silius, K., Tervakari, A.M.: An evaluation of the usefulness of web-based learning environments: The evaluation tool into the portal of Finnish Virtual University. In: Pearrocha, V., et al. (eds.) Proceedings of mENU 2003—Int. Conf. University Networks and E-learning, Valencia, Spain, May 8-9, 2003, pp. 8–9 (2003)
8. Freeman, L.C.: Methods of social network visualization. In: Meyers, R.A. (ed.) Encyclopedia of Complexity and Systems Science. Springer, Berlin (2009)

Cloud Computing for Genome-Wide Association Analysis

James W. Baurley, Christopher K. Edlund, and Bens Pardamean

Abstract. With the increasing availability and affordability of genome-wide genotyping and sequencing technologies, biomedical researchers are faced with increasing computational challenges in managing and analyzing large quantities of genetic
data. Previously, this data intensive research required computing and personnel resources accessible only to large institutions. Cloud computing allows researchers to
analyze their data without a local computing infrastructure. We evaluated the feasibility of cloud computing for association analysis of genome-wide data. Our approach utilized the MapReduce model which divides the analysis into independent
units and distributes the work to a computing cloud. We evaluated our approach
by modeling the relationships between genetic variants and disease in a simulated
genome-wide association study. We generated several data sets of 100,000 subjects
and various number of genetic variants, and demonstrated that our analysis approach
is scalable and provides an attractive alternative to establishing and maintaining a
local computing cluster.

1 Introduction

Many common diseases such as cancer, asthma, and diabetes are complex with
many genetic and environmental factors influencing disease risk. Historically only
a small set of genes could be investigated, often selected by prior experimental
evidence. With the development of high-throughput genotyping and sequencing
technologies, millions of genetic variants can be evaluated systematically across

James W. Baurley
Biorealm, Los Angeles, USA
e-mail: baurley@biorealmresearch.com

Christopher K. Edlund
University of Southern California, Los Angeles, USA
e-mail: chris.edlund@biorealmresearch.com

Bens Pardamean
Bina Nusantara University, Jakarta, Indonesia
e-mail: bpardamean@binus.edu

F.L. Gaol et al. (Eds.): Proc. of the 2011 2nd International Congress on CACS, AISC 144, pp. 377–383.
springerlink.com © Springer-Verlag Berlin Heidelberg 2012

the human genome. Genome-wide association studies (GWAS) are designed to uncover common genetic variations associated with disease. Study sample sizes are large with each individual genotyped for millions of DNA variations known as single nucleotide polymorphisms (SNPs). Management and analysis of these data is challenging, with existing statistical software not well suited for big data. Large institutions often design and implement custom software that utilize parallel programming and high performance computing clusters. The overhead of such systems makes analysis difficult or impossible for smaller institutions.

Cloud computing may be a solution to these challenges. In a general term, cloud computing is Internet-based computing and is "a model for enabling convenient, on-demand network access to a shared pool of configurable computing resources" that can be "rapidly provisioned and released with minimal management effort or service provider interaction" [9]. A service provider handles the management of cloud resources such as purchasing and maintenance of hardware and software. Computing resources can be simply requested, deployed, and accessed through an user or programming interface, with no knowledge of the underlying computing infrastructure. This model could be appealing to biomedical researchers because it translates to reduced computing overhead and increased accessibility to large computing resources. Cloud resources are available on-demand from an large shared pool as exemplified by the Amazon Elastic Compute Cloud (EC2) [1]. Investigators can easily manage their computing budget with the "pay as you go" billing structure which charges for resources used per unit time.

Many of the common statistical methods applied to genome-wide data are conceptually straightforward and fit into the MapReduce programming model. MapReduce provides an abstract way to automatically parallelize and distributes computations [8]. The model allows the user to focus on data analysis rather than the complexities of implementing a parallel program that distributes data and computation across many computers. This model has two operations: the *map* operation applied to each record of an input file and the *reduce* operation that aggregates or summarizes the set of values for each key produced by the *map* operation. Large input files are automatically partitioned into segments called splits and distributed to multiple computers.

We evaluate the feasibility of performing genome-wide association analysis in a cloud environment, where the computing infrastructure and parallel implementation details are hidden from the user. In section 2, we describe the modeling framework used to evaluate the relationships between genetic variants and disease and our implementation using Amazon Web Services [5]. In section 3, we simulate data files with increasing numbers of genetic variants and evaluate the cloud computing costs and performance under different compute cloud configurations. In section 4, we predict the computing resources needed for a large genome-wide association study and discussing performance tuning and extensions to our genetic association analysis framework.

2   Genome-Wide Association Analysis Framework

Our cloud-based approach is shown in Figure 1. The input file contains genotypes for N individuals for J SNPs. The input file extends lengthwise to the number of SNPs genotyped. When J is large, MapReduce is used to distribute the data analysis across computers [6]. In our application, each split represents a subset of the J SNPs, and is handled by a worker node that performs all the model fitting within that split. For each SNP in the split, a map function is called. For convenience we implemented the function in the [R] statistical language [11]. This function reads each row, models the relationship between the genetic variant and disease in a generalized linear model (glm), and returns summary data. In this application, the reduce function simply combines the results from all the map tasks.

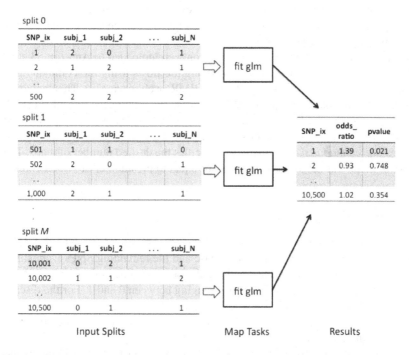

Fig. 1 Genome-wide analysis framework. Genome-wide data for N individuals is split into independent sets of SNPs, each being processed by a separate map task. The map function reads the SNP data, fits a generalized linear model (glm), and returns the estimated strength of association (odds ratio) and the p-value for a test of association between the SNP and disease. The results from all the maps are concatenated into a single results file.

In a genome-wide analysis, SNPs are tested independently for association with disease. Since the unit of analysis is a SNP, splits can take place anywhere in the genotype input file. Generalized linear models (glm) are used to model the relationship between genetic variants and disease. Consider a glm where Y denotes the outcome variable, X is a matrix of genotypes for J SNPs coded additively (i.e. 0, 1, or 2 copies of the minor allele), W is a matrix of K covariates, and β is a vector of parameters. The expected value of Y_i, the outcome variable for individual i, depends on the linear predictors through the link function g such that for each SNP j:

$$g(\mu_i) = \beta_0 + \sum_k^K \beta_k W_{ik} + \beta_{(k+1)} X_{ij}$$

where $\mu_i = E(Y_i)$.

Since our outcome is binary (an individual is either diseased or not diseased) the *logit* link function is used and called a logistic model. The maximum likelihood estimates and standard errors of the parameters are computed numerically using an iterative optimization procedure called Fisher scoring. The odds ratio is a measure of the strength of association and is given by exponentiating the effect estimate $\hat{\beta}$ for each genetic variant. The likelihood ratio test for association between the genetic variant and disease returns a p-value.

The input data files and [R] mapper function were uploaded to Amazon S3 storage [4]. We configured and launched streaming job flows using the command line interface to the Amazon Elastic MapReduce (EMR) web service [2]. This including specifying the number and type of EC2 instances, the path to the input files (SNP genotype data), cache files (disease status for all individuals), the map function script in [R], and the output directory for results. Task granularity was controlled with the *mapred.map.tasks* argument that specifies the desired number of maps. The number of maps was set to be greater than or equal to the number of virtual cores.

3 Simulations

Genotypes and outcome data were simulated for 100,000 individuals in PLINK [10]. Three SNPs had a simulated effect on disease and the remaining SNPs were simulated without an effect (null). The resulting datasets contained up to 10,000 SNPs. Each SNP had an unique allele frequency ranging between 0 and 1. For the disease associated SNPs, the odds ratio for disease was 2.0. We generated equal numbers of cases (diseased) and controls. The disease prevalence was 0.01. The simulations are summarized in Table 1.

We considered two different Amazon EC2 instance types: Large (m1.large) and High-CPU Extra Large (c1.xlarge). Both are 64-bit with high I/O performance and seven gigabytes of memory. The documentation described the c1.xlarge instance as well suited for computational intensive applications. The CPU properties and current pricing per instance-hour are summarized in Table 2. We launched MapReduce job flows with the four input data files and attempted to keep the elapsed time below two hours by varying the instance types, number of instances, and MapReduce parameters.

The elapsed time and total cost for each job are shown in Table 3. In the smaller dataset of J=1,000 SNPs, a single large instance (m1.large) was adequate to perform

Table 1 Data Simulations

Input File	J	N	File Size (MB)
1	1,000	100,000	191
2	2,000	100,000	381
3	3,000	100,000	572
4	10,000	100,000	1,843

J is the number or SNPs (rows) in the data files. N is the number of individuals (columns) in the data file. The input file size for EMR is given in megabytes (MB).

Table 2 Elastic MapReduce Configurations and Pricing

Instance Type	Cores	CU/Core	EC2 Pricing	EMR Pricing
m1.large	2	2	$0.38	$0.06
c1.xlarge	8	2.5	$0.76	$0.12

CU/Core is the number of EC2 Compute Units per virtual core which is equivalent to a 1.0-1.2 GHz processor. The EC2 and EMR pricing is per hour in U.S. dollars (see [3]).

Table 3 Job Flows Properties

Input	J	Type	Instances	Maps	Time	Total Cost
1	1,000	m1.large	1	2	0:56	$0.44
2	2,000	m1.large	1	2	1:50	$0.88
2	2,000	m1.large	1	4	1:51	$0.88
3	3,000	m1.large	1	4	2:44	$1.32
3	3,000	m1.large	2	6	2:20	$2.46
3	3,000	m1.large	3	6	1:27	$2.40
3	3,000	c1.xlarge	1	8	0:38	$0.88
4	10,000	c1.xlarge	3	24	1:22	$4.92
4	10,000	c1.xlarge	3	200	1:05	$4.92
4	10,000	c1.xlarge	4	500	0:44	$3.16

The m1.large instance has two virtual cores with two EC2 Compute Units per core. The c1.xlarge is designed for more computationally intensive tasks and has eight virtual cores with 2.5 EC2 Compute Units per core. Maps is the number of MapReduce mapping tasks. Costs are in U.S. dollars.

the job in less than an hour with only two maps (500 SNPs per map). When we doubled the dataset to $J=2,000$ and kept computation resources constant, the size of the two splits increased and the elapse time nearly doubled. We experimented with setting the argument *mapred.map.tasks* to four, but this had little impact on the elapsed time under this instance configuration. As expected, when the dataset was increased to 3,000 SNPs, the elapsed time increased to two hours and forty-four minutes. To reduce the work per map, the number of m1.large instances was increased. Two large instances decreased the elapsed time by 22 minutes but increased the costs by $1.14. Three instances of m1.large actually used less time and money. When the instance type was changed to eight virtual cores and more computational power per core (c1.xlarge), the elapse time dropped to 38 minutes and reduced costs to 88 cents. The input data file was then increased to $J=10,000$ SNPs. When the job flow was configured with three instances of c1.xlarge, 24 map tasks (416 SNPs per map) were performed and completed in 82 minutes. When *mapred.map.tasks* was increased to 200, the job completed in less time (65 minutes) indicating better load balancing. When an additional c1.xlarge instance was added and the map tasks increased to 500, the job completed in 44 minutes and saved $1.76.

4   Conclusion

We demonstrated that it is possible to run genome-wide association analysis on a compute cloud with good performance and reasonable expense. Cloud computing and the MapReduce programming model allows the user to focus on the data analysis by specifying functions in an statistical language without the complexities of maintaining a local cluster and implementing multiprocessor code.

For our application, the c1.xlarge EC2 instance had value over the m1.large instance in all but the smallest cases. The elapsed job time appears to decreases linearly with the number of instances meaning our approach should scale to large genetic datasets. MapReduce performance was sensitive to task granularity so we recommend application specific benchmarking to find the optimum number of EC2 instances and maps.

SNP genotyping technology continues to improve and the recent SNP genotyping platforms from Illumina can assay nearly 5 million SNPs per sample [7]. Based on the results in Table 3, we can estimate the compute cloud resources for a large hypothetical GWAS of 100,000 individuals genotyped for 2.5 million SNPs. The simulations showed that 10,000 SNPs could be processed is less an hour using four c1.xlarge instances. The larger dataset could be run in less than an hour using 1,000 c1.xlarge instances, costing approximately $760. This is rather inexpensive when compared to the overhead in building and operating a supercomputing cluster and implementing code to handle parallelism, data distribution, and node failures.

One limitation to all cloud computing solutions is the expense of intermediate runs. Model diagnostics and troubleshooting should be done on a thinned dataset to ensure that the full run is successful and captures the necessary information. Computation needs will vary by application and benchmarking will be needed for every genome-wide task. Optimizing the performance of MapReduce may decrease costs substantially. It is also worthwhile to compare the performance of *map* and *reduce* in a variety of languages.

This framework could be extended for many genome-wide quality control and analysis tasks such as imputation of un-typed genetic variants, estimation of SNP call rates, and tests of Hardy-Weinberg Equilibrium. The process would be similar for these applications, that is, using small job flows to evaluate the optimal configuration of EC2 instances and MapReduce.

References

1. Amazon elastic compute cloud, http://aws.amazon.com/ec2/
2. Amazon elastic mapreduce, http://aws.amazon.com/elasticmapreduce/
3. Amazon elastic mapreduce pricing, http://aws.amazon.com/elasticmapreduce/pricing/
4. Amazon simple storage service, http://aws.amazon.com/s3/
5. Amazon web services, http://aws.amazon.com
6. Apache hadoop, http://hadoop.apache.org/
7. Illumina snp genotyping, http://www.illumina.com/applications/detail/snp_genotyping_and_cnv_analysis/whole_genome_genotyping_and_copy_number_variation_analysis.ilmn
8. Dean, J.: Mapreduce: Simplified data processing on large clusters. Usenix SDI (2004), http://www.eecs.umich.edu/~klefevre/eecs584/Handouts/mapreduce.pdf
9. Mell, P.: The nist definition of cloud computing. National Institute of Standards and Technology (2009), http://www.mendeley.com/research/nist-definition-cloud-computing-v15/
10. Purcell, S., Neale, B., Todd-Brown, K., Thomas, L., Ferreira, M.A.R., Bender, D., Maller, J., Sklar, P., de Bakker, P.I.W., Daly, M.J., Sham, P.C.: Plink: a tool set for whole-genome association and population-based linkage analyses. Am. J. Hum. Genet. 81(3), 559–575 (2007), doi:10.1086/519795
11. R Development Core Team: R: A Language and Environment for Statistical Computing. R Foundation for Statistical Computing, Vienna, Austria (2010) ISBN 3-900051-07-0, http://www.R-project.org

Finding Protein Binding Sites Using Volunteer Computing Grids

Travis Desell, Lee A. Newberg, Malik Magdon-Ismail, Boleslaw K. Szymanski, and William Thompson

Abstract. This paper describes initial work in the development of the DNA@Home volunteer computing project, which aims to use Gibbs sampling for the identification and location of DNA control signals on full genome scale data sets. Most current research involving sequence analysis for these control signals involve significantly smaller data sets, however volunteer computing can provide the necessary computational power to make full genome analysis feasible. A fault tolerant and asynchronous implementation of Gibbs sampling using the Berkeley Open Infrastructure for Network Computing (BOINC) is presented, which is currently being used to analyze the intergenic regions of the *Mycobacterium tuberculosis* genome. In only three months of limited operation, the project has had over 1,800 volunteered computing hosts participate and obtains a number of samples required for analysis over 400 times faster than an average computing host for the *Mycobacterium tuberculosis* dataset. We feel that the preliminary results for this project provide a strong argument for the feasibility and public interest of a volunteer computing project for this type of bioinformatics.

1 Introduction

Cutting edge computational science is requiring larger and larger computing systems as the size and complexity of scientific data continues to increase far faster than advances in processor speeds. This is particularly true in the area of computational biology, as biologists are gathering data on the full genomes of many different

Travis Desell
University of North Dakota, Grand Forks, North Dakota, USA
e-mail: tdesell@cs.und.edu

Lee A. Newberg · Malik Magdon-Ismail · Boleslaw K. Szymanski
RPI, Troy, New York, USA
e-mail: [leen,magdon,szymansk]@cs.rpi.edu

William Thompson
Center for Computational Molecular Biology, Department of Applied Mathematics, Brown University, Providence, Rhode Island, USA
e-mail: william_thompson_1@brown.edu

F.L. Gaol et al. (Eds.): Proc. of the 2011 2nd International Congress on CACS, AISC 144, pp. 385–393.
springerlink.com   © Springer-Verlag Berlin Heidelberg 2012

species. In the domain of biological sequence analysis, high-dimensional integration allows the identification and location of DNA control signals (*cis*-regulatory elements such as protein binding sites) in analyses of dozens of co-regulated genes at a time. Identifying and locating these *cis*-regulatory elements at the genome and multi-genome scales continues to be too complex for today's clusters and clouds. However, emerging peta-scale systems (those providing petaFLOPS, millions of billions of FLoating point OPerations per Second, of computational power) can enable this computationally challenging research, which will add significantly to understanding of the cellular processes of a diverse set of organisms, including organisms for disease, biofuel production, and environmental bioremediation.

Apart from a few of the world's fastest supercomputers[1], volunteer computing projects such as Stanford's Folding@HOME [1] and the Berkeley Open Infrastructure for Network Computing (BOINC) [2] have also reached peta-scale levels of computing power. BOINC in particular is a very interesting computing environment for new scientific projects, in that it provides an open source environment for developing projects and has a highly active social community which actively seeks out new projects to participate in. Further, there is very low overhead to starting a BOINC project, as purchasing and maintaining a server machine is a fraction of cost of purchasing and maintaining supercomputer, and many volunteers actively upgrade and purchase new equipment while the hardware in a supercomputer remains fixed and degrades over time.

However, traditional numerical integration methods are highly sequential, relying on variations of Monte-Carlo Markov Chains (MCMC), which makes them not well suited to volunteer computing environments, which are inherently heterogeneous and can be extremely volatile with volunteered hosts frequently joining and leaving, and potentially returning malicious results. Current parallel approaches to MCMC are not fault tolerant and cannot scale to the number processors used by peta-scale systems. Additionally, many peta-scale systems, including the world's fastest supercomputer and many BOINC computing projects, are accomplishing these speeds by using graphical processing units (GPUs). However, utilizing GPUs for full genome analysis proves problematic, as performing computation requiring full genome scale data may not fit into a GPUs limited memory.

This paper presents preliminary work done in the development of DNA@Home, a BOINC volunteer computing project started with the goal of performing sequence analysis at the full genome level. The project extends the Gibbs Sampling algorithm, an MCMC approach for finding transcription factors [3, 4, 5, 6] which also has uses in other scientific disciplines [7, 8, 9, 10], by adding parallelism and fault tolerance making it suitable for use on a volunteer computing grid. This asynchronous Gibbs Sampling approach is currently being used by DNA@Home to analyze the full genome of *Mycobacterium tuberculosis* consisting of 2,066 intergenic regions (segments of DNA between known genes) totaling 350,825 nucleotides. Volunteered hosts report a set of samples at a rate of approximately 0.6 every second, with a sampling task taking on average 22.4 minutes – which equates to a 896 times speedup

---

[1] http://www.top500.org/

over a single average speed CPU. After only three months of operation, more than 1,800 volunteered hosts have participated in the project with over 1,600 staying active since joining, which highlights the potential for this approach to reach the scales required to look for transcription factors using the full genomes of complex prokaryotes (bacteria) and eukaryotes (even humans).

The paper proceeds as follows. Section 2 describes Gibbs Sampling and the challenges involved in using it to analyze large scale data. Section 3 describes how Gibbs Sampling was extended for use on volunteer computing grids. Section 4 provides preliminary results describing the performance and activity of DNA@Home. Finally, concluding remarks and future work are discussed in Section 5.

2 Gibbs Sampling for Finding Transcription Factors

Gibbs sampling is used by computational biologists to find transcription factor binding sites or gene regulatory elements – sites where proteins bind to DNA [11, 12]. It is a variant of Markov Chain Monte-Carlo (MCMC), where every step of the random walk must satisfy the following criteria:

$$P_i * R_{i,j} = P_j * R_{j,i} \tag{1}$$

where P_i is the probability of state i being a solution, and P_j is the probability of state j being a solution. It is sufficient for P_i and P_j to be relatively correct, as opposed to the exact probability, as this is typically unknown, while the relative probabilities of two states can be calculated more easily. $R_{i,j}$ and $R_{j,i}$ are the *transition probabilities*, or the probability that the state will move from i to j and vice versa. Fulfilling this detailed balance equation (Equation 1) ensures that a long enough Markov chain will consist of states sampled in proportion to the probability distribution of the function being integrated.

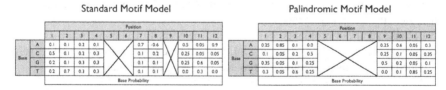

Fig. 1 *Motif models* for different types of transcription factor binding sites (sites where proteins bind to DNA). Transcription factors are non-exact which makes them computationally demanding to find. These binding sites need to be modeled probabilistically. Additionally, certain positions within the binding site may be non-binding (represented as 'X's in the figures). One goal of this research is to develop an asynchronous Gibbs sampling algorithm that can use massive scale cyber-infrastructure to find these binding sites within multiple full genomes, something not possible with current cyber-infrastructure.

The problem state of finding transcription factor binding sites consists of multiple *motif models*, probabilistic representations of DNA patterns, of varying lengths (see Figure 1). A transcription factor can bind to different sequences of DNA, so binding sites are non-exact. A motif model represents the different probabilities of each position within the motif being a DNA letter. Additionally, motif models can have skipped positions (represented by 'X's in Figure 1) and be palindromic, where tail of the motif model is the reverse complement of the front.

There are two main computational challenges involved in using Gibbs sampling to find transcription factors binding sites. First, determining the next state in the random walk can be computationally demanding. For prokaryotes, typical parameter sizes computable by current cyber infrastructure consist of less than five motifs, 12 to 24 nucleotides wide, being sampled from within 3 to 30 intergenic regions of less than 500 nucleotides, usually from one species. Eukaryotes have shorter motif sizes, 6 to 12 base pairs, however the intergenic regions are larger, 1,000 to 10,000 nucleotides. Each step requires that for every intergenic region, all the motifs must be regenerated from the samples within the other intergenic regions, and then these new motif models are used to resample within the excluded intergenic region. Other parameters that are modified during the random walks are the length of the motifs and positions of the skipped positions. Because of this, increasing the motif size, the number of motifs, intergenic regions and the region size will all increase the computational complexity.

Second, there is no known way to start the random walk from an unbiased initial position. To avoid bias in the samples taken by the random walk, Gibbs sampling performs a *burn-in* period, where the states visited by the random walk are not used as samples. Following the burn-in period, a sample can be taken after every step in the random walk. There are two main problems within the burn-in and sampling periods. First, the length of the burn-in period required to eliminate any bias in the selection of the starting state is unknown, as is the number of samples required to adequately capture the problem space. There have been a few approaches to address this issue [13, 14], but this is still an open problem. Further, as the problem size increases, the increase in burn-in period and number of samples required also increases super-linearly, at the very least.

Because of these problems, being able to perform full genome scale analysis requires massive amounts of computing power, which we feel can be effectively provided by a volunteer computing project such as DNA@Home.

3   Gibbs Sampling on Volunteer Computing Grids

Effectively distributing Gibbs sampling is non-trivial, as Markov Chain random walks are inherently sequential. This makes it difficult to effectively utilize the computing power offered by massive scale computing systems, like volunteer computing grids. The application used for computation on the volunteered computing hosts is a modification of the Gibbs sampling algorithm described in Section 2 that performs partial random walks, which are chained by the server-side *parallel walk sampling* software. Section 3.1 describes how the standard Gibbs sampling algorithm has been extended to use asynchronous communication and to provide fault tolerance and error checking. This *parallel walk sampling* enables the computational biologists participating in DNA@Home to utilize these massive scale computing environments. Section 3.2 gives details on how this was implemented and optimized for use with BOINC's open source software.

3.1   Parallel Walk Sampling

While sampling using parallel walks does not reduce the burn-in time, it is an effective way of reducing the sampling time, providing linear speedup in the number of samples generated after the walks have completed their burn-in period. In parallel walk sampling, each Gibbs sampling walk starts from an independent state, then after they have completed the burn-in period to eliminate the initial bias, samples are collected in parallel. For full genome scale data, an extremely large number of samples is required to gather an accurate picture of the sampling space to find these transcription factors, so having a large number of parallel walks being computing by volunteer computing hosts is an effective way of gathering enough samples in a reasonable amount of time.

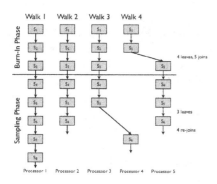

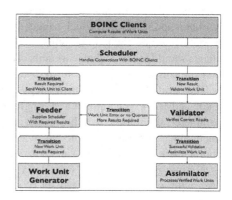

Fig. 2 A strategy for Gibbs sampling using parallel walks. Each arrow represents a *workunit*, or processing job, where a processor receives an initial state with depth x, S_x, and reports a final state with depth y, S_y. Each workunit has a fixed length walk (in this case 1). After each walk completes its burn-in period, samples can be taken. Processors can join and leave, restarting from walks of previously left processors.

Fig. 3 The BOINC architecture consists of multiple server-side daemons which handle work generation, validation and scheduling of work to clients. Tasks that clients will be or are currently computing are called *workunits* and clients report *results* to those workunits. A transitioner daemon updates the state of the workunits and results in the database, which triggers the other daemons to act on them.

However, at the massive computing scale simply running the Gibbs sampling walks in parallel is not sufficient as any processor failure will lose the burn-in steps and any samples collected, wasting all that work. It also does not take into account processors dynamically joining and leaving in the case of a volunteer project. Because of this, DNA@Home uses an asynchronous approach to perform the parallel walks, as shown in Figure 2. The BOINC server stores the last reported position of each walk along with any reported samples. Host processors perform partial walks, by contacting the BOINC server and requesting an initial starting state. The server will also specify a starting seed for randomization and the length of the walk the host will compute. This will allow the server to send the same partial walk to multiple hosts for DNA@Home so results can be compared from different hosts for validation. In this way, individual processors can fail or leave and the server can send that particular processors partial walk to another host; significantly reducing the impact of failures.

3.2   BOINC Implementation Details

The BOINC architecture (see Figure 3) consists of a set of server-side daemons which control the generation, scheduling and validation of tasks, or *workunits*, and the individual *results* from clients. After a workunit is generated by a *work generator* daemon, the *transitioner* daemon updates the database with state changes for that workunit and its results; which trigger the other daemons. The *feeder* controls what workunits are available to be sent to clients, while the *scheduler* determines which clients should receive what workunits. The *validator* compares results to ensure that they are correct, after which the *assimilator* handles the processing of valid results.

The parallel walk sampling strategy used by DNA@Home combines the assimilator and work generator into a single daemon. A database is used to store the parameters to multiple searches which can be performed concurrently, each consisting of any number of motif models of any given size and type (*e.g.* standard, reverse complement or palindromic) and a given dataset (*e.g. Mycobacterium tuberculosis* or *Yersinia pestis*). When a search is started, a fixed number of walks are created, each with a randomly selected set of sampled positions within the intergenic sequences. While burn-in is being performed on a particular walk, the results of that workunit will report the final positions walked to by the Gibbs sampling algorithm. When the results for a workunit are validated (if two or more separate hosts report the same final sample positions those results are considered valid), a new workunit is generated which includes those final positions as a new starting position. The current position of that walk is updated in the database, and its burn-in depth is increased.

After a walk has completed its burn-in period, a flag is set telling the workunits to additionally report every site sampled during its random walk. When the results for one of these sampling workunits are validated (when the final positions and accumulated samples of two or more results are the same), the server saves the accumulated samples and generates a new sampling workunit from the new end position, updating the number of samples taken by that walk in the database.

Various flags have also been specified for these workunits to enable more reliability in the turnaround time for any given workunit. While the number of results required for validation (or *quorum*) is two, BOINC's `target_nresults` flag has been set to three for every workunit. This causes the scheduler to send out the same task to three hosts initially (instead of two), which reduces the impact of one of those hosts returning an erroneous result, or from leaving the project and never returning a result.

4 Preliminary Results

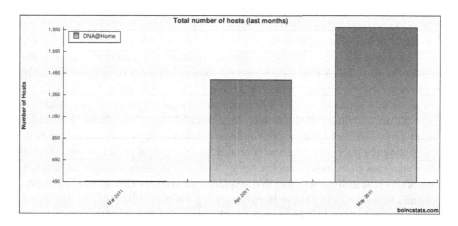

Fig. 4 The number of hosts that have participated in DNA@Home since it began sending work to volunteers in March, 2011, as recorded by the boincstats.com project tracking website. In only three months, DNA@Home has had over 1,600 hosts participate, with new hosts joining daily.

After only three months of limited operation, with small batches of workunits sent out periodically, DNA@Home has already had over 1,800 hosts participate and return valid results to the project (Figure 4 shows the increase in host participation since DNA@Home started sending out workunits)[2]. Binary versions of the application run on volunteered hosts is provided for both 32 and 64 bit versions of Windows, Mac OS X and Linux, which allows any type of computer to participate in the project. These volunteers were only 'recruited' through word of mouth and DNA@Home's listing on various BOINC project tracking websites. We feel that this provides a significant example of the public interest in this type of project, and its ability to achieve the required computational power for full genome scale analysis.

---

[2] http://boincstats.com/stats/project_graph.php?prdna&view=hosts

DNA@Home is currently using a dataset consisting of the intergenic regions of the *Mycobacterium tuberculosis* genome. In this dataset there are 2,066 intergenic regions, consisting of a total of 350,825 nucleotides. With approximately 1,600 active volunteers, it takes less than a week to gather 30,000,000 samples after a burn-in of 1,000,000 steps using 3,000 parallel walks, using a reverse complement motif model of length 16. A workunit consists of 10,000 steps, which over the set of all volunteers takes on average 22.4 minutes. Approximately 0.6 results are reported every second, for a speedup of 896 times an average processor. Alternately, to perform a burn-in of 1,000,000 steps and 30,000,000 samples on an average processor would take 2,893 days to complete, so DNA@Home can obtain a useful amount of samples over 400 times faster.

5   Conclusions and Future Work

This paper presents an asynchronous and fault tolerant implementation of parallel Gibbs sampling for use on the DNA@Home volunteer computing project. Preliminary results show DNA@Home gathering a required number of samples for analysis of the intergenic regions of the full *Mycobacterium tuberculosis* genome in less than a week, a speedup of over 400 times an average single processor. Additionally, with only limited operation over three months, the DNA@Home project has already grown to over 1,600 active volunteers, contributing 1,800 computing hosts, highlighting the public interest and potential for this type of project to analyze even larger genomes from more complex bacteria or even humans. Further, the implications of this research go beyond DNA sequence analysis, as Gibbs sampling is popular and efficient way to sample from complex probability distributions that are not easily reduced to the common tractable probability distributions. It thus has general applicability in statistics where it allows one to sample from a (Bayesian) posterior probability distribution. It also has immediate, and fairly generally applicable, uses in numerical integration where it is almost always significantly more efficient than approaches that approximate a multi-dimensional integral by evaluating the integrand at evenly spaced points or at points chosen uniformly at random.

Future research for the DNA@Home project involves development of a graphical processing unit (GPU) version of the Gibbs sampling application, which has been attempted by other groups [15, 16], as these provide a significant amount of the computational power of the BOINC computing system. Additionally, the development of a web-based interface for participating biologists would ease the use of the system for large number of researchers. Finally, we wish to investigate asynchronous methods for decreasing the time to complete the burn-in period for Gibbs sampling, which could dramatically reduce the time required to gather the number of samples required for this type of analysis.

Acknowledgment. The authors would like to thank the many volunteers at the DNA@Home volunteer computing project for making this research possible. This material is based upon work supported by the National Science Foundation under Grant No 0947637 and by the Department of Energy under Grant DE-FG02-09ER64756.

References

1. Pande, V., et al.: Atomistic protein folding simulations on the submillisecond timescale using worldwide distributed computing. Biopolymers 68(1), 91–109 (2002), peter Kollman Memorial Issue
2. Anderson, D.P., Korpela, E., Walton, R.: High-performance task distribution for volunteer computing. In: e-Science, pp. 196–203. IEEE Computer Society Press (2005)
3. Lawrence, C., Altschul, S., Boguski, M., Liu, J., Neuwald, A., Wootton, J.: Detecting subtle sequence signals: a gibbs sampling strategy for multiple alignment. Science 262(5131), 208–214 (1993)
4. Bais, A.S., Kaminski, N., Benos, P.V.: Finding subtypes of transcription factor motif pairs with distinct regulatory roles. Nucleic Acids Research (2011)
5. Stormo, G.D.: Motif discovery using expectation maximization and gibbs sampling. In: Ladunga, I. (ed.) Computational Biology of Transcription Factor Binding. Methods in Molecular Biology, vol. 674, pp. 85–95. Humana Press (2010)
6. Challa, S., Thulasiraman, P.: Protein Sequence Motif Discovery on Distributed Supercomputer. In: Wu, S., Yang, L.T., Xu, T.L. (eds.) GPC 2008. LNCS, vol. 5036, pp. 232–243. Springer, Heidelberg (2008)
7. Zhang, X.: Automatic feature learning and parameter estimation for hidden markov models using mce and gibbs sampling. Ph.D. dissertation, University of Florida (2009)
8. Finkel, J.R., Grenager, T., Manning, C.: Incorporating non-local information into information extraction systems by gibbs sampling. In: Proceedings of the 43rd Annual Meeting on Association for Computational Linguistics, ACL 2005, pp. 363–370. Association for Computational Linguistics, Stroudsburg (2005)
9. Tan, X., Xi, W., Baras, J.S.: Decentralized coordination of autonomous swarms using parallel gibbs sampling. Automatica 46(12), 2068–2076 (2010)
10. Salas-Gonzalez, D., Kuruoglu, E.E., Ruiz, D.P.: Modelling with mixture of symmetric stable distributions using gibbs sampling. Signal Processing 90(3), 774–783 (2010)
11. Newberg, L.A., Thompson, W.A., Conlan, S., Smith, T.M., McCue, L.A., Lawrence, C.E.: A phylogenetic gibbs sampler that yields centroid solutions for cis-regulatory site prediction. Bioinformatics 23, 1718–1727 (2007)
12. Thompson, W.A., Newberg, L.A., Conlan, S., McCue, L.A., Lawrence, C.E.: The gibbs centroid sampler. Nucleic Acids Research 35(Web-Server-Issue), 232–237 (2007)
13. Lartillot, N.: Conjugate gibbs sampling for bayesian phylogenetic models. Journal of Computational Biology 13(10), 1701–1722 (2006)
14. Gelman, A., Rubin, D.: Inference from iterative simulation using multiple sequences. Statistical Science 7, 457–511 (1992)
15. Yu, L., Xu, Y.: A parallel gibbs sampling algorithm for motif finding on gpu. In: 2009 IEEE International Symposium on Parallel and Distributed Processing with Applications, pp. 555–558 (2009)
16. Kuttippurathu, L., Hsing, M., Liu, Y., Schmidt, B., Maskell, D.L., Lee, K., He, A., Pu, W.T., Kong, S.W.: Decgpu: distributed error correction on massively parallel graphics processing units using cuda and mpi. BMC Bioinformatics 12(85) (2011)

Wood of Near Infrared Spectral Information Extraction Method Research

Xue-shun Wang[*], Ting Yang, Zhong-jie Lin, Fengyan Yang, Kaixin Lu, and Songyue Qiao

Abstract. In order to improve near infrared spectrum analysis precision, it is need to extract information from spectrum data. Derivative can not only eliminate spectrum background disturb , baseline drift and other factors influence, but also improve spectrum resolution ratio. However, at the same time derivative strengthen signal noise, reduce spectrum SNR (signal-to-noise ratio). Smoothing can effectively smooth high frequency noise, heighten spectrum SNR. In use of smoothing window's size has a great effect on spectrum information extraction. This paper take near infrared spectrum of eucalyptus to research NIR spectrum information extraction methods, pay more attention on smoothing window's size influence on spectrum information extraction. Then combine with multiple scatter correction and standardized variables to build eucalyptus lignin PLS model. The results showed that, using 1st derivative to combine with one of 19 point smoothing, multiple scatter correction, standardized variables to treat spectrum can all get good modeling result. In one word, spectrum information extraction can dislodge unfavorable factors influence and enhance spectrum analysis precision.

1 Introduction

Near Infrared Spectrum analytical method is a new analytical method which can carry out efficient and accurate non-destructive tests upon the physical, chemical and mechanical properties of solid, liquid or powdered organic samples.

In NIR analysis, original spectrum contains various interferences including electrical noises, vagabond rays and background interferences of the samples. All these interferences not only directly impact the accuracy and limit of measurements, but also make it more complex and difficult to pick up woods' spectrum information

Xue-shun Wang
College of Science, Beijing Forestry University, Beijing, China
e-mail: wangxueshun6688@sina.com

Ting Yang
College of Science, Beijing Forestry University, Beijing, China
e-mail: yangt624@gmail.com

[*] Corresponding author.

F.L. Gaol et al. (Eds.): Proc. of the 2011 2nd International Congress on CACS, AISC 144, pp. 395–401.
springerlink.com © Springer-Verlag Berlin Heidelberg 2012

from overlapping near infrared spectrum. Besides, near infrared spectrum is weak information. So the spectrum information should be strengthened to improve its analysing ability. What's more, in order to eliminate noise interferences and to optimize spectrum information, the characteristics of the target spectrum should be identified. Thus, the extraction of near infrared spectrum should cover the strengthening of the spectrum information, the elimination of the noises, data screening and the elimination of the impacts of other factors.

In the NIR analysis of woods, useful information should be picked up from the near infrared spectrum of woods and strenthened as high-signal-to-noise-ratio and low-background-interference information. The physical, mechanical and chemical characteristics of woods are predicted through chemometric methods. During the measurement of the spectrum, the spectrum of the samples would be easily influenced by the conditions of the samples and measurement, likely resulting in baseline drift, random noises of the instruments and background noises. So, before the development of spectrum prediction, the spectrum data should be preprocessed to eliminate the impacts of various interference factors and to improve the accuracy of the analysis and prediction of spectrum.

This article takes the near infrared spectrum of eucalyptus as the information source and introduces researches on the extraction methods of the information characteristics of near infrared spectrum. It develops the prediction model of the lignin content of eucalyptus woods with Partial Least Squares and compares the influences of different extraction methods of spectrum information upon the model

2   Experiment and Spectrum Collecting

The eucalyptus are from the Beihai Forest in Guangxi. Four eucalyptus trees are collected in different altitudes. Starting from the chest level, four round sections with a thickness of 6 cm are collected very 2 meters. 64 representative samples are chosen, shattered and screened. 40 to 60 samples of the 64 ones are chosen as the sources of analysis and near infrared spectrum collection. The spectrometer Field spec of the ASC company in U.S. Is chosen as the near infrared spectrometer. The wavelength range of measurement is 350-2500nm.

Both the near infrared spectrometer and spectrum collection are conducted indoor under a constant temperature at $(20\pm2)°C$ controlled by air conditioners. Two bifurcate optical fiber probes are used to collect the near infrared spectrum of the surfaces of the samples. After 30 times of full spectrum scanning, all the data are combined into one spectrum and saved. The spectrum consists of 2150 data points in the whole wavelength range, as shown in Figure 1.

The measurement of lignin content in eucalyptus woods is conducted respectively under CNS GB2677.10-81 and CNS GB/T 2677.8-94. The arithmetic mean value of the results under these two standards will be the final result.

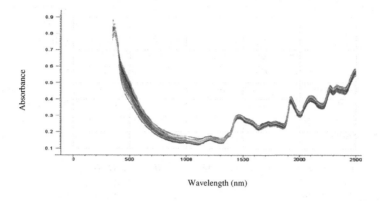

Wavelength (nm)

Fig. 1 NIR spectruim of samples of poplar

3   Principles and Methods

A   *Spectrum Derivative*

Derivative spectrum can not only eliminate baseline drift and reduce the impacts factors like background., but also provide higher resolution and more clear outline changes than original spectrum. In NIR analysis, we mainly apply derivatives of the first and second order:

$$\frac{dA}{d\lambda} = \frac{A_{i+1} - A_i}{\Delta\lambda} \qquad \frac{d^2A}{d\lambda^2} = \frac{A_{i+1} - 2A_i + A_{i-1}}{\Delta\lambda^2}$$

Usually they are also recorded as dA、d^2A in short. $A(\lambda)$ refers to spectrum absorbency. Figure 2 is the first order derivative spectrum chart of wood samples.

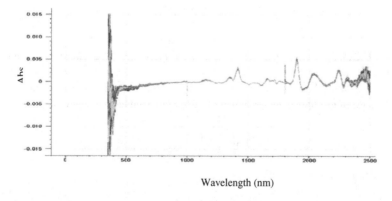

Wavelength (nm)

Fig. 2 First derivative NIR spectrum of the sample No.10

As shown in Figure 2, Derivative spectrum narrows spectrum absorption peaks and strengthen spectrum characteristics. Derivative spectrum is also conducive to determine the accurate locations of absorption peaks. It improves the quality of spectrum and strengthens spectrum information. Thus derivatives can be applied to efficiently purify the information of the spectrum figures. However, while derivative spectrum strengthens spectrum information, it also strengthens signal noises and thus lowers the signal-to-noise ratio of spectrum. In order to improve the resolution of spectrum and the accuracy of spectrum analysis, derivative spectrum should be denoised.

B   Moving Windows Average Smoothing Method

Smoothing is a common method applied to denoise signals in near infrared spectrum. The premise of its application is to assume that the noises spectrum contains are White Noise (The mean value is 0). Spectrum smoothing aims to eliminate high-frequency random noises. Its basic method is to pick up several points around smoothing point and then do "Average" or "Fitting" to get the best estimate value of the smoothing point and to eliminate random noises. The common methods of smoothing are Moving Window Average Method, Fold Smoothing and Adjacent Point Comparison Method.

This article takes Moving Windows Average Method to do denoising research on spectrum data. For a spectrum containing noises(n data points), if we average the Kth point, r points before it and r points after it, which makes the number of involved points: , the calculation formula of Moving Windows Average Method is:

$$y_k^* = \frac{1}{N} \sum_{i=k-r}^{k+r} y_i$$

y_k refers to the value of the Kth point before smoothing.
y_k^* refers to the value of the Kth point after smoothing.

Applying Moving Average Windows Smoothing, the value of the windows should be determined. Larger processing windows can improve signal-to-noise ratio. However, they might also results in signal distortion. So the principle of Moving Windows Average Smoothing is that spectrum noises should be efficiently removed while useful information of spectrum are mostly saved.

As for the evaluation of the effects of smoothing, This article takes the root mean square error of spectrum signals (RMSE) and the quadratic sum of spectrum (SUMSQ) as standards. If   refers to spectrum signals containing noises and n refers to the length of spectrum signals, the RMSE of signals is defined as:

$$RMSE = \sqrt{\frac{1}{n} \sum_{i=1}^{n} (s_i(n) - \hat{s}_i(n))^2}$$

The SUMSQ of signals is defined as:

$$SUMSQ = \sum_{i=1}^{n} \hat{s}_i^2(n)$$

The denoising of spectrum signals aims to efficiently eliminate noises while original spectrum information can be saved as much as possible. So the smaller the data squares of denoised signals and the RMS of original signals and denoised signals are, the stronger the denoising capability will be and the closer the denoised signals will be to the original signals, which means better denoising effects. This article applies the second order derivative spectrum of spectrum data to research on the denoising effects of spectrum smoothing.

Figure 3 and Figure 4 reflect the relationship between the number of moving average smoothing points and the error of data square and RMS of spectrum second order derivative.

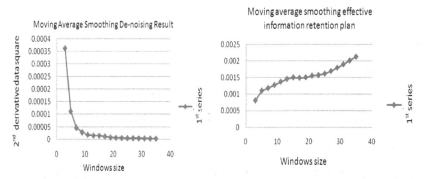

Fig. 3 Moving average smoothing denoising result effective information

Fig. 4 Moving average smoothing retention plan

As shown in Figure 3, as the number of smoothing points increases, data square gradually decreases, which means denoising effects are constantly strengthened. As shown in Figure 4, as the number of smoothing points increases, error of RMS keeps increasing, which means the similarity between the denoised signals and the original signals decreases. That is, as the number of smoothing points increases, while the average smoothing of spectrum removes noises, it also smooths much useful information of spectrum. As shown in the figures, denoising reaches the best effects when the number of moving average smoothing points is 15, 17 or 19.

C   *Multiple Scattering Correction and Variable Standardization*

Multiple scattering correction is a common method applied to eliminate spectrum drift caused by different characteristics of sample surfaces and sizes of particles. Variable standardization is a method applied to remove the impacts of changes of measuring optical path caused by sample sizes, particle sizes, uniformity and sur-face scattering on near infrared spectrum. Multiple scattering correction is appli-cable to conditions where linear relationship between spectrum and material's concentration is good and the chemical characteristics are similar. Variable stan-dardization is applied to remove the impacts of the linear drift of sample spectrum.

4   Results and Analysis

The 64 groups of data of sample spectrum are preprocessed with first order derivative, second order derivative, spectrum smoothing, multiple scattering correction
and variable standardization. Based on them, we develop a correction model of
lignin ratio of eucalyptus woods partial least-squares and fully interactive
validation. Then we evaluate the model with correlation coefficient R (absolute
coefficient R2)and calibration standard deviation (SEC). The results of models of
different preprocessing and prediction models are shown in table 1.

Table 1 Calibration data

Data preprocessing method	PCs	R	R^2	SEC
Original data	7	0.9023	0.8141	0.0533
1st Derivative	4	0.9249	0.8555	0.047
19 points smoothing	7	0.8928	0.7972	0.0557
1st derivative and 19 points smoothing	5	0.9895	0.9791	0.0036
Multivariate Scattering Correction	11	0.8925	0.7967	0.0558
1st derivative and multivariate scattering correction	6	0.9825	0.9654	0.0038
Data Standarization	8	0.9124	0.8326	0.0506
1st derivative and data standardization	*6*	*0.9836*	*0.9674*	*0.0039*

As shown by the statistics in Table 1, in the correction models of lignin ratio of
eucalyptus woods which apply different information extraction technologies to near
infrared spectrum data, the effects of the models which apply first derivative and
19-point smoothing, first derivative and multiple scattering correction, first derivative and variable standardization are better. The correlation coefficient Rs of these
models are all above 98% and the calibration standard deviation (SEC) are all below
0.03%. This indicates that spectrum information extraction technology can improve
the quality of spectrum and the accuracy of spectrum analysis. On the contrary, the
effects of the correction models based on derivative spectrum are worse, which
indicates that while derivative spectrum strengthens spectrum information, it also
strengthens spectrum noises and thus reduces the accuracy of prediction of the
models.

5   Conclusion and Discussion

The near infrared spectrum of samples measured by near infrared spectrometer always contain interference information which is irrelevant to the characteristics of target samples. So feature extraction should be conducted on spectrum to remove the impact of other factors on spectrum information. Methods like spectrum derivative, spectrum data drift, multiple scattering correction and variable standardization can all efficiently extract spectrum information. And in the quantitative analysis with near infrared spectrum, the denoising effects should be further judged by the stability and accuracy of prediction of models developed.

Smoothing precessing can efficiently smooth high frequency noises and improve signal-to-noise ratio. In application, the choosing of window's width is very important. The more the number of data points within the window is, the more the spectrum resolution will decrease and the worse the spectrum distortion will be. However, if the number is too small, the effects of smoothing denoising will be poor. Thus, the width of windows should be determined according to practical conditions.

Acknowledgment. Thank you very much to the subsidize project The National 11th five-year scientific support project 2006BAD32B03.

References

[1] Schimieck, L.R., Anthony, J.M., Carolyn, A.R.: Appita Journal 53(4), 318–322 (2000); 05, 41(4), 177

[2] Kelley, S., Hames, B., Meglen, R.: 5th International Biomass Conference of the Americas (2001)

[3] Feldhoff, R., Huth-Fehre, T., Cammann, K.: Journal of Near Infrared Sectroscopy 6(A), 171–173 (1998)

[4] Yan, Y.: Near Infrared Spectral Analysis Basic and Application. China Light Industry Press (2005)

[5] Huang, A.: Artificial Forest of Chinese Fir Near Infrared Spectrum Charactoristic Forcast. PHD thesism Chinese Academy of Forestry (July 2006)

[6] Yang, Z., Jiang, Z., Fei, B., Liu, J.: Application of Near Infrared Spectroscopy to Wood science 20183

Electromagnetic Energy Harvesting by Spatially Varying the Magnetic Field

Rathishchandra R. Gatti and Ian M. Howard

Abstract. The need for self-powering of low power electronics has accelerated the research of energy harvesters. Electromagnetic energy harvesting poses a relatively inexpensive alternative amongst other methods but has impediments like low voltage and difficulty to scale for micro size level. This study is intended to explore the possibility of increasing the voltage by spatial variation of the magnetic field. An energy harvester design similar to a microphone is considered. A novel methodology of analytically determining the spatial variation and deducing the analytical voltage expression is discussed. This is followed by development of the theoretical model and validation by experiment. The results of the study establish that the theoretical model is not only helpful in readily predicting the output characteristics of the energy harvester but also with high accuracy. This study is helpful for developing future electromagnetic energy harvesters by spatially varying the magnetic field.

1 Introduction

The growths of low power electronics and low/self-powered wireless sensor technology have fuelled the rapid advancement of energy harvesters. Traditionally low power devices and wireless sensor nodes used to be powered by batteries which needed constant replenishment which necessitated this new research of energy harvesting. Piezoelectric, Capacitive, Magnetostrictive and electromagnetic methods are commonly used to generate power from ambient machinery vibrations [1]. Compared to the other three methods, Electromagnetic vibration energy harvesting (EMVEH) is relatively inexpensive, easy to design and harvests vibration energy over a wide range of frequencies and amplitudes [2,3]. The main impediments to widespread use of EMVEHs are low voltage and the difficulty to fabricate them at MEMS/NEMS level. However, recent advances like direct AC-DC booster technology, [4] wherein the front end rectifier based two-stage DC-DC conversion is removed offers promising trends for low voltage generators like

Rathishchandra R Gatti
Curtin University, Perth, Australia

Ian M Howard
Curtin University, Perth, Australia

F.L. Gaol et al. (Eds.): Proc. of the 2011 2nd International Congress on CACS, AISC 144, pp. 403–409.
springerlink.com   © Springer-Verlag Berlin Heidelberg 2012

EMVEHs. Not only that, recently, NdFeB based MEMS magnetic films as thick as 10μm [5] were possible to be micro structured using high power plasma etching thus showing the possibility of MEMS based EMVEHs. The reverse loudspeaker and microphone designs follow the generator principle where the conductor is moving in a constant magnetic field and an EMF is generated. On the counter application, the transformer generates its EMF by a stationary conductor but varying magnetic field. This study was undertaken to ascertain the effect of having a moving coil with a time varying magnetic field (transformer combined with generator) to determine any resulting advantage in designing more efficient EMVEHs. While the transformer has magnetic field as function of time, this model considers the magnetic field, B as a function of distance x where x is a function of time due to the coil's motion, given by $\partial \mathbf{B} / \partial t = (\partial \mathbf{B}/\partial x)(\partial x/\partial t)$. This novel approach considers spatial variation of the magnetic field to achieve the desirable $\partial \mathbf{B}/\partial x$ and the use of high permeable magnetic materials like Iron to obtain the desired $\partial \mathbf{B}/\partial t$.

It is imperative to explore how one can achieve significant variation of $\mathbf{B_x}$ (B as function of x) without changing the magnet geometry thereby maintaining the size. We shall employ the design of moving coil microphone (or the reverse of loudspeaker) [6, 7] that consists of a central iron core encapsulated by a ring magnet (Fig.3) to achieve a sinusoidal variation of Bx in the intended region of vibration. While employing only the ring magnet causes a constant magnetic field in the direction parallel to the coil motion, having a central iron core distorts the magnetic field and bends it in a direction perpendicular to the coil motion thereby increasing the Lorentz forces ($\mathbf{F} = BIl \, sin\theta = \mathbf{B}Il$ where I=Current , l = length). We shall consider an extensive region of $\mathbf{B_x}$ variation as compared to the relatively constant $\mathbf{B_x}$ variation regions as considered in the loudspeaker/microphone designs. This is to utilize the large variation of Bx along the x-axis.

2 Methodology

Most of the electromagnetic designs generally use numerical analysis [6, 8] for determining the magnetic flux distribution of the complex electromagnetic structures due to limitations in analytical approach. This novel approach uses an analytical method for modeling the B_x curve of the energy harvester to have robust understanding of the system behavior and for comparison with the numerical and experimentally determined B_x curve. This is also advantageous from the design point of view because one can predict the system response of the EMVEH just like the commercial transducers available today thus making the design adaptive to the ambient vibration available for the overall system (Wireless sensor node). The current design involves a combined approach of direct computation by considering the FEA results of the spatial variation along the axis of motion of the coil (assumed to be the x-axis in this paper) as well as experimentally ascertaining the same using a Gauss meter. We later use the sinusoidal region in the analytical approach to derive a model for the theoretical voltage generated by EMVEH. This theoretical model is then validated by experiment.

3 Magnetostatic Analysis

Magnetostatic analysis using Vizimag software was performed to numerically determine the varying magnetic field (Fig.1). From the contour plot, it was observed that the physics of the design was ascertained with the flux lines being perpendicular to the coil in the operable range from x=3mm to x=7mm. To cross verify the analysis, the experimental values were measured using the gauss-meter at 2mm intervals. It was observed that the approximate experimental Bx values are close to the numerically obtained curve thus showing that modeling results are consistent with the experimental and numerical work as illustrated in Fig.2.

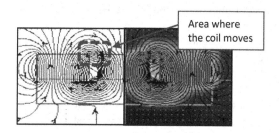

Fig. 1 Magnetic flux distribution of ring magnet and iron core assembly

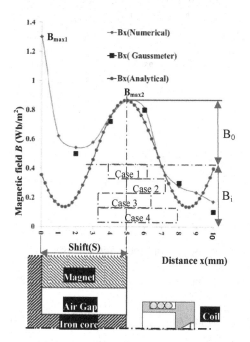

Fig. 2 Determination of spatial variation by numerical and experimental $\mathbf{B_x}$ values

It was observed that the maximum **Bx** values occur at x=0 ($\mathbf{B_{max1}}$) and x=5mm ($\mathbf{B_{max2}}$). The coil movement was designed such that it can use only x > 1.5mm to prevent the coil from hitting the iron core at its maximum forward position. Thus only the second maximum $\mathbf{B}_{max2}$ at x=5mm can be chosen which falls under the region of coil operability. It is observed that the numerically determined Bx curve is sinusoidal at the operable range surrounding the $\mathbf{B}_{max2}$ point. A curve fitting analysis was used to fit a curve found to be

$$\mathbf{B}_x = \mathbf{B}_0 \sin(gx_n) + \mathbf{B}_i , \tag{1}$$

where $\mathbf{B}_i$ =0.5T, $\mathbf{B}_0$=$\mathbf{B}_{max2}$−$\mathbf{B}_i$ =0.865-0.5=0.365T, and   is the spread of the sine wave and x_n=distance function of the nth coil (2).From the $\mathbf{B}_x$ curve, 4 cases were identified to test the robustness of the theoretical model.

Case1: Movement of coil in constant magnetic field. Where $\mathbf{B}_x$ is relatively constant and tends towards Zero (x=3 to 7 mm).
Case2: Movement of coil in varying Magnetic field. Where   is very high (x=5 to 7 mm). We chose to keep same amplitude (= .75 mm) and number of coils (=3) for both cases 1 and 2 for comparison.
Case3: Movement of coil within the sinusoidal region wherein the entire coil never goes out of this region. For this, we have amplitude =.875 mm and the number of coils possible =7. This is mainly intended to study the energy harvesting for small amplitude and higher frequency.
Case4: Movement of coil in and out of the sinusoidal region wherein the entire coil goes out of this region. For this, we have amplitude = 3 mm and no of coils possible =14. This case is also intended to study the energy harvesting for large amplitude and lower frequency.
Since the $\mathbf{B}_x$ curve is not exactly sinusoidal, we considered use of the shift factor (S) corrections in x for cases 1 and 2 to improve modeling accuracy.

4   Analytical Modeling

Consider a coil of N turns vibrating linearly at angular frequency ω and amplitude A in a magnetic field that varies with distance along the direction of motion of the coil. At any given instant of time t, the nth turn of the coil is at the position given by

$$x_n = A\sin(\omega t) + d_w(n-1) + S , \tag{2}$$

where shift S = 5mm is the shift of the sine peak from the reference (see Fig. 3) and d_w = diameter of the wire.

Considering the transformer theory, Faraday's law for generated EMF at the nth turn can be written as

$$V_n = -\partial \mathbf{\Phi}_x^n / \partial t = -(\partial \mathbf{\Phi}_x^n / \partial x)(\partial x/\partial t) . \tag{3}$$

The flux at the nth coil of radius r_c at any time t depends on its position x at that instant and hence is given by

$$\Phi_x^n = \int_s \mathbf{B}_x^n . ds == \pi r_c^2 (B_0 \sin(gx_n) + B_i) \cdot \tag{4}$$

Considering (4) for all N turns, the entire coil yields the total theoretical voltage (TTV) given by

$$V = \sum_{n=1}^{N} V_n = \sum_{n=1}^{N} \pi r_c^2 gA\omega B_0 \cos(g(A\sin(\omega t) + d_w(n-1) + S))\cos(\omega t) \tag{5}$$

5 Experimental Setup

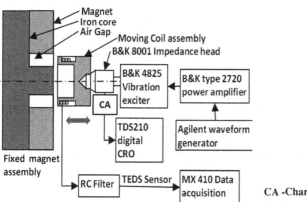

The schematic of experimental setup is as shown in Figure 4. The experiment was conducted in four sets for all the four cases separately by varying the number of coils and amplitudes. The coil assembly was made to vibrate back and forth inside the fixed magnet-Iron-core assembly and the voltage generated was measured in HBM data acquisition device. Each set of experiment was conducted for frequencies 7Hz-31Hz in steps of 1Hz. The .xls output file of the voltage from the HBM was read into a MATLAB program that also simulates the theoretical voltages for the same conditions.

6 Results and Discussion

The total uncertainty in the voltage is determined by applying Taylor's 1st order equation to generated voltage parameters [9] in (6) as

$$\frac{Uv}{V} = U_{r_c}\frac{\partial V}{\partial r_c} + U_g\frac{\partial V}{\partial g} + U_A\frac{\partial V}{\partial A} + U\omega\frac{\partial V}{\partial \omega} + U_{B_0}\frac{\partial V}{\partial B_0} + U_{x_n}\frac{\partial V}{\partial x_n} + U_t\frac{\partial V}{\partial t} \tag{6}$$

The experimental error at worst case analysis was found to be 8.69%, of which the distance error was the highest contributing factor (7.39%) mainly due to i) limited control of establishing initial static position of coil ii) mechanical design deviation iii) hand wound coil (form factor).

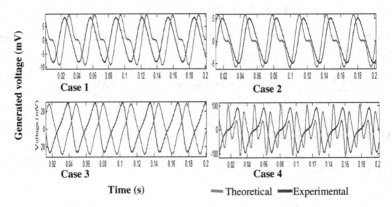

Fig. 4 Signal analyses of theoretical and experimental voltages for all the 4 cases

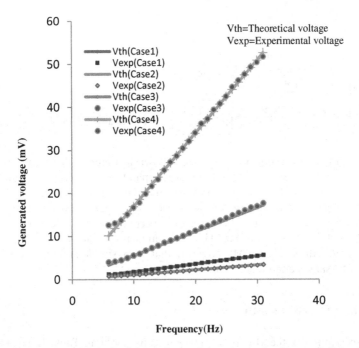

Fig. 5 RMS voltage analysis of theoretical and experimental voltages of all the 4 cases

The signals of the theoretical and experimental voltages were compared for all 4 cases(Fig.4). Cases 1, 2 and 3 were found to have approximately the same shape and Peak values. It was observed that the experimental signal was skewed towards one side and had two sine curves of the form SinA+SinB that validates (6). It can also be seen that Case 4 is the outlier with no match of theoretical and experimental voltage signals since it vibrates outside the boundary of the Bx sine curve considered. This again was expected and validates the theoretical model.

The theoretical and experimental root mean square voltages were plotted in RMS voltage plots (Fig.5) and observed that the theoretical voltage values match very well with the experimental voltage values thus validating the model within the low frequency range of 7-31Hz considered.

7 Conclusions

Since the theoretical RMS voltages matches very well with the experimental RMS voltages, the theoretical model approach is confirmed. This novel approach can be used to design many new energy harvesters by taking advantage of the B_x variation in different configurations.

Case 1, 2 and 3 was conducted for lower amplitudes and Case 4 for higher amplitudes of motion. This can be used to design an energy harvester for broad range of amplitudes (modular design) by taking advantage of the $B(x)$ profiles of the coil magnet configurations.

One of the best approaches to design the spatially varying magnetic field base EMVEH is to design it using a theoretical model over its operable range so that one can straight away include its TTV in the overall design for smart sensor applications.

References

[1] Priya, S.: Energy Harvesting Technologies (2009)
[2] Wang, L.: Vibration energy harvesting by magnetostrictive material. Smart Materials and Structures 17(4), 045009 (2008)
[3] Jonnalagadda, A.S.: Magnetic Induction systems to Harvest energy from mechanical Vibrations. In Department of Mechanical Engineering, Massachusetts Institute of Technology, p. 108 (2007)
[4] Dwari, S.: Efficient direct ac-to-dc converters for vibration-based low voltage energy harvesting. In: 34th Annual Conference of IEEE Industrial Electronics, 2320 (2008)
[5] Jiang, Y., et al.: Micro-structuring of thick NdFeB films using high-power plasma etching for magnetic MEMS application. Journal of Micromechanics and Microengineering 21(4), 045011 (2011)
[6] Hwang, S.M., Sang-Moon, H.: New development of combined permanent-magnet type microspeakers used for cellular phones. IEEE Transactions on Magnetics 41(5), 2000 (2005)
[7] Sinclair, I.R. (ed.): Sinclair, Newnes audio and hi-fi handbook (2000)
[8] Cepnik, C., et al.: Effective optimization of electromagnetic energy harvesters through di-rect computation of the electromagnetic coupling. Sensors and Actuators A: Physical 167(2), 416–421 (2011)
[9] Beckwith, T.G., Marangoni, R.D., Lienhard, J.H.: Mechanical Measurements, 6th edn. Pearson Prentice Hall (2007)

The Inrush Current Eliminator of Transformer

Li-Cheng Wu and Chih-Wen Liu

Abstract. In the power industry, transformers are key components for electrical energy transfer. However, the transformer always is affected by inrush current problems of itself. The inrush currents will cause power system network with voltage dip, harmonic, over-voltage, sympathetic current and mal-trip of protective relay and so on. In the past, there are no total solutions for all of these problems. Therefore, this paper will develop an inrush current eliminator to eliminate inrush currents of transformer, so all of the above mentioned problems of transformer will be solved.

1 Introduction

The phenomenon of transformer inrush current has been discussed in many papers since 1958 [1-9]. However, almost all those papers focus on inrush current for protective relay problems, inrush current impact power system problems, and simulation of inrush current and so on. A few papers have proposed an elimination method of inrush current with the circuit breaker (C.B.) of three-phase cut off load. Here, to limit inrush currents of transformer with energizing unloaded conditions, some methods are proposed [2-6], such as the controlled switching scheme[2]-[3], a sequential phase energizing method[4-5], improvement of transformer's winding configuration[6], and pre-insertion of series resistors method[7] and so on. In the references [2]-[3], they use controlled switching to select the optimum closing point, but the point is that the data value of the residual flux pattern, residual flux magnitudes and performances of circuit breaker should be measured in advance. The reference [4] also can get the results of inrush currents restrained, but this method needs independent pole breaker control and neutral resistor. In the reference [6], they use a new viewpoint on the design of transformer to restrain the inrush current. The reference [7] uses pre-insertion of series resistors to reduce inrush currents. This method will increase equipment costs and require a lot of maintenance. In this paper, therefore, we will show a new method of simple and low cost that no any complex information (Like: parameters of transformer, pre-insertion resistors, independent pole C.B. and power system construct) should be needed.

Li-Cheng Wu · Chih-Wen Liu
Electrical Engineering, National Taiwan University, Taipei, Taiwan

F.L. Gaol et al. (Eds.): Proc. of the 2011 2nd International Congress on CACS, AISC 144, pp. 411–419.
springerlink.com   © Springer-Verlag Berlin Heidelberg 2012

2 Theory of Inrush Current Eliminator

The equivalent circuit of transformer model shown as Fig.1 consists of an ideal transformer of ratio $N_1 : N_2$ and parameter of elements. The model takes into account of the winding resistances (R_p, R_s), the leakage inductances (L_p, L_s) and the excitation characteristics of iron core. According to Fig. 1, we can write equations as follows:

$$V_p = R_p I_p + L_p \frac{dI_p}{dt} + N_1 \frac{d\varphi_m}{dt} \tag{1}$$

$$V_s = R_s I_s + L_s \frac{dI_s}{dt} + N_2 \frac{d\varphi_m}{dt} \tag{2}$$

When the transformer is energized in no-load (Is=0), the equation (2) can be expressed by:

$$\varphi_m = \frac{1}{N_2} \int V_s dt \tag{3}$$

Due to the fact that the number of turns of primary N_1 is larger than the item of $\int R_p I_p dt$ and $L_p I_p$, we can modify the equation (3) as follows:

$$\varphi_m = \frac{1}{N_1} \left[\int (V_p - R_p I_p) dt - L_p I_p \right] \cong \frac{1}{N_1} \int V_p dt \tag{4}$$

Here, we set $V_p = V_m \cdot \sin(\omega t)$ and assuming steady-state, then the formula (4) becomes

$$\varphi_m \cong \frac{1}{N_1} \int V_m \cdot \sin(\omega t) dt \cong \frac{-V_m}{N_1 \omega} \cdot \cos(\omega t) + k \tag{5}$$

Therefore, we can know the residual flux $\varphi_{m,res}$ is from formula (5) when the circuit breaker of transformer is opened at the t_{open} moment. We can get the residual flux $\varphi_{m,res}$ as formula (6) after transformer de-energized

$$\varphi_{m,res} = \frac{-V_m}{N_1 \omega} \cdot \cos(\omega t_{open}) = \varphi_{max} \cdot \cos(\omega t_{open}) \tag{6}$$

Here $\varphi_{max} = \frac{-V_m}{N_1 \omega}$, as Fig.2 we know, if the transformer is given a random energized, the inrush currents will often occur. The main reason is that the excitation flux φ_m of transformer will tend to saturation. Therefore, our work is keeping the excitation flux φ_m of transformer not from entering the saturation zone when the transformer is energized. The task can be easily achieved by controlling C.B. of transformer. Here, we only control the closing and opening angles of C.B. at the optimal instant of transformer's energized. The optimal closing time ($t_{optimal}$) of C.B. is used to reduce the residual flux of transformer. From formula (5), we can write an equation of the excitation flux φ_m of transformer with C.B. closing:

$$\varphi_m \cong \frac{1}{N_1} \int_{t_{close}} V_m \cdot \sin(\omega t) dt + \varphi_{m,res}$$

$$\cong \varphi_{max} \cdot \cos(\omega t) - \varphi_{max} \cdot \cos(\omega t_{close}) + \varphi_{m,res} \qquad (7)$$

Using equation (6) and (7), we can control the excitation flux φ_m of transformer to make it toward steady-state flux ($\varphi_{max} \cdot \cos(\omega t)$) as Fig.3 shows when the C.B. closing time (t_{close}) is equal to the C.B. opening time (t_{open}) then the transient flux will be zero ($\varphi_{transient} = -\varphi_{max} \cdot \cos(\omega t_{close}) + \varphi_{m,res} = 0$). This condition can easily be gotten by measuring the voltage of transformer terminal. The real development and test will be introduced in section 3.

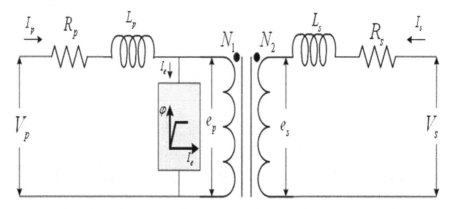

Fig. 1 Equivalent circuit of a two-winding transformer

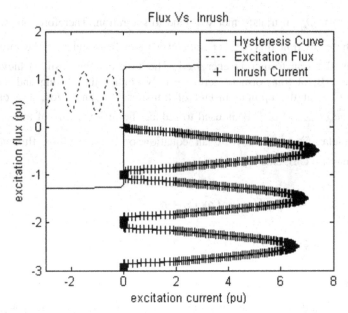

Fig. 2 Excitation flux and inrush current

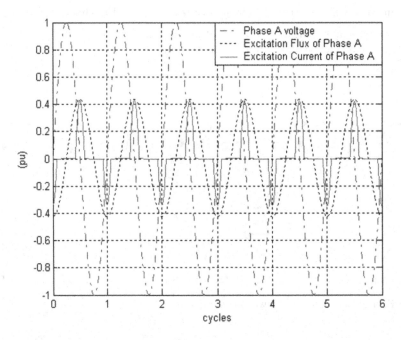

Fig. 3 The phase angle relation of voltage, excitation flux and excitation current

3   Performance Verification of Inrush Current Eliminator

According to the theory of section 2, we develop a test system of inrush current eliminator as Fig.4 shows. The prototype of inrush current eliminator has been developed in laboratory as shown in Fig.5. The inrush current eliminator consists of DSP controller(Texas Instruments TMS320F2812), ADC*16(12 Bit), 32K words on board data RAM, 32K words on board program RAM, on board 30 MHz crystal, I/O functions and voltage sensor. The transformer's parameters are shown in table 1. The C.B. is replaced by solid state relay(S.S.R) which can cut off inductance load current about three-phase 100A/600V at the same time. The waveform recorder uses Nicolet(Odyssey OD-100) with 16 channels 100kS/s.

In order to evaluate the performances of inrush current eliminator, we compare a random energized with an inrush eliminator scheme of no-load transformers. Fig.6 shows that the large inrush current will occur in no-load transformer energized without inrush current eliminator. The voltages and currents of transformers are indicated in Fig.6 (a) and Fig6.(b) respectively. The transformer with no-load are energized twice occurred in this case. Here, we can find the first stage of the largest inrush current about -35.5A in phase C(Fig.6 (b-1)) and the second stage of the largest inrush current about 37A in phase A(Fig.6 (b-2)). The Fig.7 shows the test results of transformer's energized with inrush current eliminator. The test results show the transformer can directly enter steady-state operation as Fig.7(b). Therefore, we can not find no inrush currents appearing on the Fig.7(b-1) and Fig.7(b-2). From Fig.6 and Fig.7, we can clearly know the inrush current eliminator can get best solution for transformer's energized. In addition, all of the test results will be shown in Fig.8.

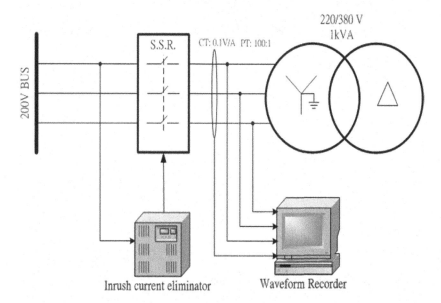

Fig. 4 The performance test system of inrush current eliminator

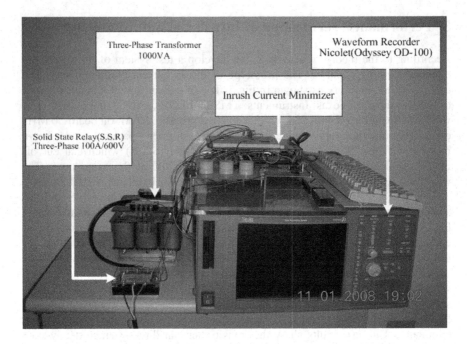

Fig. 5 The prototype of inrush current eliminator

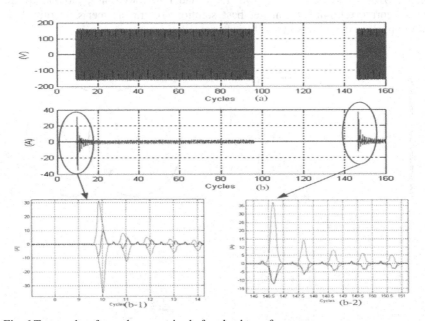

Fig. 6 Test results of a random energized of no-load transformer

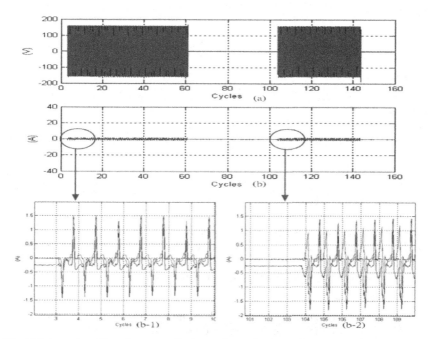

Fig. 7 Test results of inrush current eliminator for no-load transformer energized

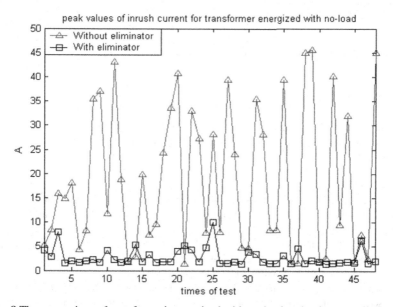

Fig. 8 The comparison of transformer's energized with no-load under the controlled and the random situations

Table 1 Transformer Parameters

Items	Values
Rating	1000VA
Voltage Ratio	220/380(Vrms)
Frequency	60(Hz)
Inductance A/B/C	0.61/0.59/0.65 (H)
Connection	Yn/Delta

4 Conclusions

This paper has presented a simple method to eliminate inrush current by transformer energized with no-load. The method has been developed by a prototype of inrush current eliminator in laboratory. All of the test results indicate that the inrush current eliminator can work effectively for transformer energized with no-load. Apart from limiting inrush current, the method of inrush current eliminator includes many advantages as follows:

A. Simplest theory;
B. Lower cost(For examples: support C.B. of three-phase together, no need added potential transformer and no need pre-resistor ect.);
C. With its simple structure, it's easy to carry out the expected result in field; and
D. It can be used in energized of transformers, cables and capacitors.

However, this inrush current eliminator has only one fault - if the transformer has never been energized in the power system before, inrush currents can not be eliminated when the transformer is energized for the first time.

References

1. Sonnemann, W.K., Wagner, C.L., Rockefeller, G.D.: Magnetizing Inrush Phenomena in Transformer Banks. AIEE Transactions, Part III 77 (October 1958)
2. John, H.B., Fröhlich, K.J.: Elimination of transformer inrush currents by controlled switching—Part I: Theoretical considerations. Transactions on Power Delivery 16(2), 276–280 (2001)
3. John, H.B., Fröhlich, K.J.: Elimination of transformer inrush currents by controlled switching—Part II: Application and performance considerations. Transactions on Power Delivery 16(2), 281–285 (2001)
4. Cui, Y., Abdulsalam, S.G., Chen, S., Xu, W.: A sequential phase energization technique for transformer inrush current reduction—Part I: Simulation and experimental results. Transactions on Power Delivery 20(2), 943–949 (2005)
5. Cipcigan, L., Xu, W., Dinavahi, V.: A New Technique to Mitigate Inrush Current Caused by Transformer Energization. In: IEEE Power Engineering Society Summer Meeting, July 2002, vol. 1(25), pp. 570–574 (2002)

6. Chen, J.F., Liang, T.J., Cheng, C.K., Lin, R.L., Yang, W.H.: Asymmetrical winding configuration to reduce inrush current with appropriate short-circuit current in transformer. In: IEE Proc.-Electr. Power Appl., May 2005, vol. 152(3), pp. 605–611 (2005)

7. Rioual, M., Martinez, M.: Energization of two transformers in series through long lines: Correlation between fluxes in both transformers, and determination of the efficiency of palliative solutions. In: IPST, June 2007, pp. 605–611 (2007)

8. Sidhu, T.S., Gill, H.S., Sachdev, M.S.: A power transformer protection technique with stability during current transformer saturation and ratio-mismatch condition. IEEE Transactions on Power Delivery 14(3), 798–804 (1999)

9. Wu, L.-C., Liu, C.-W., Chien, S.-E., Chen, C.-S.: The Effect of Inrush Current on Transformer Protection. In: IEEE/PES The 38th North American Power Symposium, September 2006, pp. 100–108 (2006)

Promoting Professional IT Training Environment Construction with Virtual Machine Technology

Liu Qingyu and Zheng Mengze

Abstract. IT specialty of Modern Vocational Education is faced with many environmental problems, of which a considerable part of the practice teaching training program requires a better environment and the biggest problem is multi-machines per student; multiple network segment environment and destructive experiments of data and system. The prevailing problem in all schools is a large investment and difficulty in environmental equipment maintenance, judging from construction cost and training environment upkeep. A virtual machine technology environment might be a good solution to it. This way is to put a machine in use for effects of multi-machines for the purpose of saving money and reducing maintenance cost.

Keywords: virtual machine, virtual technology environment, IT specialty, VMware.

1 Existing Problems

Computer practice teaching is an important component of teaching IT professionals[1,2]. To master computer science, the learning of theoretical knowledge alone is far from enough and computer science is a specialty focusing on practice. All subjects related to computer science require that skills and mastery of operations can only be obtained through practice. The objectives of practice teaching are to test and verify practical hands-on operations and theoretical knowledge learned with practice. Enhance memory is an extension and replenishment of theoretical teaching.

It is necessary to further improve the training environment[5] as the expansion of enrollment and acceptance of more and more students. Due to lack of funds,

Liu Qingyu
Ningbo Dahongying University, Ningbo, Zhejiang Province, China
e-mail: lqy7979@gmail.com

Zheng Mengze
Ningbo Dahongying University, Ningbo, Zhejiang Province, China
e-mail: zhengmengze@gmail.com

F.L. Gaol et al. (Eds.): Proc. of the 2011 2nd International Congress on CACS, AISC 144, pp. 421–427.
springerlink.com
© Springer-Verlag Berlin Heidelberg 2012

some colleges and universities belonging to the IT category are poor in laboratory environment, unitary in device type, scanty in equipment, out-of-date in instrument quantity, and the practice teaching can not meet the needs of the students' practice demands.

Meanwhile, some of the computer system maintenance experiment[3], such as disk partitions, format, operating system installment, system backup and recovery, due to destructive experiments on the disk, it may damage the hardware device or damage to data, teachers will only adopt "lecturing instead of training" approach for teaching. Computer network experiment[4,5,7], such as the local network, network configuration and management, Telnet, FTP, IIS configuration of various network services, remote access and routing setting experiments need multiple computers and multiple devices, the limited available experimental teaching conditions lead to inability of experiment to solve practical problems. Network security experiment, such as the use of scanner, sniffer monitors, Trojan horses and other hacker tools and web server security configuration, lacking experimental environment, we can only take on classroom teaching and demonstration, and the students can not operate with hand, and can not do actually image study, so it is difficult for students to understand, therefore they lost interest in study.

Comparing the experimental data, we found that the use of existing computer combined with virtual machine technology, the above-mentioned problems can be solved, not only improving experimental teaching effectiveness, but also saving a lot of money for purchasing expensive equipment.

2   A Brief Introduction to Virtual Machine [1-5]

A Virtual machine refers to the software simulation of hardware systems with full functionality, running on a completely isolated environment of complete computer systems. Virtual machine technology, one of the technologies popular in the last two years, has roused great attention among enterprises and media. By virtual machine software, you can simulate one on a physical computer or more virtual computers. These virtual machines can do work as well as on a real computer, for example, you can install the operating system, install the application and access to network resources and so on. As far as your purposes are concerned, it is only a running application on your physical computer, but in running applications on a virtual machine, it does well as the real computer. Therefore, when doing software testing in a virtual machine, the system might collapse, but the collapse is but a virtual machine operating system rather than the physical computer's operating system, use virtual machine "Undo" (recovery) function, you can immediately resume the virtual machine to the state before installing the software. In addition, virtual computer system is saved as a file, you can copy to any location, imported to a virtual machine and the virtue machine can be directly used as well as the original, so that the use is greatly facilitated.

Virtual machine can bring facilities at least as follows:

1) Demonstration environment, a variety of presentation environments, all sorts of easy examples can be installed;

2) Ensure that the host is fast running, reduce unnecessary waste installation, occasional procedure or test procedures that run in a virtual machine;

3) Avoid re-installing every time and online banking tool procedures, less often uses, and requires better confidentiality, a separate operation can be performed in an environment;

4) Test of unfamiliar application in a virtual machine can be installed and completely removed easily;

5) Experience different versions of operating systems, such as Linux, Mac, etc.;

6) Carry out all kinds of destructive experiments on the system, such as hackers, software virus detection and data recovery experiments.

Among various kinds of virtual machine software, VMware is better in performance. It can simulate a number of machines on a computer. There's no remarkable difference between a virtual machine and a physical host. They need partition, format, operating system installation and applications, as well as a real computer.

The most important feature of virtual machine VMware can run multiple operating systems without rebooting a computer. It is different from installing multiple operating systems in the same PC. Multiple operating systems installed, only one of them can run at any moment. If you want to switch to other systems, you must restart the machine. In the virtual machine environment, you can run multiple operating systems synchronically. You can switch back and forth between multiple operating systems without having to reboot the machine. The most noteworthiness is that powerful network functions of VMware can link multiple virtual machines to form one or more local area network, the use of this network is fully consistent with the real network without care about network card or damage to the switch because these are also virtual which can be added or deleted at any time.

3 Process of Project Research

In this study, our study focused on environment construction of network of IT education, and the VMware software is selected for the virtual experiment, studying the network server configuration, representative of the environmental requirements

DHCP, FTP, IIS (Web), DNS, WINS, Active Directory, Certification Authority, VPN, mail server, streaming media server are chosen as the research object. These servers are the most widely used ones in enterprises and Internet servers. They are highly applicable and practical in IT education.

3.1 The Support of Virtual Machine Environment to the Content of That IT Professional Training

Following comparison of the virtual machine environment with the real environment, we examined the actual content of configurations of IT professional training under virtual machine environment[9] to know the operation of the virtual machine and to see whether the real training requirements can be satisfied. After investigation training room configuration of more than 20 colleges and universities, integrated configuration of their computer configurations and found their computer configurations are under the middle level. we took their computer configuration environment for example and studied various server configuration and their hardware environment demands.

Table 1 Machine configuration test

Machine Configuration Test	
CPU	Celeron IV 1.7G
Memory	1G DDR
Audio Card	Built-in
Hard drive	60G

For this configuration, a detailed experiment, record the services used in the experiment of memory and CPU consumption.

Table 2 Virtual machine test data

Operating system::windows server 2003;Physical Machine memory capacity:1000MB;Virtual machine memory:256MB					
Name of Server	Start-up time	Service start time	Virtual machine file size	Running virtual machine memory	CPU usage
DHCP	73S	3S	1.96GB	146MB	5%
DNS	74S	3S	1.96GB	130MB	5%
CA	90S	45S	2.07GB	136MB	10%
AD	103S	50S	2.07GB	170MB	15%
VPN	94S	34S	2.06GB	135MB	13%
WEB &FTP	96S	40S	2.04GB	140MB	13%

Note: All server only used for authentication, service pressure is very small, and memory and CPU occupancy are very small.

These data are obtained after several tests derived from the mean. From the data, we can see, when memory is 1G, 256MB virtual machine memory is set, all services are properly installed and only part of the call to open files to service, more time is consumed.

Test data obtained are in accordance with the current mainstream configuration of the computer, all the necessary requirements can be easily reached. In the actual teaching process the machine configuration used is generally higher than the test machine configuration, therefore, there is no problem for the virtual machine to run.

3.2 Comparison between Physical Demands and Virtual Machine Environmental Needs

(1) Statistics of real environmental needs (each student needs).

Table 3 Real statistics of environment needs (per student)

Real environment needs	Real cost (to maximize the demand for statistics, $410/per machine
One Server & Three Clients	4*$410=$1640
Computer Configuration：Lenovo E2300,(CPU: INTEL E2300 (Dual Core 1.8G); Memory:2G DDR2; HD:320G SATA2; Integrated lan &HD Audio;17'LCD	
Note: The test performance demand is less on the server, so the same configuration and verification of machine can be used. April 2011 quoted market price of the machine.	

(2) Statistics for virtual machine environment demand (per student).

Table 4 Statistics of virtual machine environment needs (per student)

Virtual Machine Environment	Real cost (to maximize the demand for statistics)
Lenovo E2300（Run 2-3units of virtual machine)	$410
The same Computer Configuration as Table3. Note: Experimental training demand is less on the server performance, so the same configuration and verification of machine can be used. April 2011 quoted market price of machine.	

(3) Total cost Comparison of real environment and virtual machine environment.

Table 5 Cost comparison

	Real Environment	**Virtual machine environment**
Each student's need	3	1
Class size	45	45
Price of each machine	$410	$410
Total cost	3*45*$410=$55350.00	1*45*$410=$18450.00
Virtual machine environment saves	$55350.00-$18450.00=$36900.00	

(4) Comparison of real environment and virtual machine environment (other aspects)[9].

Table 6 Other comparisons

	Real Environment	**Virtual machine environment**
Total machines required	135	45
Classroom occupancy	50 computers * 3 rooms	50 computers * 1 room
Difficulty of management	hard	easy
Concrete situation of Course Teaching	The situation required more than one computer, running back and forth, fail to observe classroom disciplines	Only operating on one computer

Through the above comparison, we can draw the following conclusions: Using a virtual machine environment to build IT training courses will save a lot of money (to save $36900.00) and the site (save 2/3 of the classroom space), and management and students performance become easer. These are of great advantage.

4 Summary

In this study, we implemented the use of 50 computers to create 150 virtual machines running training machine environment which allows a student's need of using 3-4 computer training operation projects and greatly enhanced the student practical hands-on problem-solving abilities.

In finance, the virtual machine environment saves environment than real machine by nearly $37000; as for area, the virtual machine environment saves environment than real machine by nearly 2/3; as for configuration flexibility, virtual machine training environment is based on need, at any time the number of virtual machines can be added as long as there is sufficient memory and hard disk space, open multiple virtual machines simultaneously test is not a problem. In addition, network training programs using virtual machine can build more LANs, thus a physical machine on the purpose of building multiple logical networks is achieved, further saving to buy network switch environment building costs.

Application and popularization of virtual machine applied to all training courses will greatly enhance experiment course and training results, enhancing the capacity of students to master skills, and save lot of money and space.

From the study results using virtual machine for computer specialty training environment to improve experiment is feasible and effective. It also saves money, saves floor space. Configuration is flexible and easy to manage. Its construction method is simple, low in cost and effective. It combines with computer technology that can help students and teachers at any time to operate computer experiments, to compensate the limitation of the real laboratory. Once it is put to really flexible use, it can test and help IT specialty environment construction and improvement training environment.

References

[1] Shu, Y., Zheng, W.: Vmware Virtual Computer-based Experimental System. Research and Exploration 25(9), 1087–1088 (2006)
[2] Zhuang, X.: VMware Workstation on the Computer Teaching. Xi'an Aviation and Technical College 27(3), 76–78 (2009)
[3] Wang, J.: VMWare Virtual Machine Technology in the Computer Room Management and Application. Science and Technology Information 1, 96–38 (2009)
[4] Shen, J.: Realization of Single Virtual Network-based Experimental Platform. Computer Knowledge and Technology (Academic Exchange) (17) (2007)
[5] Huang, Z.: Multi Operating System Environment Based on VMware and Virtual Network. Computer Knowledge and Technology 4(3), 729–730 (2008)
[6] Xie, S.: Constructing Virtual Software Environment for Teaching Experiment. Financial Computer (3), 37–39 (2009)
[7] Xing, Z., Lin, P.: The Role of Virtual Machine for Network Experiment Course. Research and Exploration 26(12), 304–306 (2007)
[8] Jia, L., Lin, P.: Application for Computer Education Based on VMware Virtual Machine Technology. China's Modern Education Equipment 2(60), 15–17 (2008)
[9] Chen, J., He, Z.: On VPN-based Virtual Machine Experiment Environment Construction. Research and Exploration 29(1), 59–61 (2010)

The Study of Parametric Modeling and Finite Element Analysis System of Skirt Support

Su Liu and Changming Yang

Abstract. Combining APDL in ANSYS with VC, the Parametric Modeling and Finite Element Analysis system of skirt support is developed. In the system, VC is used to establish dialog interfaces and encapsulate the process of ANSYS finite element analysis, the system itself will produce the APDL program and run ANSYS automatically through inputting the needed parameters, which realizes the parameterized finite element analysis of the skirt support. A case study is used to prove the feasibility of this system.

1   Introduction

Skirt support is one of the main high stress areas in pressure vessel, it has harsh operating conditions and strict design requirements. In traditional design, it usually uses a simplified structure model to do a rough calculation and analysis, or even estimates results with experience, so that the obtained result is quite unreliable.

Some finite element analysis software are widely used to do modeling and analysis, which can get relatively ideal results, but they are quite ineffective in design changes or modification. if some size or data of object changes, It will brings a lot of repetition work and human errors.

Therefore, the Parametric Modeling and Finite Element Analysis system of skirt support is developed based on Visual C + +6.0, ANSYS10.0, which can generate finite element model and grids of skirt support automatically to realize the finite element analysis of the skirt support by just inputting the structural parameters.

2   Basic Function Principle of the System

The system builds the parametric design interface of hydrogenation reactor with VC++6.0 tool, which can produce the APDL command stream in ANSYS when its structural parameters are changed.

Su Liu
College of Mechanical Engineering Nanjing University of Aeronautics and Astronautics
e-mail: meesuliue@nuaa.edu.cn

Changming Yang
College of Mechanical Engineering Nanjing University of Aeronautics and Astronautics
e-mail: 258227931@qq.com

F.L. Gaol et al. (Eds.): Proc. of the 2011 2nd International Congress on CACS, AISC 144, pp. 429–435.
springerlink.com   © Springer-Verlag Berlin Heidelberg 2012

By modifying ANSYS startup files, APDL command stream generated by VC can be directly read when ANSYS is started, so that it can obtain the stress analysis of skirt support. Flow chart of system development is shown in Figure 1.

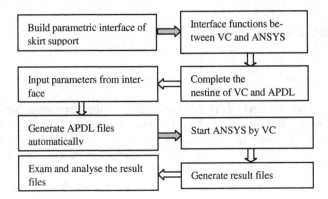

Fig. 1 Flow chart of system development

3　Parametric Modeling and Finite Element Analysis System of Skirt Support

Skirt support is usually "h" type forged piece, as Figure 2, the upper part of "h" is connecting with cylinder, and the bottom part is joint with Seal Cap and skirt

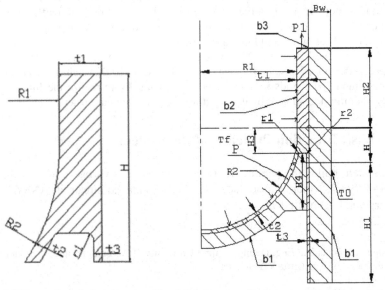

Fig. 2 "h" type forged piece　　　**Fig. 3** Total stress calculation model with insulation layer

support. Generally, "h" type forged piece has three basic thickness designs, they are cylinder (R1), skirt support (R2) and   the thickness of skirt support (T3).

The hydrogenation reactor, working in high temperature and pressure environment, not only has a large mechanical stress, but also possibly has a great thermal stress, so the stress status of support area may deteriorate. Therefore it is necessary to do a comprehensive analysis of mechanical stress and thermal stress. Figure 3 shows the total stress calculation model with insulation layer.

The analysis adopts axisymmetric model, in which the length of cylinder and skirt support in "h" type forged piece is much 2.5 times longer than the length of the edge stress decay .

Figure 3 shows the Loads and constraints on model. P1 is the surface force on the top of cylinder, which is converted from gravity. P2 is the surface force that simulates force is placed on the top of a closed cylinder:

$$P_2 = \frac{PR_1^2}{(R_1 + t_1)^2 - R_1^2}$$ (1)

ANSYS offers two methods of combined stress analysis to calculate thermal stress.

One is the direct method. In this method, the coupling units of temperature and displacement free degrees are used to carry out both thermal analysis and stress analysis.

The other is the indirect method. In this method, the first is to do temperature field analysis, read the results, and add the cell node temperature on the model as body load. Then do the thermal analysis. Finally, the boundary displacement constraints, internal pressure load, and gravity load are added, thus the analysis of combined stress is completed.

Of the two methods, the indirect method can use all the functions of thermal analysis and structural analysis, so we choose it as our analysis method.

The flow chart of combined stress analysis is shown in Figure 4.

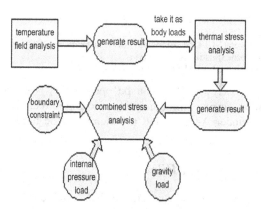

Fig. 4 Combined stress analysis process

4   The Nesting of VC++ and ANSYS

4.1   APDL Code of Finite Element Analysis of the Skirt Support

Parametric APDL command code of finite element analysis of skirt support primarily consists of ten parts: ①Definition of model cell type; ② Definition of heat transfer coefficient; ③Definition of section model; ④Definition of the main body and the insulation layer; ⑤Definition of thermal load; ⑥Thermal analysis of temperature field; ⑦ Definition of mechanical loads; ⑧definition of material property; ⑨definition of solving; ⑩ definition of post-processing.

The method obtaining parametric APDL code is as follows: firstly, do finite element analyses of skirt support in ANSYS, then define parameter names to represent relevant values of variable. After analysis is completes, select "File \ Write DB LogFile" menu to output APDL batch text, then modify and simplify it, so that we can get the parametric APDL code needed by the program.

4.2   Interface between VC and ANSYS

There are several functions that can be used to start other applications in VC++6.0, such as WinExec, ShellExecute, CreateProcess and so on. Of these functions, CreateProcess can specify the security attributes inheritance and class priority.of the process, so CreateProcess is chosen to start ANSYS.

Ansys10.0 provides a format of batch processing, "ansys100-bi inputname-o outputname", in which "inputname" and "outputname" are respectively input and output file name. A menu can be set in this program, which calls Create Process function to make the finite element analysis program of Ansys batch mode run automatically. Code is as follows:

```
STARTUPINFO si ;   // recorder the original state of process
PROCESS_INFORMATION pi;   // recorder the return information of process

memset(&si,0,sizeof (si) );        // initialize the variable "si"
si.cb=sizeof (si);                 // set the value of "si"

si.wShowWindow=SW_SHOW;   // set ANSYS shown in the manner of windows
si.dwFlags=STARTF_USESHOWWINDOW;
BOOL fRet=CreatProcess(NULL, "C:\\Program Files\\Ansys Inc\\v100\\Ansys\\bin\\ in-
tel\\ansys100.exe -b -pane3flds-iE:\\ansysbat.txt-o
E:\\resultout.out",NULL,NULL,FALSE,0,NULL,NULL,&si,&pi);
```

The results file "AnsysBat.Out" can be opened by "NotePad.exe". As for the pictures obtained during the analysis process,redirect command "Redirect-Plots" ,which is under Ansys's PlotCtrls menu, can be used to output those pictures to a file.

4.3   APDL Batch Text File

The program can automatically generate APDL batch code files according to the input parameters. The process is as follows: (1) create a text file that is used to store the generated APDL batch code; (2) read and write the text file; (3) the program wirites APDL command string into the files according to the input parameters. Part of APDL batch text code is as follows:

```
fstream outfile   // define the object of "fstream"
outfile.open("ansysbat.txt",ios::out);   //     open the appointed files named "ansysbat.txt"
// ansysbat.txt    write the character string into file"ansysbat.txt"
outfile<<"FINI"<<endl;                        // clean the memory, start a new round analysis
outfile<<"/CLEAR"<<endl;
outfile<<"/TITLE, THERMAL STRESS ANALYSIS OF SKIRT SUPPORTING ZONE OF
        HYDROGENATION REACTOR "<<endl; // define the name of analysis
outfile<<"/REPLOT"<<endl;
                                   // nter pre-processing
outfile<<"/PREP7"<<endl;
outfile <<"ET,1,PLANE183"<<endl;       // define cell type
outfile<<"KEYOPT,1,3,1"<<endl;    // define cell is axisymmetric
// build the section model of skirt support
outfile<<"K,1,0,0"<<endl;                   // create the center of ball shell
outfile<<"K,2,"<<TONGT_DI/2<<"+"<<TONGT_T<<",0"<<endl;  // create points on the top
of"h"type forged piece
.......................................................
                                   // make meshes
outfile<<"LREFINE,4,,5"<<endl;     // coarctate the meshes of outside arc
.......................................................
// temperature analysis and solving
outfile<<"SFL,ALL,CONV,"<<HAIR<<",,"<<TAIR<<endl;    // add air convection boundary
.......................................................
// reenter pre-process to define elastic modulus, poisson ratio, and thermal expansion coefficient
outfile<<"/PREP7"<<endl;
outfile<<"ETCHG,TTS"<<endl;                // convert thermal cell into structure cell
.......................................................
outfile<<"/SOLU"<<endl;
outfile<<"LDREAD,TEMP,,,,,,RTH"<<endl;    // read the temperature distribution
outfile<<"SFL,6,PRES,"<<endl;                  // add stress on inside surface
.......................................................
// post-processing
```

Before the above code performs, a Pre-processing command # Include "Fstream.h" must be added to the header files. The batch file AnsysBat.txt is default saved in the running directory of program. After the batch code text AnsysBat.txt is created, it can use Notepad to open the text file with CreateProcess function, and check the correction of the generated APDL command code. It can also select "File \ Save As "in the menu of the window, to save the file to a designated directory.

5   The Result of the Program and Conclusion

A case study is used to prove the feasibility of this system. It takes a hydrotreating reactor, which annually yields 600,000 tons of gasoline and diesel, as an example to show how the system works. Firstly, run the program and input parameters from the interface shown in Figure 5. After that, the program automatically generates APDL code. Then click Ansys menu item and start Ansys batch processing to make APDL batch command file "ansysbat.txt" run automatically. Results are shown in Figure 6.

The maximum stress of skirt support, which is 306.345Mpa, occurs in the transition roundness of h-type forgings piece. It has been verified, with the same material and load, the result of this system is as same as that calculated from ANSYS10.0, which proves the system is correct and reliable . According to the above example, we can get the following conclusions:

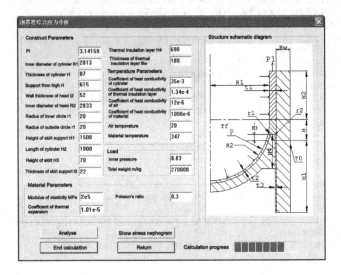

Fig. 5 Input interface

- The system successfully uses VC to package ANSYS APDL command stream. Combining VC interface with ANSYS finite element analysis, it also achieves ANSYS read APDL command stream automatically.
- The system realizes parametric modeling and finite element analysis of skirt support, avoiding information loss during the conversion process between the modeling and analysis of skirt support, and improving the efficiency and reliability of modeling and finite element analysis of skirt support.
- The system has simple operation, high robustness, which is quite easy and convenient for people who are not familiar with ANSYS to complete the parameters modeling and finite element analysis of skirt support. So it has great practical value in engineering applications.

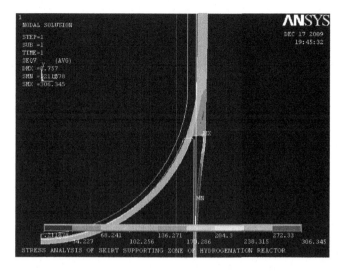

Fig. 6 The result generated by program

References

1. Yu, W., Gao, B.J.: The Application of ANSYS in the Mechanical and Chemical equipment. China WaterPower Press, Beijing (2005)
2. Zhou, N.: Application and Further Development of ANSYS-APDL. China WaterPower Press, Beijing (2007)
3. Tian, H.F., Zhang, J.F.: The Parametric Finite Element Analysis based on VC and ANSYS. Traffic and Computer (22), 116–119 (2004)
4. Li, H.J., Lu, P.: The Parametric Modeling and Finite Element Analysis System of Valve based on VC and APDL. Journal of Wuhan University of Technology (31), 325–327 (2007)

A Quantitative Research into the Usability Evaluation of China's E-Government Websites[*]

Zhou Guimei and Yan Taowei

Abstract. This thesis is devoted to a research into the usability of China's E-Government websites. According to Nielsen's usability principle, a theoretical model of usability is initially constructed. Based on this model, from the perspective of the user's needs, this thesis evaluates the weight of the indicators that indicate the usability. Concerning the evaluation of the importance of the indicators, this thesis mainly adopts the approach of Analytic Hierarchy Process . From the perspective of the market research, this thesis evaluates the weight of every indicator in the whole system so as to learn about the performance of the indicators indicating the usability of China's e-Government websites and rationalize the evaluation result.

1   Introduction

Through its previous work, the project on the research into the usability of China's E-Government has completed the task of evaluating the four selected indicators and their sub-indicators. Since there are so many selected indicators and sub-indicators, evaluating their degree of importance can help us to make targeted improvements and strengthen the website's usability so as to satisfy the needs of the public. Therefore, determining the usability of the websites is very important. Previous studies have shown that the key point in evaluating the usability is to determining the importance of every indicator from the perspective of the user's needs. Evaluating the importance is equal to weighing every indicator in the whole system based on the market research.

According to Nielsen's usability principles, based on preliminary studies, this thesis mainly adopts AHP to determine the weight off the indicators.

Zhou Guimei
Business School of Shandong University at Weihai, Weihai, China
e-mail: guimeizhou@.sdu.edu.cn

Yan Taowei
Scientific Research Department of Shandong University at Weihai, China
e-mail: yantw@163.com

[*] Fund: Natural Science Foundation of Shandong Province (Grant No.: Y2007H20).

F.L. Gaol et al. (Eds.): Proc. of the 2011 2nd International Congress on CACS, AISC 144, pp. 437–442.
springerlink.com

2   The Theoretical Model of Evaluating the Usability of the Websites

E-government websites, as a window and application platform, is a man-computer interaction system. And its usability study follows [1] Nielsen (1993) usability principles Including the following elements: User-friendly. That is, whether it is easy for the users to master the use of the product; Interaction efficiency: how efficient it is for the user to complete specific by using the products; Easy to remember. To put products into use again after some time, whether the user still remembers how to operate; Error frequency and severity: whether the error frequency is high and the error is serious or not; User's satisfaction: Whether the users are satisfied with the products or not. Based on the principles above, the model for China's e-government websites is constructed as illustrated in Fig. 1.

Fig. 1 This model consists of the target layer, rule layer and program layers.[2]According to different indicators, the system is divided into several hierarchical levels as shown in Figure 1: The index is illustrated in the form of trees of indicators which is hierarchical. The usability of the E-government website is the first layer (the target layer, assumed weight is W), the second layer is called the Layer G (layer criteria, assumed weight Wi, i = 1,2 ... n), the third layer is called the Layer C(program level, assumed weight WIJ, i = 1,2 ... n; j = 1,2 ... m). The importance of Layer G targets is weighed each by Wi (i = 1,2 ... n), we call it weight; the importance of Layer C is measured by Wij (i = 1,2 ... n; j = 1,2 ... m). To weigh the indicators of Layer C, a criteria of five levels A, B, C, D, E is set. The selection of indicators is performed as the following

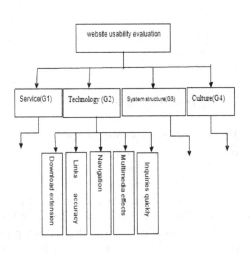

First:The indicator measuring the usability of the service. Usability of the service refers to the usability of the content of the websites and the service function of the websites. It should be the ideal measure used to evaluate the usability of e-government website. It is also applied to evaluate the design of the websites. [3] The usability of the service refers to its reliability; responsiveness; accuracy; communication tools and emotional investment.

Second:The indicator indicating technical usability. The technical usability ensures the proper performance of the system by providing software and hardware.

It is very important in materializing the construction and usability of the websites. We have selected the following indicators, namely, the accuracy of the link; website download delay; navigation; quick and accurate inquiries and multimedia effects demonstration.

Third:The indicator indicating the usability of the system structure. The structure of the system is a depiction of the design style and pattern of the system. Whether it is reasonable or not is an important indicator to evaluate the usability of the websites. [6,7] The usability of the system structure is weighed by the following indicators: a reasonable layout of the website; a proper organization of the content and simplicity in the design of the website.

Four:The indicator indicating the usability of the culture. The usability of the culture refers to the perceived quality of service under different cultural backgrounds. It is a key measure weighing the quality of the transmission of the culture while being transmitted. It is weighed by the following four indicators: public relations; the detailed content of the service provided by the service; the evaluation of the public and the clear statistics and accounting.

3   Determining the Weight of the Indicators Evaluating the Usability of the E-Government Websites

A   *Approaches to the Analysis*

In determining the target weight, due attention should be paid to the following three aspects:

First, the approach evaluating the usability of the e-government website is of vital importance. Owing to the subjective and objective factors in evaluating the usability of the e-government websites, many requirements of the content and expectations of the evaluation are vague. Thus relying solely on a mathematical model to solve the problem is often not feasible. We should find solutions by doing operational research. So Analytical Hierarchy Process (AHP method) is applied to figure out the result of the evaluation.

Second, on Analytical Hierarchy Process (AHP method). [9] Analytical Hierarchy Process (AHP method) was first proposed In the mid-1970s by Saaty T. L, an expert on Operations Research from University of Pittsburgh. The basic principle of AHP is to study complex problems as a large system, through the analysis of a number of factors in the system; various interrelated factors are distinctly arranged into orderly hierarchy. Then we make an objective evaluation of the factors in every hierarchy and finally work out the weight of every factor (Weight is equal to importance). This approach is people-oriented. By comparing the factors one by one, we aim to find out theindividual psychological perception and difference of the users. Therefore the results are more objective and effective.

Third, the approach of AHP involves the issue of checking the consistency. Using the AHP constituted by the weight ratio of pair-wise comparison matrix, the importance of the matrix is judged by people. Since people can not be completely consistent in a pair-wise comparison while dealing with complicated things, the estimation errors are inevitable. The estimation error should be controlled within reasonable limits. If the error exceeds the reasonable limit, the estimation will be ineffective. Therefore, we should check the consistency of the estimation and work out the weight according to the result of the checking.

B   The Quantitative Analysis of the Data of the Evaluation Model

By constructing a comparison matrix and matrix operations, we aim to determine the level of an element on a case and the elements associated with the hierarchy in order of importance and relative weight. Specific process: first the general theorem is given based on an analysis, and then work out the data according to the theorem and reasoning.

First, we mark each factor in this way: G1: service; G2: technology; G3: system structure; G4: culture. According to the relevant research, we select five levels in conducting qualitative comparison in pairs. The five levels are: the same, slightly stronger, strong, much stronger and absolutely strong. Therefore, we can use the following table 1 to 9 scale to quantify the indicators..

Table 1 The Significance of Determining the Scale

Significant of scale (qualitative results)	Scale value (quantitative results)
Gi and Gj of the same effect	Gi : Gj = 1:1
The effect of Gi is Slightly stronger than that of Gj	Gi : Gj = 3:1
The effect of Gi is stronger than that of Gj j	Gi : Gj = 5:1
The effect of Gi is much stronger than that of Gj	Gi : Gj = 7:1
The effect of Gi is absolutely stronger than that of Gj	Gi : Gj = 9:1
the impact of Gi and Gj is between these two levels	Gi : Gj = 2,4,6,8:1
Gi and Gj with the opposite effects	Gi : Gj = 1:1,2,...,9

Suppose Gi: Gj = fij, we usually put the above results in matrix form, this is called pair-wise comparison matrix. Based on the table we assume that the relationship between the importance is: the effect of G1is stronger than that of G2; it is much stronger than the effect of G3 and G4; The effect of G2 is slightly stronger than that of G3. the effect of G2 is stronger than that of G4; The effect of G3 is slightly stronger than that of G4. Then we compare the quantities in pairs, and the results are shown as the following.G1 : G1=1:1   G1 : G2=5:1   G1 : G3=6:1   G1 : G4=7:1,G2 : G1=1:5   G2 : G2=1:1   G2 : G3=5:1   G2 : G4=5:1,G3 : G1=1:6   G3 : G2=1:3   G3 : G3=1:1   G3 : G4=2:1,G4 : G1=1:7   G4 : G2=1:5   G4 : G3=1:2   G4 : G4=1:1. Have symmetric matrix:

$$A = \begin{pmatrix} 1 & 5 & 6 & 7 \\ 1/5 & 1 & 5 & 5 \\ 1/6 & 1/3 & 1 & 2 \\ 1/7 & 1/5 & 1/2 & 1 \end{pmatrix} \begin{pmatrix} w_1 \\ w_2 \\ w_3 \\ w_4 \end{pmatrix} = n \begin{pmatrix} w_1 \\ w_2 \\ w_3 \\ w_4 \end{pmatrix} \tag{1}$$

In order to facilitate mathematical treatment, suppose there are n pieces of objects: A1, A2, ..., An, their weights were: w1, w2, ..., wn. If we compare their weight in pairs, the ratio (relative weight) is in the form an n × n pair-wise comparison matrix:

$$AW = \begin{pmatrix} a_{1,1} & a_{1,2} & \cdots & a_{1,n} \\ a_{2,1} & a_{2,2} & \cdots & a_{2,n} \\ \cdots & \cdots & \cdots & \cdots \\ a_{n,1} & a_{n,2} & \cdots & a_{n,n} \end{pmatrix} \begin{pmatrix} w_1 \\ w_2 \\ w_3 \\ w_4 \end{pmatrix} = n \begin{pmatrix} w_1 \\ w_2 \\ w_3 \\ w_4 \end{pmatrix} = nW \tag{2}$$

After careful observation, we found that pair-wise comparison matrix of each row is exactly proportional to the weight vector W = (w1, w2, ..., wn) T According to analogy, we assume that the importance of factors, vectors and pair-wise comparison matrix (1) have the same relationship. As a result, we can get the importance of vector elements of the matrix (1):

$$W = \begin{pmatrix} w_{11} \\ w_{21} \\ w_{31} \\ w_{41} \end{pmatrix} \propto \begin{pmatrix} 1 \\ 1/5 \\ 1/6 \\ 1/7 \end{pmatrix} + \begin{pmatrix} 5 \\ 1 \\ 1/3 \\ 1/5 \end{pmatrix} + \begin{pmatrix} 6 \\ 5 \\ 1 \\ 1/2 \end{pmatrix} + \begin{pmatrix} 7 \\ 5 \\ 2 \\ 1 \end{pmatrix} = \begin{pmatrix} 19 \\ 56/5 \\ 7/2 \\ 129/70 \end{pmatrix} \tag{3}$$

$$W = \begin{pmatrix} w_{11} \\ w_{21} \\ w_{31} \\ w_{41} \end{pmatrix} = \begin{pmatrix} 0.535 \\ 0.315 \\ 0.098 \\ 0.052 \end{pmatrix} \tag{4}$$

For the sake of convenience in application, we can choose an appropriate scale factor so as to make the value of the weight of each factor equal to 1 (this process is called normalization, normalized importance factor values as weights, the importance of vector called weight vector), so a weight vector can be achieved as the following.

Namely: W = (0.535, 0.315, 0.098, 0.052)T,In Formula (4), the degree of the weight determines the overall order of the major elements.

In the same way single-level sort weight can be calculated from the above. That is, the same level in the corresponding element on one level for a factor in the relative order of weight. Applying the principles of Table 1, we assume that each sub-index layer and the importance of the relationship between the more forward in pairs, according to the matrix (1), (2), (3), (4) and so we can obtain the weights of Layer G and sub-indicators (Layer C).

The weight vector of the different options to improve service in the form of a pair-wise comparison matrix is shown as the following:

W1= (0.382, 0.286, 0.209, 0.080, 0.043)T

The weight vector of the different options to improve technology in the form of a pair-wise comparison matrix is shown as the following:

W2= (0.432, 0.284, 0.172, 0.072, 0.040)T

The weight vector of the different options to improve the system structure in the form of a pair-wise comparison matrix is shown as the following:

$$W3= (0.571, 0.286, 0.143, 0, \quad 0)T$$

The weight vector of the different options to improve the cultural element in the form of a pair-wise comparison matrix is shown as the following:

$$W4= (0.478, 0.338, 0.135, 0.048, \quad 0)T$$

4   Conclusion

Since people can not be completely consistent in a pair-wise comparison of the weight ratio while dealing with complicated matters, the estimation errors are inevitable. Therefore, the results listed above are subjective to consistency checking whether they are arranged in overall order or the single order.

It ha been proved in theory that for the consistent pair-wise comparison matrix, the maximum feature value is n; Vice versa, if the biggest feature value is n, then surely there is consistency between the two elements.We use "n" to stand for the maximum feature value with a deviation, So the size of the difference between n' and n reflects the extent of inconsistency. Taking into account the effect of the number of the factors will, Saaty defines CI = (n'-n) / (n-1) as the indicator of consistency.   When CI = 0, the pair-wise comparison matrix, Matrix is exactly consistent. Otherwise there are inconsistencies. The greater CI is, the greater the degree of inconsistency will be. Usually it is required that   CI ≤ 0.1, while CI ≤ 0.15 is acceptable only to the overall order arrangement or the order arrangement of higher level.   In this thesis, for the overall order arrangement, through the characteristic equation (1), we can work out the maximum feature value n '= 4.178, CI = (n'-n) / (n-1) = 0.0593. it is clear that the matrix to determine the extent of the inconsistency is acceptable.

References

[1] Nielsen, J.: Usability engineering, pp. 13–16. Academic Press, Boston (1993)
[2] Cheng, X.: Management Information System, vol. 4, pp. 314–315. Tsinghua University Press, Beijing (2003)
[3] Chaffee, D., Meyer, R., et al.: Internet marketing strategy, implementation and practice, vol. 6, p. 165. Mechanical Industry Press, Beijing (2004)
[4] NIST.Webmetrics.Technicaloverview (EB/OL) (January 2, 2003), http://zing.ncsl.nist.gov/WebTools/
[5] Hoff, R.C., McWilliams, G., Saveri, G.: The 'click here' economy. Business
[6] Constantine, L.L., Lockwood, L.A.D.: For the use of software design, vol. 5, pp. 17–18. China Machine Press, Beijing (2004)
[7] Daft, R., Lengel, R.: Organizational information requirements, media richness and structural design. Management Sci. 32, 554–571 (1986)
[8] Wang, L., Xvshubo: AHP Introduction. China Renmin University Press, Beijing (1990)

Experimental Power Harvesting from a Pipe Using a Macro Fiber Composite (MFC)

Eziwarman, G.L. Forbes, and I.M. Howard

Abstract. Piezoelectric material can be used to transform ambient vibration energy into small amounts of electrical power. As piezoelectric materials have been developing, such as lead-zirconate-titanate (PZT), Quick Pack (QP), macro fiber composite (MFC) and polyvinylidineflouride (PVDF), many researchers have been investigating their applications. This research investigates the possibility of electrical energy being generated by a macro fiber composite (MFC) patch mounted onto the surface of a pipe structure. In addition to the application of MFC, this research investigates the effect of varying load impedance to; the resonance frequency, voltage and electrical power generated by the structure. The results presented are for a large structure where most previous research on power harvesting has been undertaken on much smaller structures.

Keywords: energy harvesting, vibration, MFC, piezoelectric harvesting, pipe vibration.

1 Introduction

Research into power harvesting using piezoelectric materials and components has been developed rapidly. A review of energy harvesting literature using piezoelectric materials can be found in [1-3]. Power harvesting produced by an embedded piezoelectric system on the human body and mechanical motion was studied by [4, 5]. Additionally, power harvesting in macro systems was explored in [6]. The study of the extraction of power from low level vibration sources has included vibration from car structures, house hold devices, clothes, HVAC vents, windows and notebook CD devices as presented in [7]. Parasitic shoes with different kinds of harvesting devices were implanted and investigated in [8] for extracting electrical energy during walking. Power harvesting from back pack straps was also studied in [9].

Eziwarman
Department of Mechanical Engineering, Curtin University Perth,
Australia & University of Bung Hatta, West Sumatran, Indonesia

G.L. Forbes · I.M. Howard
Department of Mechanical Engineering Curtin University Perth, Australia

F.L. Gaol et al. (Eds.): Proc. of the 2011 2nd International Congress on CACS, AISC 144, pp. 443–449.
springerlink.com © Springer-Verlag Berlin Heidelberg 2012

Power harvesting from beam vibration has been widely researched, [10, 11], using the smaller bimorph piezoelectric beams that have become readily available. Many methods relating to beam modelling have been presented including analytical, simulation and also experimental investigation. Research using the Euler-Bernoulli beam theory can be found in [10], along with the application of the weak Hamiltonian principle in [11]. In addition, the work in [10] modelled a piezoelectric bimorph beam with translational base excitation and small rotations, while base excitation with translation and longitudinal motion was presented in [11]. Previous research has shown the investigation of resonant frequency, voltage and power changes as a function of resistive load impedance.

One of the main technological advances in recent years has been the development of low profile piezo devices, [3]. This has included the development of the Active Fiber Composite (AFC), Macro Fiber Composite (MFC), Radial Field Diaphragm (RFD) and Quick Pack (QP). The advantage of these types of piezoelectric components is their lighter weight, flexibility, ease of mounting and low frequency operation.

A comparison of piezoelectric energy harvesting for recharging batteries was conducted using three different types of piezoelectric components, [12]. The three types of piezoelectric materials included the lead-zirconate-titanate (PZT), macrofiber composite (MFC) and Quick Pack (QP). Structural vibration sources for the analysis included that found in an engine compartment of an automobile. A PCB accelerometer was initially placed at various locations within the compartment to understand the dominant frequency components present for various vehicle speeds. The corresponding frequency ranges were then used in a laboratory scale test. The experimental test results over the frequency range from 0 – 1 kHz showed that the PZT material performed better than the other piezoelectric components resulting in greater output power and consequentially shorter recharging time for the battery.

Even though many researchers have investigated piezo power harvesting, most have only presented tests on small structures, typically modeling a bimorph beam as a source of mechanical power. This research investigates the use of a larger cylindrical pipe structure. A Copper pipe with diameter 101.6 mm and length 1400 mm was chosen for this experimental investigation. This paper presents the results of the investigation including the vibration and voltage frequency response measurements and the voltage and output power response as a function of resistive load impedance and circumferential mode shape during vibration.

2   Piezoelectric Energy Harvesting Device

Macro Fiber Composite materials (MFC) were first developed by the NASA Langley Research Center and can be used as a sensor, an actuator or a power harvesting device. MFC's are marketed by Smart Material Corporation [13] and are designed to operate in either d_{33} or d_{31} mode. The first subscript indicates the direction of polarization generated in direction 3 (z-axis), and the second subscript denotes the direction of applied force or induced strain per unit stress applied in direction 3 (z-axis), or direction 1 (x-axis) perpendicular to the direction in which the material is polarized.

The MFC type M-8528-P2 was used in this experimental study operating in the d_{31}-mode. Fig. 1 shows a photo of the prototyped power harvesting device which was used in this experimental study. Table 1 also presents the material properties and structural parameters of the power harvesting device. The MFC patch was glued to the surface of the Copper pipe using epoxy glue. The piezoelectric patch was located one quarter of the length along the pipe. Two wires were attached to the piezoelectric element and connected to a potentiometer in parallel. The Bruel & Kjaer accelerometer type 4507 B was placed close to the bonded piezoelectric patch as shown in Fig. 1. The accelerometer measured the frequency response of the structure during the test with different load resistances.

Table 1 Materials properties

Structure Material: Cooper Tube-As 1432 DN 100	
Composition	Alloy C12200 Copper=99.9% min
	Phosphorus =0.015-0.040%
Melting Point	1083^0 C
Density	8.94×10^3 kg/m^3
Modulus Elasticity	17,000 MPa
Outer Diameter	101.60 mm
Thickness	1.63 mm
Length	1400 mm
Piezoelectric material: MFC M-8528-P2 (d_{31} effect actuators)	
Dimension: Active length	85 x 28 mm
Overall length	106 x 34 mm
Capacitance	172 nF
Free strain	-820 ppm
Blocking force	-205 N
Low field operation d_{31}	-1.7E+02 pC/N
Constant electric field	15.857 GPa

3 Experimental Set-Up and Setting

An experimental investigation was performed on the flexible piezoelectric material and the mating Copper pipe structure using the Bruel & Kjaer Pulse analyser. The experimental set up is shown in Fig. 1. The Copper pipe was lightly supported at both ends using rubber bands to approximate the free-free boundary condition.

The B&K exciter type 4809 was suspended by a crane above the pipe with a B&K impedance head type 8001 connected to the pipe and exciter via an adapter stud glued onto the surface of the Copper pipe providing the input excitation. Random input signal excitation was obtained from an Agilent 33120A waveform generator. The voltage waveform from the signal generator was then transferred to a power amplifier B&K type 2706 and then from the power amplifier, the output voltage was connected to the exciter.

Fig. 1 Piezoelectric patch to copper pipe and experimental set up

The measured excitation force from the impedance head was transferred to a charge amplifier prior to connection to the Pulse analyser. The B&K accelerometer was used to measure the vibration frequency response of the Copper pipe during the tests. The voltage response from the MFC piezo material was measured using a simple parallel resistor circuit with variable potentiometer with load resistance ranging from 0-10 kΩ. The potentiometer enabled the easy adjustment of the load resistance during the test. The resistance and circuit voltage was measured using a multimeter and was also subsequently connected to the Pulse analyser for further frequency response analysis.

4 Experimental Results

4.1 Modal Analysis

A modal analysis study was undertaken and the frequency response, mode shapes and damping ratio of the Copper pipe was obtained for each resonance in the frequency band from 0 – 800 Hz. This analysis was essential to understand the circumferential and bending mode shapes of the pipe structure. The results of the modal analysis tests are presented in table 2 showing the 7 dominant modes of vibration within this frequency range.

Table 2 Circumferential and Bending Modes

Frequency (Hz)	Mode (m, n)[a]	Damping ratio (%)
225	1, 1	0.299
296	0, 2	0.272
308	1, 2	0.281
364	2, 2	0.201
493	3, 2	0.160
586	2, 1	0.136
696	4, 2	0.101

[a] m= axial bending number, n=circumferential bending number.

4.2 Power Harvesting Energy

Fig. 2 shows the voltage frequency response result in units of dB/1.00 V/N from the MFC piezo element glued onto the pipe structure with random force excitation and eleven resistance values from 0.1 kΩ to 10 kΩ. Note that the noise harmonics of 50 Hz in the result should be ignored. The voltage response at the resonant frequency of 296.375 Hz changes significantly with increasing load resistance. Closer examination of the voltage frequency response of the second resonance at 296.375 Hz is shown in Fig. 3 as a function of load resistance. The result shows that the resonance frequency appears to shift slightly lower due to the change of load impedance however the frequency change is at the limit of the frequency resolution of 0.125 Hz. This result appears to indicate that the resonance frequency at 296.375 Hz does not change significantly due to varying load impedance. This is an important result showing that the MFC patch has negligible effect on the larger structural resonance.

Fig. 4 plots the voltage response at the resonant frequency of 296.375 Hz as a function of load resistance. It can be observed that the output voltage increases as the load resistance increases. For the resistance of 1 kΩ at resonance, the absolute voltage output was 1.325 V/N. At the load resistance of 5kΩ the voltage output was 3.715 V/N and then reached 4.557 V/N at the maximum resistance of 10 kΩ.

The output power result as a function of load resistance is also shown in Fig. 4, where the output power was dissipated across the resistive potentiometer. The results show that the output power falls after reaching a peak at 3.14 mW/N^2 and this maximum output power was obtained at a resistance load of 4 kΩ from the resonance frequency at 296.375 Hz. The variation of voltage and power response as a function of load impedance reported here is similar in nature to that shown previously from the much smaller bimorph beam models, [10 – 11].

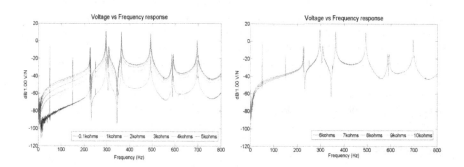

Fig. 2 Piezoelectric voltage frequency response function with varying resistance 0.1-10kΩ

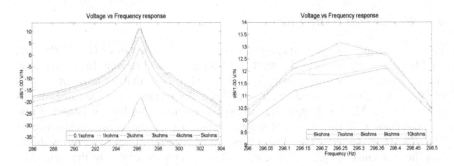

Fig. 3 Voltage with varying resistance at mode shape (0, 1) frequency resonance 296.375 Hz

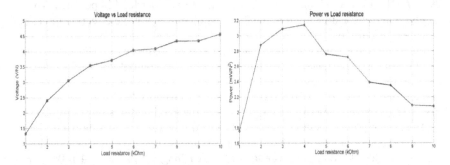

Fig. 4 Voltage and Power with varying load resistance at resonance frequency 296.375 Hz

5 Conclusion

This research has been undertaken to investigate energy harvesting by using macro fiber composite (MFC) patches on the surface of a copper pipe, with a diameter 101.6 mm and length 1.4 m. Output power was measured as a function of load resistance by using a potentiometer with variable load from 0-10 kΩ. The maximum output power measured at the resonance of 296.375 Hz was found to be 3.14 mW/N^2, when the resistive load of 4 kΩ was applied. This resonance represents the second circumferential mode of the pipe structure which appeared to have zero or insignificant bending behaviour.

References

1. Sodano, H.A., Inman, D.J., Park, G.: A review of Power Harvesting from Vibration Using Piezoelectric Material. The Shock and Vibration Digest 36, 197–205 (2004)
2. Anton, S.R., Sodano, H.A.: A review of Power Harvesting Using Piezoelectric Materials (2003-2006). Smart Material and Structures 16, R1–R21 (2007)
3. Priya, S.: Advance in Energy Harvesting Using Low Profile Piezoelectric Transducers. Electroceram. 19, 165–182 (2007)

4. Chalasani, S., Conrad, J.M.: A survey of Energy Harvesting Sources for Embedded System. In: Proc. IEEE Southeastcon, Huntsville, AL, April 2008, pp. 442–447 (2008)
5. Mitcheson, P.D., Yetaman, E.M., Rao, G.K., Holmes, A.S., Green, T.C.: Energy Harvesting form Human and Machine Motion for Wireless Electronic Device. In: Proc. IEEE, September 2008, pp. 1457–1486 (2008)
6. Beeby, S.P., Tudor, M.J., White, N.M.: Energy Harvesting Vibration Sources for Microsystems Applications. Measurement Science and Technology 17, R175–R195 (2006)
7. Roundy, S., Wright, P.K., Rabaey, J.: A study of Low Level Vibration as a Power Source for Wireless Sensor Nodes. Computer Communication 26, 1131–1144 (2003)
8. Kysmissis, J., Kendall, C., Paradiso, J., Gershenfeld, N.: Parasitic Power Harvesting in Shoes. In: Proc. IEEE Internationational Symposium on Wearable Computers, October 1998, pp. 132–139. IEEE Press, Pittsburg (1998)
9. Feenstra, J., Granstrom, J., Sodano, H.A.: Energy Harvesting Through a Backpack Employing a mechanically Amplified Piezoelectric Stack. Mechanical System and Signal Processing 22, 721–734 (2007)
10. Erturk, A., Inman, D.J.: An Experimentally Validated Bimorph Cantilever Model for Piezolectric Energy Harvesting from Base Excitation. Smart Material and Structures 18, 18 p (2009)
11. Lumentut, M.F., Howard, I.M.: An Analytical Method for Vibration Modelling of Piezoelectric Bimorph Beam for Power Harvesting. In: Proc. ASME 2009 Conference on Smart Materials Adaptive Structure and Inteligent Systems, Oxnard, California, USA (September 2009)
12. Sodano, H.A., Inman, D.J., Park, G.: Comparison of Piezoelectric Energy Harvesting Device for Recharging Batteries. Intelligent Material Systems and Structure 16, 799 (2005)
13. http://www.smart-material.com

Fast Access Security on Cloud Computing: Ubuntu Enterprise Server and Cloud with Face and Fingerprint Identification

Bao Rong Chang*, Hsiu Fen Tsai, Chien-Feng Huang,
Zih-Yao Lin, and Chi-Ming Chen

Abstract. This study employed an open source Ubuntu Enterprise Server together with Ubuntu Enterprise Cloud to construct a "Private Small-Cloud Computing" (PSCC) in which a cloud controller (CLC) has been linked to some cluster controllers (CC), where the cloud computing initiates the services like SaaS, PaaS, and/or IaaS. A cloud controller (CLC) may connect remote mobile devices or thin clients through the wired or wireless networking, for example, Ethernet, WiFi or 3G. We also made mobile device or thin client to be a low-capacity Linux embedded platform where JamVM virtual machine is utilized to form the J2ME environment and GNU Classpath is viewed as Java Class Libraries. Finally, the rapid identification fulfills access security within 2.2 seconds in PSCC in order to verify the cloud system effectiveness and efficiency in access security.

Keywords: Ubuntu Enterprise Server, Ubuntu Enterprise Cloud, Access Security, Fingerprint identification, Face Recognition, Linux Embedded Platform, JamVM Virtual Machine.

1 Introduction

Cloud computing is divided into two categories, namely "cloud services" and "cloud technology". "Cloud services" is achieved through the network connection to the remote service. Such services provide users installation and use a variety of operating systems, for example Amazon Web Services (CE2 and S3) services. This type of cloud computing can be viewed as the concepts: "Infrastructure as a

Bao Rong Chang · Chien-Feng Huang · Zih-Yao Lin · Chi-Ming Chen
Department of Computer Science and Information Engineering
National University of Kaohsiung, Kaohsiung, Taiwan
e-mail: {brchang,cfhuang15}@nuk.edu.tw, qazwsxee3118@xuite.net,
gcsul0725@hotmail.com

Hsiu Fen Tsai
Department of Marketing Management
Shu Te University, Kaohsiung, Taiwan
e-mail: soenfen@stu.edu.tw

* Corresponding author.

F.L. Gaol et al. (Eds.): Proc. of the 2011 2nd International Congress on CACS, AISC 144, pp. 451–457.
springerlink.com © Springer-Verlag Berlin Heidelberg 2012

Service" (IaaS) "Storage as a service" (StaaS), respectively. Both of them are derived from the concept of "Software as a Service" (SaaS) that is the biggest area for cloud services in demand, while "Platform as a Service" (PaaS) concept is an alternative for cloud computing service. Using these services, users can even simply to rely on a cell phone or thin client to do many of things that can only be done on a personal computer in the past, which means that cloud computing is universal. The "cloud technology" is aimed at the use of virtualization and automation technologies to create and spread computer in a variety of computing resources. This type can be considered as traditional data centers (Data Center) extension; it does not require external resources provided by third parties and can be utilized throughout the company's internal systems, indicating that cloud computing also has the specific expertise. Currently on the market the most popular cloud computing services are divided into public clouds, private clouds, community/open clouds, and hybrid clouds, where Goggle App Eng [1], Amazon Web Services [2], Microsoft Azure [3] - the public cloud; IBM Blue Cloud [4] - the private cloud; Open Nebula [5], Eucalyptus [6], Yahoo Hadoop [7] and the NCDM Sector/Sphere [8] - open cloud; IBM Blue Cloud [4] - hybrid cloud.

2   Motivation and Objective

The main purpose of this study is to build a "Private Small-Cloud Computing" (PSCC). The idea of private small-cloud computing is based on three concepts: small clusters, virtualization, and general graphics processor [9]. This cloud system will include the use of Xen [10] virtualization technology, VMGL [11] planning to use general-purpose graphics processors, Open Nebula [5] Management of cluster structure, Eucalyptus [6] implementation of the cloud controller.

In many applications, embedded devices often require huge computing power and storage space, just the opposite of the hardware of embedded devices. Thus the only way to achieve this goal is that it must be structured in the "cloud computing" and operated in "cloud services". The idea is how to use the limited resources of embedded devices to achieve the "cloud computing", in addition to using the wired Ethernet connection, and further use of wireless mobile devices IEEE802.3b / g or 3G to connect, as shown in Fig. 1.

First, we use the standard J2ME [12] environment for embedded devices, where JamVM [13] virtual machine is employed to achieve J2ME environment and GNU Classpath [14] is used as the Java Class Libraries. In order to reduce the amount of data transmission, the acquisition of information processed is done slightly at the front-end embedded devices and then processed data through the network is uploaded to the back-end, private small-cloud computing. After the processing at the back-end is completed, the results sent back to the front-end embedded devices. As shown in Fig.2, an open source package, Ubuntu Enterprise Server [15] & Ubuntu Enterprise Cloud [16], is utilized to establish the private cloud computing easily; in such a way that we can focus on installing the back-end cluster controllers and cloud controller in order to build a private small-cloud computing.

An embedded platform in conjunction with a cloud computing environment is applied to testing the capabilities of fingerprint identification and facial recognition served as the access security system. The basic structure of Ubuntu cloud computing is developed and has been deployed as well. We will then test the performance of the embedded platform operating in cloud computing to check whether or not it can achieve immediate and effective response to required functions. Meanwhile, we continue to monitor the online operation and evaluate system performance in statistics, such as the number of files, file size, the total process of MB, the number of tasks on each node, and throughput. In a cluster implementation of cloud computing, the statistical assessment by the size of each node is listed. According to the analysis of the results, we will adjust the system functions if changes are required.

Fig. 1 Cloud controller (CLC) architecture. **Fig. 2** Node controller (nc) structure. **Fig. 3** A complete structure of CLC + nc private small-cloud computing.

3 Access Security on Cloud Computing

3.1 Deploying Ubuntu Cloud Computing

A private small-cloud computing (PSCC) is built by packages Xen, OpenNebula, Eucalyptus, Euca2ools [17] and so on. For the purpose of simplicity, Ubuntu Enterprise Server [18] includes all of packages we need to install a private small-cloud computing. Cloud Controller (CLC) structure as shown in Fig. 3 in which each Cluster Controller (CC) has its own OpenNebula, Zen, VMGL, Lustre [19] and hardware resources, through an unified CLC to manage all of CC. Node Controller (nc) structure as shown in Fig. 4. nc hardware resources will determine the cloud service capabilities; the more powerful nc hardware (more CPU core and more memory), the more virtual machine resources. Furthermore, according to the above Fig. 3 and Fig. 4 they provide a way to establish CLC and nc, and we can completely describe the structure of private small-cloud computing, as shown in Fig. 5. The storage server in conjunction to Eucalyptus is Walrus [20] that is a compatible storage interface like Amazon S3 storage system and can be managed through the web interface to modify it. In addition, a control unit managing the storage server is called the storage controller (sc) [21] as shown in Fig. 6. In Fig. 7, PSCC can still link to the remote cloud through the Internet such that node device gets remote cloud services via private small-cloud, such as Goggle App Eng, Amazon Web Services, Yahoo Hadoop, Microsoft Azure, IBM Blue Cloud, and NCDM Sector/Sphere.

Fig. 4 Node controller (nc) structure.

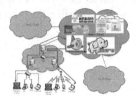

Fig. 6 Advanced Eucalyptus setup along with Walrus.

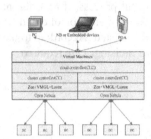

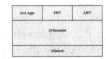

Fig. 5 A complete structure of CLC + nc private small-cloud computing.

Fig. 7 Various cloud services in PSCC and remote cloud.

3.2 Establishing Thin Client

In terms of thin client, JamVM is treated as the framework of programming development; however the virtual machine JamVM has no way to perform the drawing even through their core directly, and thus it must call other graphics library to achieve the drawing performance. Here some options we have are available, for example, GTK+DirectFB, GTK+X11, QT/Embedded, and so on, as shown in Figs. 8, 9 and 10 below. The problem we encountered is that GTK needs a few packages to work together required many steps for installation, compiling different packages to build system is also difficult, and it is often time-consuming for the integration of a few packages no guarantee to complete the work. Therefore this study has chosen QT/Embedded framework instead of GTK series, in such a way that achieves GUI interface functions. In Fig. 11, no matter SWT or AWT in JamVM they apply Java Native Interface (JNI) to communicate C- written graphics library. Afterward QT/Embedded gets through the kernel driver to achieve graphic function as shown in Fig. 12. According to the pictures shown in Fig. 11 and Fig. 12, we can string them together to be the structure of a node device as shown in Fig. 10. This part will adopt a low-cost, low-power embedded platform to realize a thin client.

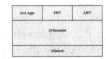

Fig. 8 Terminal node with GTK + DirectFB.

Fig. 9 Terminal node with GTK + X11.

Fig. 10 Terminal node with QT / Embedded.

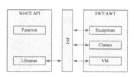

Fig. 11 Communication between SWT/AWT and QT/Embedded.

Fig. 12 QT/embedded communicates with the Linux frame buffer.

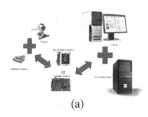

(a)

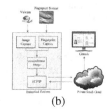

(b)

Fig. 13 System architecture.

3.3 Installing Access Security System

As a private small-cloud together with terminal node in terms of a overall structure is shown in Fig. 7, users (personal computers, laptops, embedded platforms, and smart phones) go over the wireless network IEEE802.3b/g or wired Ethernet network to connect to the private small-cloud computing, PSCC. The storage server in conjunction to Eucalyptus is Walrus [20] that is a compatible storage interface like Amazon S3 storage system and can be managed through the web interface to modify it. In addition, a control unit managing the storage server is called the storage controller (sc) [21] as shown in Fig. 8. As show When a Ubuntu cloud computing has been done, we will test the cloud employing an embedded platform in a cloud environment to perform fingerprint identification [22] and face recognition [23] to fulfill the access security system [24]. Meanwhile the development of basic structure and deployment for Ubuntu cloud computing are valid and even more we test the service performance for an embedded platform collaborated with cloud computing, checking an immediate and effective response to client. The access security system is shown in Fig. 13.

4 Experimental Results and Discussions

The host takes a lot of capacity with CPU virtualization technologies (VT) [25], a large number of CPU computing power, large memory and fast disk. In order to verify the cloud system effectiveness and efficiency in access security, the experiment on fingerprint identification and face recognition by using rapid identification in PSCC cloud computing has been done successfully within 2.2 seconds in average access identification to exactly cross-examine the subject identity. As a result the proposed PSCC cloud computing has been performed very well when it has deployed in local area.

Fig. 14 Binarization processing automatically running in program. **Fig. 15** Processing fingerprint features to reduce the amount of information. **Fig. 16** Information sent to the cloud and cloud returns the results of recognition to the consol.

Steps are as follows: (a) the operation for face recognition is quickly to open the video camera for the first, and then press the capture button, the program will execute binarization automatically as shown in Fig. 14; (b) the rapid fingerprint identification is first to turn on device, then press the deal button for feature extraction that reduces the amount of information as shown in Fig. 15; (c) at first the terminal device test the connection if Internet is working properly, and then we press the identify button and information sent to the cloud, and at last the cloud will return the identification results as shown in Fig. 16.

5   Conclusions

This study has achieved a "Private Small-Cloud Computing" (PSCC) where the cloud computing can initiate the cloud services like SaaS, PaaS, and/or IaaS. Furthermore, low-capacity Linux embedded platforms can be attached to PSCC through the wired or wireless networking, for example, Ethernet, WiFi or 3G. Finally, the rapid identification fulfills access security within 2.2 seconds in PSCC.

Acknowledgments. This work is supported by the National Science Council, Taiwan, Republic of China, under grant number NSC 99-2218-E-390-002.

References

1. Google App Engine (2010), http://groups.google.com/group/google-appengine
2. Amazon Web Services (AWS) (2010), http://aws.amazon.com/
3. Windows Azure- A Microsoft Solution to Cloud (2010),
   http://tech.cipper.com/index.php/archives/332
4. IBM Cloud Compputing (2010), http://www.ibm.com/ibm/cloud/
5. OpenNebula (2010), http://www.opennebula.org/
6. Eucalyptus (2010), http://open.eucalyptus.com/
7. Welcome to Apache Hadoop (2010), http://hadoop.apache.org/
8. Sector/Sphere, National Center for Data Mining (2009),
   http://sector.sourceforge.net/

9. General-Purpose Computation on Graphics Processing Units (2010),
   `http://gpgpu.org/`

10. Barham, P., Dragovic, B., Fraser, K., Hand, S., Harris, T., Ho, A., Neugebauery, R., Pratt, I., Warfield, A.: Xen and the Art of Virtualization. Technical report (2003),
   `http://www.cl.cam.ac.uk/research/srg/netos/papers/2003-xensosp.pdf`

11. VMGL: VMM-Independent Graphic Acceleration (2007),
   `http://www.cs.toronto.edu/~andreslc/publications/slides/Xen-Summit-2007/vmgl.pdf`

12. Java 2 Platform, Micro Edition (J2ME) (2010),
   `http://www.java.com/zh_TW/download/faq/whatis_j2me.xml`

13. JamVM – A compact Java Virtual Machine (2010),
   `http://jamvm.sourceforge.net/`

14. GNU Classpath, GNU Classpath, Essential Libraries for Java (2010),
   `http://www.gnu.org/software/classpath/`

15. Ubuntu Enterprise Server (2011),
   `http://docs.sun.com/app/docs/doc/821-1045/ggfrh?l=zh_TW&a=view`

16. Ubuntu Enterprise Cloud (UEC) (2011),
   `https://help.ubuntu.com/community/UEC`

17. Euca2ools User Guide (2010),
   `http://open.eucalyptus.com/wiki/Euca2oolsGuide_v1.1`

18. ubuntu-9.10-server-i386.iso, Ubuntu 9.10 (Karmic Koala) (2010),
   `http://releases.ubuntu.com/karmic/ubuntu-9.10-server-i386.iso`

19. Lustre a Network Clustering FS (2009),
   `http://wiki.lustre.org/index.php/Main_Page`

20. Walrus/Eucalyptus (2010),
   `http://open.eucalyptus.com/wiki/EucalyptusStorage_v1.4`

21. SC/Walrus/Eucalyptus (2010),
   `http://open.eucalyptus.com/wiki/EucalyptusAdvanced_v1.6`

22. VeriFinger SDK, Neuro Technology (2010),
   `http://www.neurotechnology.com/verifinger.html`

23. VeriLook SDK, Neuro Technology (2010),
   `http://www.neurotechnology.com/verilook.html`

24. opencv (open source) (2010),
   `http://www.opencv.org.cn/index.php?title=%E9%A6%96%E9%A1%B5&variant=zh-tw`

25. Virtualization technologies from Intel (2011),
   `http://www.intel.com/technology/virtualization/`

Reform of and Research Based on the Higher Vocational Education of "Database Principles" Teaching

Liu Shan and Cao Lijun

Abstract. In the database principles course teaching reform, we should determine teaching mode and training target according to the professional requirements and the course property, and in the reform of teaching contents and practice, we should pay attention to the integration and refining of theoretical knowledge and practical skills . Through letting the student write demand analysis report, complete database conceptual model, a logical model, the physical model design, the implementation of database, make the students master the practical skills, In the teaching process of mastering of using comprehensive assessment of the way to complete the course teaching.

1 Introduction

In the computer professional of higher vocational education, database principles is a stronger theoretical course, and often is set to professional courses. The course knowledge point not only to have an important impact on practice course of study and application, but to lay a foundation on understanding the subsequent course. According to the characteristics of curriculum and the professional requirements, puts forward the new challenge to teachers' teaching. It requires the teachers in the teaching process through the knowledge teaching, experimental, course design, teaching chain for various angles, different levels to improve the students' practical ability and finally reach the purpose of good skills. It make students in a follow-up study always adhere to the theory of knowledge as the foundation, understand and expand application of knowledge. Therefore, this paper based on the principle of database of the course of years' teaching practice discusses the teaching reform of the course and method, so it can provide teaching reference For similar courses.

Liu Shan · Cao Lijun
College of Mathematics and Information Technology, Hebei Normal University
of Science & Technology, Qinhuangdao, China
email: ls3252003@163.com, misscao6666@163.com

F.L. Gaol et al. (Eds.): Proc. of the 2011 2nd International Congress on CACS, AISC 144, pp. 459–464.
springerlink.com © Springer-Verlag Berlin Heidelberg 2012

2 Establish Teaching Mode and Training Goal According to the Curriculum Properties and Professional Requirements

At present, many higher vocational colleges and universities set the course of "principles of database", but the course in different professional setting, teaching mode and training on the establishment of the target has different focus, so in the curriculum and teaching in the process of implementation, we should fully consider the application requirements and the target of profession to prevent the teacher teaching at random and teaching at will.

For the course, the training target of the course " database principle" is through the course of study to make students understand and master the basic principle of database system, including the database of some basic concepts, the characteristics of the data model, the basic concept of relational database, standard SQL and database design theory; Master database application system design and development method; Understand the main content of the database technology and development trends. At the same time, with the experiment, the use of the database principle knowledge and practical tools is to train students to use the database technology analysis, the ability to solve practical problems in this field, to stimulate students' study and research to continue in the desire, to train high applied talents.

3 The Reform of Teaching Content and Practice

The database principles course is a relatively fixed systemic discipline, what different is that with the application of database technology and database management software, many database principles course is taught   and is joined the database management software application content   such as SQL SERVER, MYSQL, ORACLE, mainly for the verification experiment to provide help.

Generally speaking, the database principles of teaching content mainly includes three parts: one is the database theory, the second is the database design, and the third is the database application. The database principles involves database teaching content, from the knowledge of systemic and integrity of the above three angle, if the content in a semester study is completed, will need at least 90 class hours, in addition to the computer profession, many vocational colleges, information management and the industry and commerce management specialty, this course is mostly in the 50 ~ 70 hours of range, in such a short time, to achieve the teaching goal, must through teaching reform and practice, to make three parts content in a semester learned. The principles of database course teaching reform can through the following two aspects: one is the reform and integration of theory teaching; the second is the reform. and integration of   practical teaching.

3.1 The Reform and Integration of Theory Teaching

The curriculum knowledge points about database principles   are so scattered that it is hard for the beginner to grasp and   understand its points , and the higher vocational education emphasizes" theory knowledge is sufficient , pay attention to the skill training", so, In the professional education under the prerequisite of the course pays attention to skills training, three content in the teaching should be closely combined together, Breaking chapters restrictions, reflecting the integrity and the systematic features of the knowledge. In the course, the first two parts is the thread of the teaching, we will integrate the content of the database application into the other two parts of instruction, combining theory with practice, and help students understand and master the points. Such as: in the chapter about relation model of data structure, we can let the student observe commonly used relational database (such as SQL server, the access, mysql, oracle) of the data storage model, so that students can understand the relation model of data structure (two-dimensional form), thus deepening the understanding of the knowledge.

3.2   The Reform and Integration of Practical Teaching

In the course of "Principles Of Database", practical knowledge is mainly database application and database design. The two parts are integrated and combined with the database design theory knowledge to complete database design, which is   the main course practice teaching features. Through Integrating   and refining the practice teaching content of   "Principles Of Database", Sums up the following three practice skill points, the teacher can depend on each skill points to guide the student to carry on the related training.

Design scheme database and write demand analysis report. The teacher according to professional students belong to, guide the group to confirm and design difference database scheme. The core is the design requirement analysis. This skill point training is the first step of the database design and also necessary link. According to this skill points' task, The teacher can arrange students to use two classes time for exampling how to analysis requirement of database, the independent demand analysis of each group to their choice of database is arranged after class to finish, and the time is limited to giving the analysis report to the teacher. This skill points   will end with students spending a large number time, and the demand analysis report might involve some demand analysis method and is suported by the other course knowledge. Therefore, this skill point training should give students enough time, the teacher should always know   each group   how to do the training , and   effectively guide   students and provide some advice that this skill points work successfully is completed to prepare for the subsequent experiment.

Concept structure, logic structure, physical structure design,   preparation for the implementation. This part of the skill point training mainly through the group work and discussion, will the actual case analysis and computer operation, and the combination of will and the design of software design by hand, combined with the application of the database design strengthen students' skills and computer design software operating skills, enhances the student to study independently and analysis problem, problem-solving skills in the use.

The implementation of the database, finally complete database design "Printciples of Database" is involved in the course of the needs of the methods of analysis and design method, a self-contained course of the main skills training content. The database in the implementation stage main complete specific relation database design, students can choose familiar database management system, such as SQL server, mysql, oracle. In most schools computer specialized curriculum setup, database management system will alone as a course, teaching process in open knowledge is isolated, therefore two course should be combined, increasing scheduling, make the theory and practice of knowledge is more compact, makes it easier for students to understand and accept.

In the reform of teaching contents in the study , the theory of knowledge in the whole teaching course   should not take up too much, most school should be put in practice skill points on the study, and the two part teaching contents are not completely inalienable, the teachers should have Rich experience in the course of teaching and can effectively accumulated a good grasp of the theory of knowledge and practice skill points to join each of these two aspects, abandon   knowledge content that is not related to the content.

4   The Way of Evaluation Reform and Practice Demonstration

The examination of Principle of the database is consistent with appraisal to the course and teaching aim at teaching content and so on a series of reform process. In the process of curriculum reform, to try to replace the written exam form simply, mainly using the written examination, verification experiment, database training project development and so on varied ways.

4.1 Written

"Principles of database" includes the above three contents, the first part is suitable for the use of written, and written examination is consistent with the practical application of knowledge behind. The focus of first part of the theory of knowledge is to test students' assessment of theoretical knowledge and understanding of the administrative levels and deepness, such as for this course, some important concepts and knowledge assessment of the theory is mainly related with graduation design examples, and mainly from different angles of theory knowledge helps students distinguish and distinguish the accuracy, practice and comprehensive application for the subsequent theoretical basis. The written mainly make with points measure in final score, accounting for 50% of the course grade.

4.2 Verification Experiment

Verification examination of Principles of database mainly is around the second part-database design. First of all   the teacher assigned topics or title, and then students arrange database design, computer operation and training, and promptly to the students training assessment. Computer operation assessment of Principles of database mainly includes requirement analysis, concept structure design, logic structure design, physical structure design, database implementation five stages of the design results. Students can choose from own subject, or provided by teacher, according to the database design theory knowledge gradually completed. In the design process, students need to use proper class time searching for information and study the application of the software design skills. Verification examination accounting for 20% of the course grade.

4.3 Database Project Development

Database project development examines student assignment of the database development ability, it is a comprehensive identification for students' knowledge. In the assessment process, we shoule give students a certain class time, in the process, the course knowledge is not only involved, other course content will be involved, such as: JSP, Java, such as the use of c # and so on other development tools. The process will be finished in the form of group, and students can complete it independently choosing the familiar development tools, but The difficulty and knowledge structure of the final submit application are reasonable, a clear division of members, the design results applied effect is good, have certain innovation. Teachers should communicate with students forunderstanding the completion progress, and give more guidance. According to the operation is completed and members, the teacher gives each student a grade, the grade accounting for 30% of the course final grade.

5 Closing

At present, in my school, the teaching process, according to the different major requirements, in addition to the requirements of each part of this course, but also for normal students organized the so-called "curriculum Graduation reply", namely, the panel recommended 1-2 students to show database development results, in addition to prepare necessary PPT, but need a detailed introduction such as: main functions, characteristics, technology, innovation and other aspects. It is arranged in such a way not only to achieve the knowledge exchange and discussion, the more important thing is to improve the students' cooperation consciousness, arouse students' the interest of students to participate in the enthusiasm greatly. The part can not as a basis for evaluation, not included in the course examination total score at the end of the course. Student will achieve the curriculum knowledge summary and sublimation, and for the same kind of curriculum and curriculum reform provides a certain reference value.

References

[1] Wang, S., Sa, S.: Introduction to database systems, 4th edn. Higher Education Press, Beijing (2006), 王珊,萨师煊.数据库系统概论（第四版）.北京:高等教育出版社,
[2] Xu, L., Tang, J.: Ireland to our college teaching practice of the reform of the reference. China University Teaching (4), 50–61 (2006), 徐立臻,唐继卫.爱尔兰高等教育对我国高校教学实践环节改革的借鉴意义.中国大学教学

[3] Lei, H., Zou, H.: Teaching method reform of Database principles. Computer Know-ledge and Technology (1), 1766–1768 (2007), 雷红艳,邹汉斌.数据库原理课程教学方法改革探讨.电脑知识与技术

[4] Dong, C., Wang, R., Wang, X., Wang, J., Wang, A.: Study on Teaching Reform of Statistical Principle Course. Occupation Technology Education (29), 18–20 (2009), 董春玲,王锐,王晓宇,王金会,王爱英.《统计原理》课程教学改革研究.职业技术教育

[5] Yang, G.: Class teaching approach and the teaching strategy. Harvard University Press, Shanghai (2009), 郭景扬.课堂教学模式与教学策略.上海:学林出版社

[6] Gang, L.: Principle and application of database course teaching reform thinking. China Education Technology and Equipment (14) (2009), 柳佳刚.数据库原理与应用课程教学改革的思考.北京:中国教育技术装备

Application of Virtual Instrument Technique in Data Acquisition of Gas-Water Pulse Pipe Cleaning Experiment

Kun Yang, Chenguang Wu, Yixing Yuan, and Jingyang Yu

Abstract. Gas-water pulse pipe cleaning technology based on the theory of gas-water two-phase flow is a high-efficient, water-saving and low-cost method for cleaning fouling in the pipe wall. Pressure drop and flow rate during cleaning can. A data acquisition system based on virtual instrument is designed to acquire real-time data of pressure drop and flow rate which can express cleaning effect. The experiment shows that the data acquisition system is suitable for acquiring data of high-speed multi-channel signals and LabVIEW could decrease development time and improve the efficiency of program.

1 Introduction

During water pipeline operation all year round, pipe acts like a reactor, there are physics, chemistry, electrochemistry and microbiology reactions in the pipe, which will form growth-ring[3]. Growth-rings not only influencing water quality but also reducing ability of transportation and water pressure and increasing operational costs[6,8].Pipe cleaning is most effective method for removing the growth-ring, keeping pipes clean, reducing twice pollution and prolonging pipe life. And gas-water pulse pipe cleaning based on gas-liquid two-phase flow theory is a high frequency and high energy instantly quickly transfers momentum into impulse, high shear stresses act on growth-rings on the pipe wall and wash them down [3].

This paper describes a gas-water pulse pipe cleaning experiment to find the law of pressure drop and flow rate. The most difficulty of this experiment is data acquisition and model construction due to the pressure and flow rate flow in the pipe as high-speed waves. Therefore, virtual instruction is introduced to gas-water pulse pipe cleaning experiment.

Kun Yang · Chenguang Wu
School of Municipal and Environment Engineering, Harbin Institute of Technology, Harbin, China
e-mail: yklgd@163.com, wu.cg@126.com

F.L. Gaol et al. (Eds.): Proc. of the 2011 2nd International Congress on CACS, AISC 144, pp. 465–470.
springerlink.com © Springer-Verlag Berlin Heidelberg 2012

2 Virtual Instruments

Virtual instruments (VI) are computer programs that interact with real world objects by means of sensors and actuators and implement functions of real or imaginary instruments[7].LabVIEW (short for Laboratory Virtual Instruments Engineering Workbench) is a Graphical Programming Language developed in 1986 by National Instruments.

Each function or routine is stored as a virtual instrument having three main parts: (1) the Front Panel is the interactive environment of a VI. (2) the Block Diagram which defines the actual data flow between the inputs and outputs. (3) the connector terminals determine, where the inputs and outputs on the icon must be wired. They correspond to the controls and indicators on the Front Panel of the VI[5].

3 Application of VI

3.1 Experiment Platform

According to the gas-water pulse pipe cleaning technology, an experiment platform was set up (see Fig. 2 for a schematic).The experimental system can be divided into two sections, including air-water loop and test section, consists of a DN40mm, 16m long horizontal plexiglass pipe, a $0.6m^3$ water tank, a water pump with pressure gage, an air compressor, an air tank with pressure-stabilization valve and gas-water pulse generating device, five pressure transmitters, a differential pressure sensor, a flow sensor, a data acquisition system and a computer.

The water pumped from water tank was metered using a flow meter, the air from air compressor was pulsed by air-water pulse generating device, and then mixed in point a (see Fig.2). The air-water pulse flushed through the test pipe, finally drained into the water tank, air was released, and water could be reused.

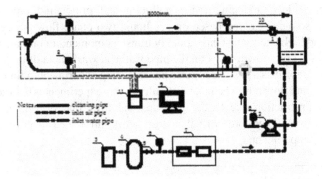

Fig. 1 Schematic diagram of the experimental air-water pulse pipe cleaning system 1.water tank 2.water pump 3. air compressor 4.air tank 5. pressure- stabilization valve 6. pressure gage 7. air-water pulse generating device 8. pressure sensor 9. computer 10. flow sensor 11. data acquisition card

The transient of pressure drop and flow rate were measured through changing aerated and interval time to change the cleaning process.

3.2 Pressure Drop Data Acquisition

The data acquisition system includes a data acquisition module NI USB-6251 and 5 pressure transmitters. The NI USB-6251 is a USB high-speed M Series multifunction data acquisition (DAQ) module optimized for superior accuracy at fast sampling rates, has 16 analog inputs (16-bit,1.25 MS/s single-channel), 2 analog outputs (16-bit, 2.8 MS/s), 24 digital I/O (8 clocked) and 32-bit counters.

Signal sampling mode, sampling rate, signal input range, et. can be set in the computer. The signal can be shown in Fig.3.

Open LabVIEW and create a New VI. Switch to the block diagram. Place the DAQ Assistant on the block diagram by dragging and dropping it from the Functions palette. The Assistant should automatically launch. On the first screen, select Acquire Signals and then Analog Input for Measurement Type. Next, select Voltage (pressure transmitters and flow sensor transmit voltage signal). To select n channels in the next screen, hold down the Ctrl button while clicking on the channel names (5 pressure channels and 1 flow channel in this experiment). Then, configure channel-specific settings such as scaling, input limits, and terminal configuration. Also configure task-specific settings such as timing and triggering if need. In these experiments, use the default values of 10 for Max and -10 for Min, select Differential as the terminal configuration used for the signal, select N Samples for the Acquisition Mode. Enter 100 for Samples To Read, and enter 1000 for Rate (Hz) (if you know the theoretical limits for the signal you are measuring, use it). DAQ assistant can set n channels to sample continuously and will acquire n-dimension dynamic waveform data which are difficult to analyze. Therefore, put the DAQ assistant and Index Array function together to transfer n-dimension dynamic waveform data to data set which are more convenient for analyzing and operating. From the Functions palette, select Index Array from Programming»Array and place Index Array function next to the DAQ assistant indicator. Wire the array input of the Index Array function to the data output of the most recent DAQ Assistant Express VI. The Index Array will have n index terminals. 1-dimension data can be extracted from n-dimension dynamic waveform data by setting index terminals. Connect data of every channel plus sampling time to a cluster can collect all data of a condition in certain period [2].

Fig.3 shows the block diagram of this experiment. There are 6 channels in this experiment include 5 pressure channels and 1 flow rate channel. The acquired data store in Excel files for the date and time as file name.

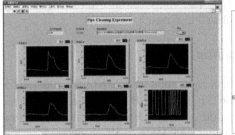
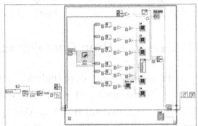

Fig. 2 The front panel of the application of these   **Fig. 3** The block diagram of this experiment
experiments.

3.3   Flow Rate Data Acquisition

In this experiment a modified water meter was used to obtain the real-time recording
of flow rate in the pipeline. The common needle was changed to a magnetic needle.
A magnetic switch was installed on the meter panel. When the needle turned to the
switch, the switch was closed. Once the needle turns away, the switch was open. The
conversion happened once when the needle moved in a round. The volume of the
needle moves in a round V divided by the time dt is the flow rate Q in the pipeline.In
order to get dt, a virtual voltage output channel was set in the data acquisition module
(output voltage is 5V). This channel connected in series with water meter and a
10KΩresistor. The voltage across the resistor was 5V when the magnetic switch was
closed, and was 0V when the switch was open. A signal of square wave was received
in the computer, as the last graph seen in Fig.2. The period of square wave was as
same as the time of the needle moved in a round.

3.4   Data Analysis

The effects of aerated time and interval time on pressure drop in straight and curve
pipes were investigated. Fig.4 showed the pressure variation of 2 # pressure point at
different interval times when aerated time was 11s. From this chart we could see
pressure drop increase first and then stay steady when interval time was more than
10s, and the maximum pressure drop was 9.78m, the minimum pressure drop was
4.53m.

   Real-time data when aerated time or interval time is changed and other conditions
can be acquired by using the data acquisition system. Virtual instrument increases
the speed and accuracy of data acquisition which provides good data base for
theoretical analysis.

Fig. 4 Pressure variation of 2 # pressure point at different interval times when aerated time is 11 s

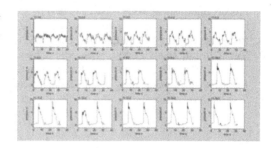

4 Summary

Virtual instrument is an advanced technology combines electronic measurement with PC observation and control[1].NI LabVIEW has proven to be a powerful tool to create both rapid prototype applications as well as an entire framework for system integration and process execution[4]. With VI being introduced into measurement of gas-water pulse pipe cleaning, it becomes more simple and convenient to acquire pressure and flow rate data of multithreading and long-distance. The interface of the data acquisition system is simple and human-computer interaction is friendly, not only save labour costs but also reduce the factitious errors. Virtual instruments displace hardware-based instruments to reduce the cost, moreover, the acquisition speed will increase with the development of computer technology. This system also can be used in many other hydraulic experiments by changing pipe length, pipe diameter, sensor type, et.

Acknowledgment. Research presented in this paper has been supported through National Major Projects of the Polluted Water Control and Management in the 11th Five year Plan of China (2009ZX07424-001) and State Key Laboratory of Urban Water Resource and Environment (Harbin Institute of Technology) (No 2010TS02).

References

1. Elliott, C., Vijayakumar, V., Zink, W., Hansen, R.: National Instruments LabVIEW: A Programming Environment for Laboratory Automation and Measurement. Technical Brief (2007), doi:10.1016/j.jala.2006.07.012
2. Wu, C., Xu, C., Yu, J.: Application of virtual instrument technique in acquiring hydraulic experiment changing transient data. Journal of Harbin University of Commerce (Natural Sciences Edition) 25(2), 210–213 (2009)
3. Zhao, H., Li, X., Zhao, M.: Water Supply Pipeline Hygienics. M. China Architecture & Building Press, Beijing (2008)
4. Cao, J., Liu, S.: Development and Prospect of Virtual Instrumental Technology. Automation & Instrumentation 1, 1–5 (2003)

5. Product Information: What is NI LabVIEW?,
   http://www.ni.com/labview/whatis/
6. Bott, T.R.: Fouling of Heat Exchangers, pp. 357–405. Elseviwer (1995)
7. Virtual Instruments, http://www.abc.chemistry.bsu.by/vi/vi.htm
8. Yuan, Y., Zhao, H., Zhao, M.: The Research on Growing-ring of Water-supply Network. Journal of Harbin University of C.E. & Architecture 31(1), 72–76 (1998)

Quick Formulas Based on DDA Outputs for Calculating Raindrop Scattering Properties at Common Weather Radar Wavelengths

Wang Zhenhui, Guo Lijun, Dong Huijie, and Zhang Peichang

Abstract. Quick Formulas based on DDA outputs is a set of partitioned polynomials aiming at easy-coding and time-saving during calculating extinction and backscattering properties for oblate spheroid raindrops. The Quick Formulas for the common radar wave bands of S, C, X and Ka have been obtained. It has been shown that the accuracy of Quick Formulas is more than 95%, that is, the error is less than 5%, and the time consumed is negligible as compared with DDA computation. The resulting codes can be inserted into the software for radar reflectivity simulation and real-time attenuation correction of rainfall estimation with radar.

1 Introduction

Raindrop scattering calculations play an increasingly important role in radar quantitative measurement of precipitation, millimeter wave technology and polarization technique. The discrete dipole approximation (DDA) technique [1] for non-spherical particle scattering has become the most popular algorithm but the computing is very time-consuming and the code is difficult to be embedded into other softwares. It is necessary to put up Quick Formulas which is equivalent to DDA in accuracy for calculating scattering properties of non-spherical raindrop at common radar wave bands (10.7 cm, 5.6 cm, 3.2 cm, 8.6 mm) but quick and flexible for practice.

2 Brief Introduction of Quick Formulas Equivalent to DDA

To maintain the complex curvature in scattering features with equi-volume radius r, partitioned multinomial is used as the expression of Quick Formulas ('QF' for short) [2]:

$$Q = Ar^3 + Br^2 + Cr + D \tag{1}$$

Wang Zhenhui · Guo Lijun
Key Laboratory of Meteorological Disaster of Ministry of Education,
Nanjing University of Information Science & Technology, 210044, Nanjing, P.R. China

Wang Zhenhui · Guo Lijun · Dong Huijie · Zhang Peichang
School of Atmospheric Physics, Nanjing University of Information Science & Technology,
210044, Nanjing, P.R. China

F.L. Gaol et al. (Eds.): Proc. of the 2011 2nd International Congress on CACS, AISC 144, pp. 471–476.
springerlink.com © Springer-Verlag Berlin Heidelberg 2012

where Q stands for either extinction efficiency factor $Q_e=C_e/\pi r^2$ or backscattering efficiency factor $Q_b=C_b/\pi r^2$ under the horizontal and vertical incident polarization wave. C_e is the extinction cross section and C_b is the backscattering cross section. Letters A, B, C and D in (1) are fitting coefficients which can be obtained from fitting analysis on DDA output database. The order of the multinomial obtained may be equal to or less than three and the fitting results are expected to be more accurate than that of pure linear fitting.

3 QF at Common Radar Wave Bands

3.1 Preferences

Suppose the rotation axis for oblate raindrops orient upward and the equi-volume radius r ranges from 0.0100 cm to 0.5000 cm. The incident wave is horizontally transmitted and polarization is indicated by subscript letter h for horizontal and v for vertical, respectively. Axial ratio (the length of rotation axis divided by the length of symmetry axis, b/a) of an oblate raindrop is determined by the equivalent radius and terminal velocity of the raindrop according to reference [3,4].

3.2 Fitting Coefficients for Quick Formulas

Fitting Coefficients for Quick Formulas have been obtained for $Q_{h,e}$, $Q_{v,e}$, $Q_{h,b}$ and $Q_{v,b}$ at the all four wavelengths. Table 1, as an example, shows part of the fitting coefficients for Quick Formulas on $Q_{h,b}$ at $\lambda=3.2$ cm.

Table 1 Coefficients of Quick Formulas for calculating $Q_{h,b}$ at $\lambda=3.2$ cm

r(cm)	A	B	C	D
[0.0100,0.0130]	/	0.0461374297	-0.0007028032	0.0000029957
(0.0130,0.0200]	3.7518232266	-0.0900997893	0.0009506292	-0.0000037145
(0.0200,0.0350]	5.7299298969	-0.2068898535	0.0032193689	-0.0000181536
(0.0350,0.0600]	10.62025869	-0.71054927	0.02063213	-0.00021962
(0.0600,0.0850]	17.7604705	-2.0498862	0.1042571	-0.0019563
(0.0850,0.1200]	23.7644031	-4.0744580	0.3167006	-0.0090695
(0.1200,0.1600]	275.3712633	-100.3062407	12.5692117	-0.5282360
(0.1600,0.2000]	-154.9247003	138.9779076	-30.9197631	2.0664059
(0.2000,0.2400]	/	35.58568	-8.27691	0.43381
(0.2400,0.2900]	/	36.10415	-8.63659	0.49110
(0.2900,0.3400]	/	35.73670	-7.05832	0.06678
(0.3400,0.3900]	/	-61.67439	58.62517	-11.00325
(0.3900,0.4250]	1708.51867	-2108.08211	872.55926	-118.53266
(0.4250,0.4650]	2980.19242	- 4022.29312	1805.55735	-266.93756
(0.4650,0.4850]	24183.28212	-34549.44081	16432.78205	-2599.72026
(0.4850,0.5000]	-76487.30914	112001.90705	-54674.19043	8899.54936

3.3   Comparison of Quick Formula with DDA

The results are shown in Figs.1 and 2. It can be seen that the results from QF and DDA agree with each other quite well in general. Both Q_e and Q_b enter into the range of resonance scattering at smaller radius for shorter wavelength. The Q_e and Q_b reaches the peak values, then fluctuates after the onset of the resonance scattering. The maximum radius for Q_b at $\lambda = 8.6$ mm as shown in Fig. 2d is only 0.4300 cm, because DDA calculation for larger particles is inaccurate.

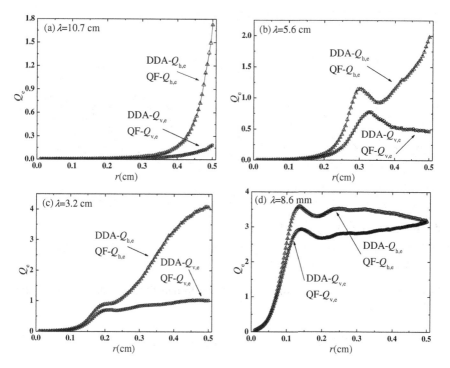

Fig. 1 Extinction efficiency factor Q_e of liquid oblate spheroids from QF calculations and DDA database at common radar wavelengths. (a)λ=10.7 cm; (b)λ=5.6 cm; (c)λ=3.2 cm; (d)λ=8.6 mm

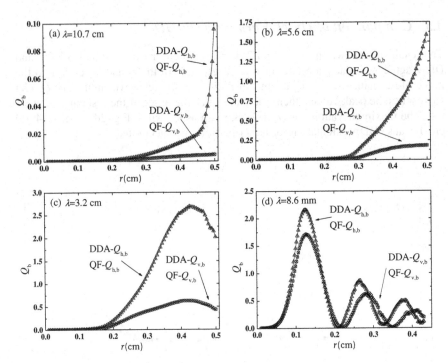

Fig. 2 The same as Fig.1 but for Q_b

4 Evaluation of Quick Formulas

4.1 Calculation Accuracy

Definition of the relative error *err* and its mean and the root mean square error *RMSE* are as follows:

$$err = \left|Q_{QF} - Q_{DDA}\right|/Q_{DDA} \times 100\% \qquad (2)$$

$$Mean\ \ err = \sum err/n \qquad (3)$$

$$RMSE = \sqrt{\sum\left(Q_{QF} - Q_{DDA}\right)^2/n} \qquad (4)$$

The independent dataset with a sample size $n=49$ is used for the above statistics and the results are shown in Fig.3 and Table 2. It can be seen that the maximum and the mean of the relative error are all less than 5% and the *RMSE*'s are also very small.

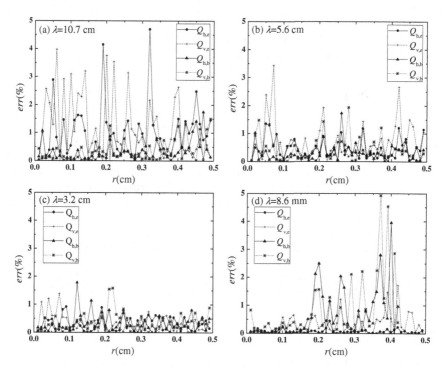

Fig. 3 Quick Formula error in Q_e and Q_b relative to DDA .(a)λ=10.7 cm; (b)λ=5.6 cm; (c)λ=3.2 cm; (d)λ=8.6 mm

Table 2 Error statistics for Quick Formulas

λ (cm)	Mean err				RMSE			
	$Q_{h,e}$	$Q_{v,e}$	$Q_{h,b}$	$Q_{v,b}$	$Q_{h,e}$	$Q_{v,e}$	$Q_{h,b}$	$Q_{v,b}$
10.7	0.99%	1.21%	0.31%	0.38%	0.00418	0.00034	0.00009	0.00002
5.6	0.45%	0.72%	0.28%	0.40%	0.00520	0.00356	0.00157	0.00036
3.2	0.29%	0.33%	0.34%	0.39%	0.00700	0.00235	0.00582	0.00159
0.86	0.08%	0.32%	0.71%	0.83%	0.00242	0.00345	0.00343	0.00283

4.2 Computational Efficiency

The time consumed for calculation has been counted. Take λ=10.7 cm for example, it takes about 56 hours for DDA to complete the computation to obtain 99 sets of data (each set includes $Q_{h,e}$, $Q_{v,e}$, $Q_{h,b}$ and $Q_{v,b}$) but Quick Formulas requires only 0.031 seconds. The ratio is about 6500000:1. Therefore the Quick Formulas improve the computational efficiency greatly.

5 Conclusion

Quick formulas in form of partitioned polynomials have been set up based on DDA outputs. The formulas are easy-coding and time-saving during calculating extinction and backscattering properties for oblate spheroid raindrops. The Quick Formulas for the common radar wave bands such as S, C, X and Ka have been obtained. It has been shown that the accuracy of Quick Formulas is more than 95%, that is, the error is less than 5%, and the time consumed is negligible as compared with DDA.

Acknowledgments. This work is jointly supported by National Natural Science Foundation of China (40875016), National High Technology Research & Development Program of China (863 Program, 2007AA061901) and Doctoral Program of Higher Education Research Foundation (20060300002).

References

1. Draine, B.T., Flatau, P.J.: Discrete-dipole approximation for scattering calculations. J. Opt. Soc. Am. A 11, 1491–1499 (1994)
2. Guo, L.J., Wang, Z.H.: A study on Quick Formulas for calculating liquid oblate spheroid scattering properties (unpublished)
3. Pruppacher, H.R., Beard, K.V.: A wind tunnel investigation of the internal circulation and shape of water drops falling at terminal velocity in air. Quart. J. Roy. Met. Soc. 96, 247–256 (1970)
4. Atlas, D., Srivastava, R.C., Sekhon, R.S.: Doppler radar characteristics of precipitation at vertical incidence. Reviews of Geophysics and Space Physics 11, 1–35 (1973)

Optimization Analysis for Louver Fin Heat Exchangers

Ying-Chi Tsai and Jiin-Yuh Jang

Abstract. The present study is aimed at optimization of the geometers of the fin-and-tube heat exchanger with louver fin through numerical simulation. The optimization is carried out by using the simplified conjugate-gradient method (SCGM) is adopted for solving the optimal problem. Using the optimizer, the louver angle of louvered fin is adjusted toward the maximization of the performance of the heat exchanger. It is also shown that the maximum of area reduction ratios at the louver fin for Re_{Lp} 100 – 400 with Lp = 1 mm. For the louver pitches, the following correlations for the optimal louver angle are derived, based on Reynolds number Re_{Lp} ranging from 100 to 400. The results indicate the optimal louver angle applied in heat exchangers can effectively enhance the heat transfer performance. Thus, the correlations is derived that can be applied to the design of heat exchangers.

1   Introduction

Heat exchangers have been used in a wide variety of applications. Typical among them are the district heat stations, HVAC (heating, ventilating, air-conditioning, and refrigeration) systems, food and chemical process systems, and heat recovery systems. Especially, the fin and tube heat exchanger was frequently applied to the air-conditioning system of the automotive industry. Therefore, enhanced surfaces were often employed to effectively improve the overall performance of the fin and tube heat exchanger.

The design of louver fin has been extensively studied experimentally and, more recently, numerically with computational fluid dynamic (CFD) codes using the finite difference or finite volume methods. The louver fin surfaces are characterized by fins that have been cut and bent out into the flow stream at frequent intervals.

Ying-Chi Tsai

Candidate, Department of Mechanical Engineering, National Cheng Kung University, Tainan, Taiwan 70101

e-mail: n1897108@mail.ncku.edu.tw

Jiin-Yuh Jang

Department of Mechanical Engineering, National Cheng Kung University, Tainan, Taiwan 70101

e-mail: jangjim@mail.ncku.edu.tw

F.L. Gaol et al. (Eds.): Proc. of the 2011 2nd International Congress on CACS, AISC 144, pp. 477–482.
springerlink.com   © Springer-Verlag Berlin Heidelberg 2012

The purpose of louver is to break up the boundary layers so as to yield higher heat transfer coefficients and lower thermal resistance than which are possible with plain fins under the same flow conditions.

The first reliable published data about louver fin was by Kays and London [1] in 1950. Investigations on louvered fin heat exchangers are principally divided into three categories. Firstly, for louvered fin having flat tube configurations, extensive experimental data were reported by Davenport, [2] and Achaichia. [3] Webb and Jung [4] presented experimental data for six brazed aluminum heat exchangers. Suga et al. [5] used a rectangular flow domain filled with overlapping Cartesian meshes to compute the flow and heat transfer over a finite-thickness fin. In the 1990's several researchers developed CFD codes based on non-orthogonal, boundary-fitted meshes to compute the flow over louver fins. Jang et al. [6] numerically investigated three-dimensional convex louver finned-tube heat exchangers. The effects of different geometrical parameters, including convex louver angles (h = 15.5°, 20.0°, 24.0°), louver pitch (Lp = 0.953 mm, 1.588 mm) and fin pitch (8 fins/in., 10 fins/in., 15 fins/in.) are investigated in detail for the Reynolds number ReH (based on the fin spacing and the frontal velocity) ranging from 100 to 1100.

The optimization technique may be used in the geometrical optimization of louver fin in order to obtain optimal performance under specified design objectives within the allowable pressure drops. To reach the goal, the simplified conjugate-gradient method (SCGM) has been combined with a commercial CFD code to build an optimizer for designing the angle of louvered fin. Using the optimizer, the louver angle of louver fin is adjusted toward the maximization of the performance for the heat exchanger.

2 Mathematical Analysis

Fig.1 (a) shows the physical model and computation domain for the louvered fin heat exchanger in Cartesian coordinates. The 3D computational domain is composed of a louvered fin cross-section with entry region, five louvers on either side of the turnaround louver and outlet region as seen from Fig.1 (b). In the present study, the fluid is considered incompressible with constant properties and the flow over the louvers assumed to be laminar in the steady-state and with no viscous dissipation. The detailed geometrical for the calculation are given as follows:

In the present study, the fluid is considered incompressible with constant properties. Consequently, due to the low inlet velocity and the small fin pitch the flow in the channel of the heat exchanger is assumed to be laminar in the steady-state and with no viscous dissipation. This assumption seems to be reasonable at the considered Reynolds numbers Re_{Lp} ranging from 50 to 500. The dimensionless equations for mass, momentum (Reynold-averaged Navier-Stokes equation) and energy may be expressed in tensor form as

$$\frac{\partial U_i}{\partial X_i} = 0 \tag{1}$$

$$\frac{\partial}{\partial X_j}\left(U_iU_j\right) = -\frac{\partial P}{\partial X_i} + \frac{1}{Re_H}\left[\nabla^2 U_i\right] \tag{2}$$

$$\frac{\partial}{\partial X_j}\left(\Theta U_j\right) = \frac{1}{Re_H Pr}\left[\nabla^2 \Theta\right] \tag{3}$$

In the above equations, the dimensionless temperature is defined as Θ. Pr is the Prandtl number, which is set equal to 0.71 (for air) in the present study.

Fig. 1 Schematic of the computational domain: (a) Definition of geometrical parameters for a multi-louvered fin heat exchanger. (b) Cross-section of louvered fin geometry.

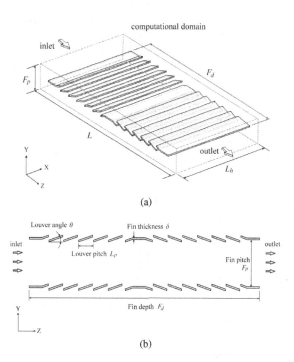

Because the governing equations are elliptic in spatial coordinates, the boundary conditions are required for all boundaries of the computation domain. For the computation domain, no-slip conditions and constant wall temperature T_w (333K) are specified at the louvered fin surface. However, one more pair of boundary conditions for the velocity and thermal fields needs to be specified on the periodic interface of fluid between the successive fins. At the upstream boundary, uniform flow with the velocity V_{fr} and temperature T_{in} (293K). On the other hand, for the boundary conditions at the exit, it is assumed that the flow becomes fully developed downstream and the outlet pressure is operational atmosphere pressure.

3 Numerical Methods

In this study, the governing equations are solved numerically using a control volume based finite difference formulation. The numerical methodology is briefly described here. Finite difference approximations are employed to discretize the transport equations on non-staggered grid mesh systems. A third-order upwind TVD (total variation diminishing) scheme is used to model the convective terms of governing equations. Second-order central difference schemes are used for the viscous and source terms. The coupling between velocity and pressure is performed with SIMPLEC algorithm.

A pressure based predictor/multi-corrector solution procedure is employed to enhance velocity-pressure coupling and continuity-satisfied flow filed. A grid system of 239×46×11 grid points is adopted typically in the computation domain. Again, the plate geometry is coarsely meshed for the convenience of the reader to visualize the flow regions. For this purpose, three grid systems, 219×40×10, 239×46×11 and 251×51×11, are tested. It is found that for V_{fr}=2.0 m/s, the relative errors in the local pressure and temperature between the solutions of 239×46×11 and 251×51×11 are less than 3 %.

However, a careful check for the grid-independence of the numerical solutions has been made to ensure the accuracy and validity of the numerical results. Computations were performed on an INTEL Core2 Q9300 2.54G personal computer and typical CPU times were about 5000s for each case.

4 Optimization Method

In the present study, the simplified conjugate-gradient method (SCGM) has been combined with the commercial CFD code as an optimizer for designing the louver angle with louvered fin. In addition, a grid generation unit is used to generate the mesh for numerical computation. Using the optimizer, the uniform angle of the louver fin is adjusted toward the maximization of the area reduction ratio compared with the fin and tube heat exchanger.

In the simplified conjugate-gradient method, the objective function (J) in conjunction with the optimization process for the angle of the louver fin is defined in the following:

$$J(X_i)=1-A_c/A_{ref} \tag{4}$$

where X denotes the parameter for optimization (referred to as search variable or design variable). In this manner, as the objective function J is approaching its maximum value in the optimization process, the area reduction ratio reaches a minimum. This leads to an optimal design having a higher performance.

5 Results and Discussion

For the optimization objective, the evaluation criteria, area reduction ratio $1-(A_c/A_{ref})$ is chosen, where the subscripts of 'c' and 'ref' refer to compared louver fin and reference plate fin, respectively. In addition, it is noted that the performance Colburn factor (j) and the friction factor (f) calculations are based on the above correlations. The optimization process for area reduction ratio is shown in Fig.2. For the case at $Re_{Lp} = 50$ and $Lp = 1.0$ mm, the area reduction ratio of the optimal design is substantially improved by 58.5%; whereas for original design at louver angle $\theta = 15.0°$, the optimal louver angle is increased to 28.953°. In optimization search, the magnitude of area reduction ratio is increased rapidly during the first 6 iterations, and the maximum is obtained in 30 iterations. Moreover, after certain generation (larger than 11 iterations), in turn the variation of louver angle is small, finally a level off value of optimal design is found. Therefore, the optimized louvered fin offers better heat transfer performance than the reference plate fin. On the other hand, the optimized louvered fin also offers better heat transfer performance than the original louvered fin.

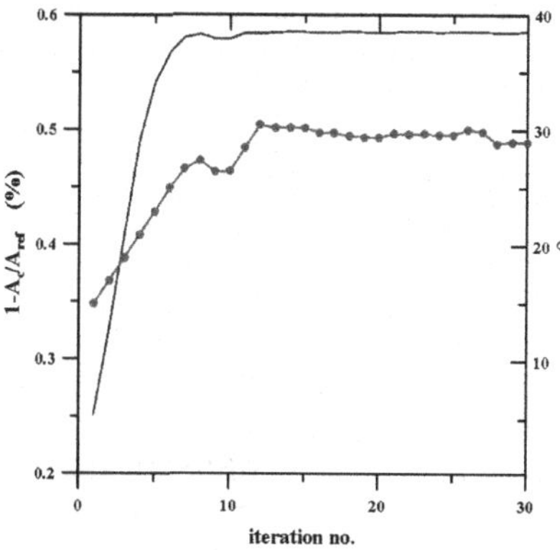

Fig. 2 Iteration process for maximum area reduction ratio and louver angle.

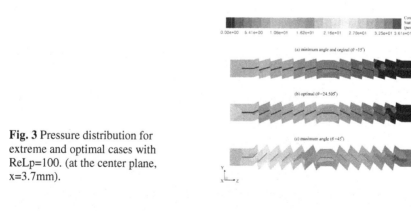

Fig. 3 Pressure distribution for extreme and optimal cases with ReLp=100. (at the center plane, x=3.7mm).

Fig. 4 Temperature distribution for extreme and optimal cases with ReLp=100. (at the center plane, x=3.7mm).

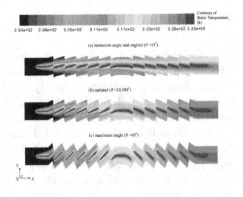

Acknowledgment. Financial support for this work was provided by the National Science Council of Taiwan, under contract NSC99-2221-E-006-141-MY2.

References

1. Kays, W.M., London, A.L.: Heat transfer and flow friction characteristics of some compact heat exchanger surfaces-Part I: Test system and procedure. Trans. ASME 72, 1075–1085 (1950)
2. Davenport, C.J.: Correlations for heat transfer and flow friction characteristics of louvred fin. In: AIChE Symposium Series, pp. 19–27 (1983)
3. Achaichia, A.: The performance of louvered tube-and-plate fin heat transfer surface. PhD thesis, Department of Mechanical and Production Engineering, Brighton Polytechnic (1987)
4. Webb, R.L., Jung, S.H.: Air side performance of enhanced brazed aluminum heat exchangers. ASHRAE Trans. 98, 391–401 (1992)
5. Suga, K., Aoki, H.: Numerical study on heat transfer and pressure drop in multilouvered fins. In: Proc. of ASME/JSME Thermal Eng. Joint Conf., vol. 4, pp. 361–368 (1991)
6. Jang, J.Y., Shieh, K.P., Ay, H.: 3-D thermal-hydraulic analysis in convex louver finned -tube heat exchangers. In: ASHRAE Annual Meeting, Cincinnati, OH, USA, June 22–27, pp. 501–509 (2001)

Numerical Investigation of Nanofluids Laminar Convective Heat Transfer through Staggered and In-Lined Tube Banks

Jun-Bo Huang* and Jiin-Yuh Jang

Abstract. Laminar forced convection of a water-based suspension of Al_2O_3 nanoparticles through in-line and staggered tube banks with constant wall temperature boundary condition have been investigated numerically. A two phase mixture model is employed to simulate the nanofluid convection, taking into account appropriate thermophysical properties. Nanoparticles are assumed spherical with a diameter equal to 50 nm. The effects of Reynolds number and nanoparticle volume concentration on the flow and heat transfer behavior are studied. Results show that convective heat transfer coefficient and pressure drop for nanofluids is greater than that of the base fluid. It is found that the heat transfer enhancement increases with increase in Reynolds number and nanoparticle volume concentration. In general, the heat transfer in a staggered array of tubes is found to be higher than that in an in-lined array of tubes.

Keywords: Nanofluids, In-line tube bank, Staggered tube bank.

1 Introduction

The phenomenon of crossflow in a tube bank occurs in a wide variety of heat exchanger applications. A better understanding of the details of flow and heat transfer in tube banks is important for the improvement of heat exchanger designs.

Numerical modeling of heat transfer of nanofluids can be performed using single phase (homogeneous) or two phase approaches. The single phase approach is simpler and requires less computational time, which assumes that the fluid phase and nanoparticles are in thermal equilibrium with zero relative velocity. In two phase model, nanoparticles and base fluid are considered as two different liquid and solid phases with different momentums respectively. Since the two phase approach considers the movement between the solid and liquid phase, it has better prediction

Jun-Bo Huang · Jiin-Yuh Jang
Department of Mechanical Engineering, National Cheng-Kung University, Tainan, Taiwan
e-mail: n1897102@mail.ncku.edu.tw

* Corresponding author.

F.L. Gaol et al. (Eds.): Proc. of the 2011 2nd International Congress on CACS, AISC 144, pp. 483–490.
springerlink.com   © Springer-Verlag Berlin Heidelberg 2012

in nanofluid study. The mixture model was successfully applied to model nanofluids convection.[1-5] Behazadmehr [3]simulated nanofluid turbulent convection in a circular tube. They showed the higher accuracy of mixture model with respect to the single phase model and the agreement with experimental correlation proposed by Xuan and Li [6].

Two types of tube array are solved numerically: in-line tube arrays and staggered tube arrays. Considerable work has been done for an in-line arrangement and for a staggered arrangement[11-17].

In this paper, a two phase mixture model was applied to study the laminar convection of water-Al$_2$O$_3$ nanofluid flow through an in-line and a staggered tube bank with constant wall temperature boundary condition. There are only a limited number of published works related to the heat transfer enhancement from a tube bank using nanofluids. The importance of this research is it shows that water-Al$_2$O$_3$ nanofluid can increase the heat transfer rate from tube banks.

2   Physical Model and Numerical Procedures

The physical model of flow around an in-lined tube bank and a staggered tube bank is shown in Fig.1. The inlet flow has a uniform velocity V_{in}. The tube diameter is 20mm. The solution domain is bounded by the inlet, the outlet, and by dotted lines KL and MN, which can be referred to as the bottom and top boundaries, respectively. The coordinate system is illustrated in the figure. The full domain included five and ten longitudinal rows of tubes for an in-lined and a staggered tube bank, respectively. The flow is assumed to be laminar, steady, 2D and exhibiting no viscous dissipation.

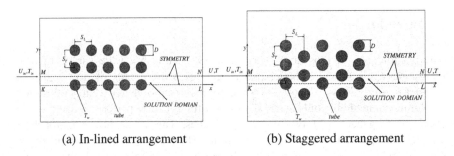

(a) In-lined arrangement   (b) Staggered arrangement

Fig. 1 Physical model and computational domain of a tube bank

2.1   Mixture Model

The mixture model is employed in the simulation by assuming the coupling between phases is strong, and particles closely follow the flow. Instead of using the governing equations of each phase separately, the continuity, momentum and fluid energy equations for the mixture are employed.

2.2 Boundary Conditions

At the inlet area, profiles of uniform inlet velocity, V_{in}, and temperature, $T_{in} = 300K$, are assumed. The set of coupled non-linear differential governing equations has been solved subjected to following boundary conditions:

Table 1 Summary of boundary conditions

Boundary Condition	V_x	V_y	T
Inlet	$V_x = V_{in}$	$V_y = 0$	$T = T_{in}$
Outlet	$\dfrac{\partial V_x}{\partial x} = 0$	$V_y = 0$	$\dfrac{\partial T}{\partial x} = 0$
Symmetry	$\dfrac{\partial V_x}{\partial y} = 0$	$V_y = 0$	$\dfrac{\partial T}{\partial y} = 0$
Tube surface	$V_x = 0$	$V_y = 0$	$T = T_w$

3 Results and Discussions

3.1 Parameters of Interest

The average heat transfer results are presented using an average Nusselt number, based on the total rate of heat transfer and a log-mean temperature difference. This parameter is defined by

$$Nu_{LM} = \frac{\bar{h}D}{k} = \frac{q}{A_s \Delta T_{LM}} \frac{D}{k} \tag{1}$$

Where q is the rate of heat transfer to the fluid and A_s is the total surface area of tubes exposed to the fluid in the solution domain. The log-mean temperature difference is defined by

$$\Delta T_{LM} = \frac{(T_w - T_{in}) - (T_w - T_{out})}{\ln\left[(T_w - T_{in})/(T_w - T_{out})\right]} \tag{2}$$

The total rate of heat transfer to the fluid was calculated as the total change in the fluid enthalpy between the inlet and the outlet.

The overall pressure drop for the tube bank is presented using an average friction, f_c, defined as

$$f_c = \frac{P_{in} - P_{out}}{2\rho \left(V_{max} \right)^2 N_L} \tag{3}$$

where N_L, the number of tube rows.

3.2 Streamlines, Pressure and Temperature Contours

Figs. 2 and 3 illustrate the numerical result for the streamline, pressure and isotherm distributions with two different nanoparticle volume concentration. For the in-lined array, the flow separates at the rear portion of a tube and reattaches at the front portion of the following tube to form a larger stationary recirculation region between the two adjacent tubes, resulting in a dead flow zone. For the staggered array, the streamline patterns have a similar character because of the repeated blockage of the staggered tube bank, there is a smaller recirculation zone behind each tube. The streamlines indicate almost fully developed flow pattern behind the third row up to the front of the last row. Pressure contours display the drop of pressure along the flow direction. It is found that the magnitude of the overall pressure drop increased slightly with increased nanoparticle volume concentration.

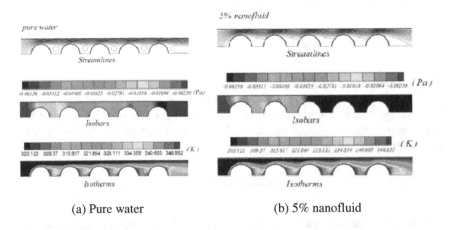

(a) Pure water   (b) 5% nanofluid

Fig. 2 The streamline, pressure and isotherm distributions for in-lined arrangements at Re=100

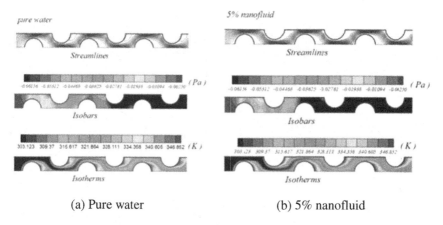

(a) Pure water (b) 5% nanofluid

Fig. 3 The streamline, pressure and isotherm distributions for staggered arrangements at Re=100

The temperature distribution of a heat transfer configuration can provide detailed information about a temperature field such that one can determine the existence of hot spots or other phenomena related to the temperature. Closely spaced isotherms on the front of the first tube indicate the higher heat transfer rate there. According to these figures, it can be seen that there is a non-uniform temperature distribution in front of the tube. The density of the temperature decreases along the flow direction.

3.3 Average Heat Transfer and Pressure Drop

The average heat transfer coefficient for the tube bank is presented in terms of Nu_{LM} in Fig. 4. In the figure, the empirical correlations of Zukauskas [11-12] are plotted.

For an in-line arrangement:

$$Nu_{LM} = 0.979\left(0.71 \cdot Re_{max}^{0.5} \cdot Pr^{0.36}\right) \tag{4}$$

For a staggered arrangement:

$$Nu_{LM} = 0.52 \cdot Re_{max}^{0.5} \cdot Pr^{0.36} \tag{5}$$

The relationships between average Nusselt number and Reynolds number for an in-lined and a staggered tube bank are shown in Fig. 4. It can be seen that the average Nusselt number for nanofluids is higher than that for water, and the presented work are in agreement with available correlations. As shown in Fig. 5, it is observed that the presence of nanofluids around tube banks also increase the overall pressure drop.

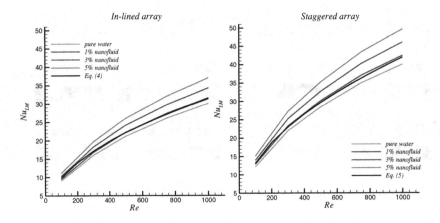

Fig. 4 Average Nusselt number for an in-lined and a staggered tube bank

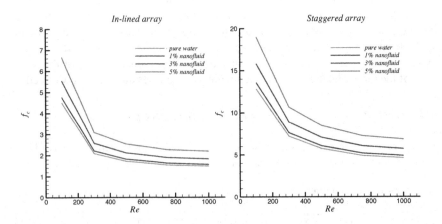

Fig. 5 Average friction factor results for an in-lined and a staggered tube bank

4   Conclusions

Laminar forced convection of a water-based suspension of Al_2O_3 nanoparticles through in-line and staggered tube banks with constant wall temperature boundary condition have been investigated numerically. The numerical results demonstrate that the heat transfer coefficient of staggered array is higher than that of in-lined array, while the pressure drop of staggered configuration is higher than that of in-lined configuration. Results show that the addition of nanoparticles produce a significantly increase of the heat transfer with respect to that of the base fluid. Heat transfer enhancement is increasing with the nanoparticle volume concentration. However, it is accompanied by increasing pressure drop values.

References

[1] Lotfi, R., Saboohi, Y., Rashidi, A.M.: Numerical study of forced convective heat transfer of Nanofluids: comparison of different approaches. Int. Commun. Heat Mass Trans. 37, 74–78 (2010)

[2] Bianco, V., Manca, O., Nardini, S.: Numerical investigation on nanofluids turbulent convection heat transfer inside a circular tube. Int. J. Therm. Sci. 29(17-18), 3632–3642 (2009)

[3] Akbarinia, A., Laur, R.: Investigation the diameter of solid particles affects on a laminar nanofluid flow in a curved tube using a two phase approach. Int. J. Heat Fluid Flow 30(4), 706–714 (2009)

[4] Behzadmehr, A., Saffar-Avval, M., Galanis, N.: Prediction of turbulent forced convection of a nanofluid in a tube with uniform heat flux using a two phase approach. Int. J. Heat Fluid Flow 28, 211–219 (2007)

[5] Mirmasoumi, S., Behzadmehr, A.: Effect of nanoparticles mean diameter on mixed convection heat transfer of a nanofluid in a horizontal tube. Int. J. Heat Fluid Flow 29, 557–566 (2008)

[6] Xuan, Y.M., Li, Q.: Investigation on convective heat transfer and flow features of nanofluids. Journal of Heat Transfer 125, 151–155 (2003)

[7] Maiga, S.E.B., Nguyen, C.T., Galanis, N., Roy, G.: Heat transfer behaviours of nanofluids in a uniformly heated tube. Superlattices and Microstructures 35, 543–557 (2004)

[8] Das, S.K., Putra, N., Thiesen, P.W., Roetzel, R.: Temperature dependence of thermal conductivity enhancement for nanofluids. J. Heat Transfer 125, 567–574 (2003)

[9] Rea, U., Mckrell, T., Hu, L.-W., Buongiorno, J.: Laminar convective heat transfer and viscous pressure loss of alumina-water and zirconia-water nanofluids. Int. J. Heat and Mass Transfer 52, 2042–2048 (2009)

[10] Hwang, K.S., Jang, S.P., Choi, S.U.S.: Flow and convective heat transfer characteristics of water-based Al2O3 nanofluids in fully developed laminar flow regime. Int. J. Heat and Mass Tranfer 52, 193–199 (2009)

[11] Zukauskas, A.: Heat transfer from tubes in crossflow. In: Hartnett, J.P., Irvine, J.T.F. (eds.) Advances in Heat Transfer, vol. 8, pp. 93–160 (1972)

[12] Zukauskas, A., Skrinska, A., Ziugzda, J., Gnielinski, V.: Single phase convective heat transfer: Banks of plain and finned tubes. In: Hewitt, G. (ed.) Heat Exchanger Design Handbook, ch. 2.5.3. Begell House, New York (1998)

[13] Chen, C.J., Wung, T.-S.: Finite analytic solution of convective heat transfer for tube arrays in crossflow: part-heat transfer analysis. J. Heat Transfer 111, 641–648 (1989)

[14] Dhaubhadel, M.N., Reddy, J.N., Telionis, D.P.: Penalty finite-element analysis of coupled fluid flow and heat transfer for in-line bundle of cylinders in cross-flow. Int. J. Non-Linear Mech. 21, 361–373 (1986)

[15] Dhaubhadel, M.N., Reddy, J.N., Telionis, D.P.: Finite-element analysis of fluid flow and heat transfer for staggered bundles of cylinders in cross flow. Int. J. Numerical Methods in Fluids 7, 1325–1342 (1987)

[16] Wang, Y.Q., Penner, L.A., Ormiston, S.J.: Analysis of laminar forced convection of air for cross flow in tube banks of staggered tubes. Numer. Heat Transfer, Part A 38(8), 819–845 (2000)

[17] EI-Shaboury, A.M.F., Ormiston, S.J.: Analysis of laminar forced convection of air for cross flow in in-line tube banks with nonsquare arrangements. Numer. Heat Transfer, Part A 48(2), 99–126 (2005)

[18] Manninen, M., Taivassalo, V., Kallio, S.: On the Mixture Model for Multiphase Flow/. Technical Research Center of Finland, VTT Publications 288, 9–18 (1996)

[19] Schiller, L., Naumann, A.: A drag coefficient correlation. Z. Ver. Deutsch. Ing. 77, 318–320 (1935)

A Kind of High Reliability On-Board Computer

Pei Luo, Guofeng Xue, Jian Zhang, and Xunfeng Zhao

Abstract. Because of the special application environment and high reliability requirements of the on-board computers, this article presents a design of on-board computer based on ERC32 with high reliability. This article also addresses the conception of 'Cumulative Single Event Upset' as well as the corresponding strategies to diminish such errors. It puts forward a variety of coping strategies for single event upset effects in space; it describes the design of the storage modules which are hard to cosmic rays briefly. The design has characteristics of high reliability and high compatibility; some ideas in it have been used in some of our earlier on-board computer systems. The long-term on orbit working performance shows that these strategies are effective to mitigate the errors caused by radiation and to improve the reliability of the payload systems.

1 Introduction

Spacecraft is a system which requires high reliability. Its central control module - the on-board computer - requires higher reliability because of its specific application environment. Thus how to improve the reliability and extend the life span of the computer system become the key issues to be discussed during the design process.

Reliability is the basic requirement for the aerospace and military system, and it has become one of the key issues to be discussed during the design process. As the control unit of the whole system, the Central Processing Unit (CPU) used in an aerospace electronic payload design is a key factor for the reliability of satellite system. Rad-hard devices are widely used in such designs to mitigate Single Event Upset effects (SEUs) caused by cosmic radiation. Traditional methods such as Triple Modular Redundancy (TMR) have been proven effective in mitigating SEUs, but triplication of CPU cannot be implemented in some designs because of the limitations of power and area. Many other special methods should be implemented to insure the system against errors and interruptions.

Some special hardware frameworks are often used in the on-board computer design, such as multiple redundancy, hot backup and so on. Besides, processors with error correction and error detection functions are specially designed for aerospace use. To further enhance the stability of the system and extend its life span,

Pei Luo · Guofeng Xue · Jian Zhang · Xunfeng Zhao
No. 1 Nanertiao, Zhongguancun, Haidian, Beijing, China
e-mail: luopei09@mails.gucas.ac.cn, xueguofeng@sccar.ac.cn

F.L. Gaol et al. (Eds.): Proc. of the 2011 2nd International Congress on CACS, AISC 144, pp. 491–496.
springerlink.com   © Springer-Verlag Berlin Heidelberg 2012

some measures including triple modular redundancy, RAM backup are introduced into this design. This article also puts forward the conception of "Accumulative Single Event Upset" which would cause uncorrectable errors, and the strategies to eliminate such errors are also proposed.

2 System Construction

Nowadays, the processors used in on-board computer are mainly 80X86, PowerPC and SPARC V7, V8 and some other series. The processors based on PowerPC structures such as RAD6000 and RAD750 are mainly used by National Aeronautics and Space Administration (NASA), and the SPARC series including TSC695 and AT697 are developed by European Space Agency (ESA) and Atmel Corporation.

The TSC695F processor used in this design belongs to SPARC family and it is produced by ATMEL. It is a kind of high integrated, high-performance 32-bit RISC embedded processor which adopts the SPARC architecture V7 specification and is able to support ESA criteria. It is a space-tailored design with on-chip concurrent transient and permanent error detection [1] [2].

This system adopts a 32K bytes PROM to store Bootloader, and another 2M bytes EEPROM to store the operating system image. Two SRAM chips are used to construct the RAM of the system, one 2M bytes, 32 bits wide SRAM used for data storage, the other 512K Byte, 8bit wide one for storing the check bits and the parity bit. Figure 1 shows the structure of the system.

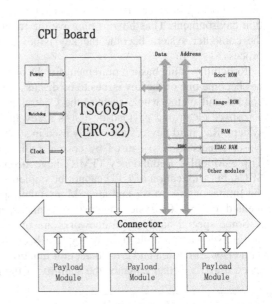

Fig. 1 Hardware construction of the system

3 Triple Modular Redundancy of VxWorks Image

Redundancy is frequently implemented in the design of on-board computers for higher reliability and stability. Triple modular redundancy, hot backup and cold backup are also often used in such designs. The key idea is to backup the critical parts and get the right result through arbitrator, so that the system can run properly even if one critical module fails in the function, and thus the system's stability is strengthened and its life span can be prolonged.

To prevent data errors of VxWorks image resulted from the SEU effects in deep space, triple modular redundancy is also used in the design of ROM. This system will store three copies of VxWorks image in the ROM separately, and get the right result by comparing the three copies of image in the initialization of the system and then write it into the RAM in order to prevent errors caused by SEU.

The system uses a PROM of 32K bytes and an EEPROM of 2M bytes as the ROM space. The 32K bytes PROM is used to store the Bootloader, and the 2M bytes EEPROM is used to store the VxWorks image. Triple modular redundancy is used in the storing of the VxWorks image, as shown in Figure 2.

Fig. 2 Triple modular redundancy of VxWorks Image

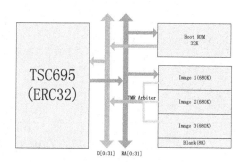

The process to realize triple modular redundancy of the VxWorks image is as following: there are three identical copies of VxWorks image in the EEPROM, when the Bootloader reads the copies of image after the initialization of the registers and the RAM, it will get the right result by comparing the three copies of data according to the formula (1):

$$data = (data\_A \& data\_B) | (data\_B \& data\_C) | (data\_C \& data\_A) \qquad (1)$$

In the equation above, data_A, data_B and data_C are the contents read from the same address of the three image copies. We can get the right result 'data' by comparing the three copies according to the formula and write it to the relevant address in the RAM. The Bootloader will jump to the initial address of the image in RAM to run the operating system after the whole VxWorks image is copied into the RAM correctly. This strategy can guarantee that it will not affect the startup of the operating system even if some single bits are wrong in the images of VxWorks.

4 Double RAM

Double RAM structure and EDAC are also implemented in the design to improve the reliability and extend the life span of the system. There are two RAM space in this design, one is located on the CPU board while the other on the external payload board. Both the two RAMs are 40 bits wide (32 bits data, 8 bits EDAC) and 2M Byte of the capacity. They can be accessed with different RAM selection signals without affecting each other.

The TSC695F provides chip selects for two redundant memory banks for replacement of faulty banks. There are eight selection signals MEMCS [0:7] used to select the RAM chip. Meanwhile, MEMCS [8:9] are used for the replacement of MEMCS [0:7]. For example, MEMCS0 will be invalid when MEMCS8 takes the place of it to use another memory space. So MEMCS0 will be inactive and MEMCS8 active instead. This strategy can effectively prevent the potential failure of the whole system caused by the failure of RAM by just replacing the main RAM with the backup one; the operating system will be copied to the second RAM and run there instead. The two RAM of the system are designed to use the same address in this system. The structure is shown in Figure 3.

Fig. 3 Structure of RAM

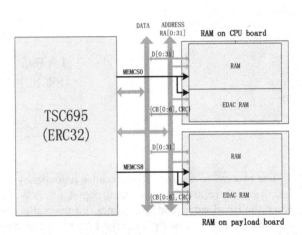

In the initialization of the whole system, Bootloader will check the two RAM blocks before the images are copied. If the Bootloader finds out that the RAM on the CPU board is damaged, MEMCS8 will take the place of MEMCS0. In other words, the RAM on CPU board will be replaced by another RAM located on the external payload board which has identical address. Then Bootloader will copy the VxWorks image to the RAM on the payload board instead. It will jump to the low address of the operating system and the VxWorks will start to run in sequence. This strategy insures that even if the RAM on the CPU board is damaged, the system will still go on working properly, so the reliability and stability of the whole system will be improved. Figure 4 shows the details of the flow diagram of Bootloader's initialization.

Fig. 4 Flow Diagram of Bootloader

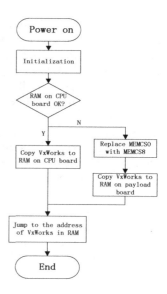

5 Eliminating Accumulative SEU Error

A key design point of on-board computer is how to protect the memory from the effect of SEU and SEL (Single Event Latch-up) [3].The TSC695F includes a 32-bit EDAC (Error Detection and Correction). Seven bits (CB [6:0]) are used as check bits over the data bus. The Data Bus Parity signal (DPAR) is used to check and generate the odd parity over the 32-bit data bus [1] [2].

The TSC695F's EDAC uses a seven bit Hamming code which detects any double bit error on the 40-bit bus as a non-correctable error. Also, the EDAC corrects all single bit data errors on the 40-bit bus. However, in order to correct any error in memory (e.g., Single Event Upset induced) the data has to be read and re-written by software because the TSC695F does not automatically write back the corrected data [1]. The EDAC of both TSC695F and AT697E use Hamming code to generate the seven bits check code [1] [4]. The algorithms in these two processors are almost the same. The one in TSC695F is shown in the Figure 5 [1].

If one single bit SEU error is not corrected in time, the effect will accumulate when SEU happens on the other bit of this data and then turn to uncorrectable error which cannot be corrected and may cause serious accident in spacecraft. We call such error CSEU (Cumulative Single Event Upset), which is caused by multiple bits of single event upset.

CSEU is caused when one single bit SEU error is not corrected in time and SEU effect accumulated in this data. Take RAM as an example, there are much data which is frequently read and written in RAM while some others are seldom visited. The CSEU may happen on such seldom visited data if it has not been visited for a long time after one SEU effect happened.

Then, one simple method will be used to eliminate such CSEU errors. One thread will read data from and write it back to the RAM to eliminate such errors when no other thread is using the CPU. The EDAC will find one single bit SEU error when reading data from RAM, and correct it and then write it back. So it can prevent CSEU errors before they occur. Taking "0x12345678" as an example, if the first bit turned from 0 to 1 because of the radiation, the data would be "0x12345679". EDAC will find this error and correct it according the check bits if we read it from the RAM before one other bit upsets too. CPU will write the corrected data back to the RAM and the chance for the happening of CSEU error will be reduced. The long-term on orbit working performance shows that this strategy can find and correct SEU errors in RAM timely, which prevents such errors from turning to uncorrectable CSEU errors, so that the possibility of the failure of the spacecraft has been reduced greatly.

Bit	Parity	D[31:0] (* indicates bit of D[31:0] used in parity calculation)																														
		31	30	29	28	27	26	25	24	23	22	21	20	19	18	17	16	15	14	13	12	11	10	9	8	7	6	5	4	3	2	1	0
DPAR	N.XOR	•	•	•	•	•	•	•	•	•	•	•	•	•	•	•	•	•	•	•	•	•	•	•	•	•	•	•	•	•	•	•	•
CB[6]	XOR	•	•	•	•	•				•	•			•	•	•	•	•				•			•			•				•	•
CB[5]	XOR	•				•	•		•		•	•		•				•	•			•	•	•	•		•	•					•
CB[4]	N.XOR		•	•		•	•	•	•		•		•		•				•		•	•	•						•	•	•		•
CB[3]	XOR	•	•			•			•	•		•				•		•	•	•	•	•	•	•	•		•						•
CB[2]	N.XOR	•					•			•			•	•		•	•	•				•		•			•	•	•	•	•	•	•
CB[1]	XOR		•		•			•	•			•				•	•	•		•	•			•	•	•	•	•		•	•		
CB[0]	XOR	•	•	•	•				•				•	•	•			•				•	•	•	•					•	•		•

Fig. 5 EDAC algorithm of TSC695F

6 Conclusions

The combination of hardware and software strategies in this design eliminates the possibility of failure of the system caused by cosmic radiation. The long-term on orbit working performance of several on-board computers shows that our strategy can find out and correct SEU errors in RAM timely, which prevents such correctable errors from turning to uncorrectable CSEU errors, and the failure possibility of the spacecraft has been eliminated greatly. Many ideas in this article have been implemented in our early on-board computer designs. These designs have improved the reliability and extended the life span of the system.

References

1. ATMEL, TSC695F SPARC 32-bit Space Processor User Manual, 4148H-AERO-12/03
2. ATMEL Web site, http://www.atmel.com/products/radhard
3. Xiong, J., Cheng, Z., et al.: On board computer Subsystem Design for the Tsinghua Nanosatellite. In: 20th AIAA International Communication Satellite Systems Conference and Exhibit
4. ATMEL, Rad-Hard 32 bit SPARC V8 Processor AT697E, 4226G–AERO–05/09

The Spatial Differentiation and Classification of the Economic Strength of Counties along the Lower Yellow River

Zhang Jinping and Qin Yaochen

Abstract. The spatial differences of economic strength of 109 counties along the Lower Yellow River in 2009 have been analyzed by weighted principal component analysis, trend analysis tools and Moran's I index of GIS. On this basis, the classification of economic strength is realized by the SOFM neural network model in MATLAB 7.0. The study shows that, the spatial concentration of the county economy is very significant as was indicated by Moran scatter plot. The level of economic strength of 109 counties can be divided into five categories with the principal component scores as the input of the SOFM network.

1 Introduction

Coordinated development of a county's economy and its dynamic equilibrium in space will help improve the overall efficiency of regional economic development [1]. Therefore the analysis on spatial differentiation and classification of economic strength is critical to clarify the developmental period of the county's economy so that it can contribute to relate decision-making [2]. The region along the Lower Yellow River is one of the major grain producing areas in China, and it is also a combination of the Efficient Eco-economic Zone in the Yellow River Delta and Zhongyuan Economic Zone, where there are significant regional differences on the county economic development. In China, the region has been defined as one of the main producing areas of agricultural products that are restricted to exploit, however, in recent years, the contradiction between rapid urbanization and environmental protection have been very conspicuous. On the whole, counties in this region are facing desire for intense industrialization and rapid economic growth. In this paper, we used spatial analysis methods to reveal the global and local spatial distribution of county economy, and take full advantage of nonlinear fitting of the SOFM neural network model to evaluate and classify the counties' economic strength in 2009 with the intention to reveal the nature of spatial differences and promote a healthy development of the regional economy [3] [4].

Zhang Jinping · Qin Yaochen
College of Environment and Planning, Henan University, Kaifeng, China
e-mail: maryzhjp@126.com, qinyc@henu.edu.cn

F.L. Gaol et al. (Eds.): Proc. of the 2011 2nd International Congress on CACS, AISC 144, pp. 497–501.
springerlink.com © Springer-Verlag Berlin Heidelberg 2012

2 Study Area and Indicators

The scope of the region along the Lower Yellow River is defined by the counties covered by the Lower Yellow River irrigation area with Taohuayu in Xingyang, Henan Province as the cut-off point of the Middle and Lower Yellow River. The total land area is 103,000 square kilometers, involving three provinces of Henan, Shandong, and Hebei, and covering 109 counties. The agricultural production conditions of the region are the best in Huang-Huai-Hai Plain in China with the arable land is 5.626 million hectares, and grain output is 42.079 million tons in 2009.

A total of ten indicators, i.e. indicators of GDP per capita (0.1410), revenue per capita (0.1197), savings deposits of residents per capita (0.0794), total investment in fixed assets per capita (0.0889), average wage of staff and workers at work (0.0972), rural per capita net income (0.1090), proportion of nonagricultural industries in GDP (0.0948), proportion of gross industrial production in GDP (0.0972), proportion of fiscal expenditure in GDP (0.0770), all-personnel labor productivity (0.0958), have been selected to evaluate economic strength of 109 counties. The number in the parenthesis means the weight of each variable obtained by scoring from 20 experts. Indicators were Z-scored before multiplying the respective weights to calculate the correlation matrix [5].

3 Spatial Differentiation of the Economic Strength of Counties

The composite score derived from weighted principal component analysis is used to measure the economic strength of county. Weighted principal component analysis can combine the advantage of the subjective weight of indicators with the objective weight of data to make the results more in line with the reality [6]. The cumulative contribution rate of the first three principal components was up to 82.56%, and could reflect the vast majority of information of the total variance of the original indicators. The composite score F was calculated by multiplying the normalized contribution rate of eigenvalues with the corresponding principal component score to characterize the development of county economy.

Trend analysis tools in ArcGIS 9.3 and global Moran's I index in GeoDA 0.9.5-i can reveal the global spatial differentiation rules of the economic strength of counties along the Lower Yellow River [7]. The three-dimensional perspective of the composite score F of 109 counties shows that, the overall features of the spatial differentiation of county economy are obvious, with a clear U-shaped trend in the southwest - northeast direction, and a clear inverted U-shaped trend in the southeast - northwest direction.

In GeoDa software, the global Moran's I index was calculated by creating space adjacency matrix by the Queen adjacent way with Z value being 9.57 and p value 0.01. The index was 0.6571, indicating that the spatial autocorrelation is significant. Namely, the phenomenon that the developed counties are adjacent to the developed ones, and the economically backward counties are adjacent to the backward ones is very prominent.

Local spatial heterogeneity of economic strength of counties can be characterized by Moran scatter plot. Economic strength of counties along the Lower Yellow River demonstrated significant characteristics of aggregation of homogeneous areas and isolation of heterogeneous ones in the local space. Scatter in the first and third quadrants is very dense, accounting for 75%, with most being located near the regression line, revealing that the counties of high economic strength and those of low economic strength have formed significant aggregation areas respectively. The number of scatter in the third quadrant is more than that in the first quadrant, and is closer to the regression line, indicating that the concentration of counties of low economic strength along the lower Yellow River is much stronger.

Local indicators of spatial association (LISA) has been utlized in this paper to reveal the autocorrelation of economic strength in the neighborhood space, and find the 'hot' and 'cold' areas of economic development. It is found that, Xingyang and counties in Dongying and Jinan have the highest potential for the economic development in the region, as the hot areas, i.e. the core area in the well-developed economic zone (Figure 1). These counties have the ability to drive the adjacent counties to improve the level of economic development. The economically most backward counties are mainly in Yudong Plain. These counties distribute contiguously and become the economically 'cold' area in the region. There is a long way to enhance the level of economic development in these counties.

Fig. 1 LISA cluster map of the economy strength of counties

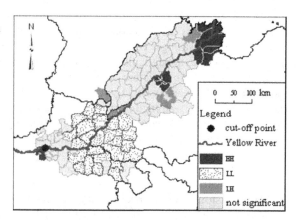

4   Classification of the Economic Strength of Counties

The three principal component scores derived from the weighted principal component analysis were normalized as the input data of the SOFM network to simplify the network structure, and therefore to avoid the instability of network solution process and the slow convergence speed [8]. In MATLAB 7.0, the function 'newsom' in neural network toolbox has been used to create a self-organizing feature map network. The classification of economic strength of 109 counties was achieved by programming. There were three input neurons in the network, and each neuron corresponded to one normalized principal component score, while the output neurons depended on categories to be distinguished.

According to the results by hierarchical clustering of gravity method, 109 counties were more appropriately to be divided into five categories by their economic strength [9]. Therefore, a five-layer neural network was designed in this study with each layer containing one neuron to represent one category. Set the maximum number of iterations as 100, 200, ..., 600 respectively to train the SOFM network designed thereby observe the distribution of the network weights after training.

The initial weights of the network were at [0.5, 0.5, 0.5], and not very well reflected the spatial distribution of input vectors. The network weights continuously moved towards the position of the input vector with the ongoing training, and the weight vectors gradually became orderly with a narrower range of neighborhood. After 400 times training, the layer of neurons has finished the adjustment of the weight vectors, so each neuron could respond to a small area that the input vector was located, and has completed the feature map of contacting 109 elements with five neurons through the neighborhood. Therefore, the network has been stable and could get a good output. Finally, use the simulation function 'sim' to train the network and turn the results into serial data using the function 'vec2ind'.

The counties were divided into five categories. Category one includes counties with the strongest economic level which are mainly located in Jinan, Zhengzhou and Dongying, where the economy is based on industry and services with a high degree of openness and obvious location advantages. Category two includes counties with the stronger economic level, mainly located in the east and west of the region, where nonagricultural industries dominate the economy and have formed the economic structure adapted to their economic development, but the industrial development is slightly less. Category three includes counties with the moderate economic strength, mainly in the western and northern Shandong Province, where industries of distinctive agricultural products have developed, and have somewhat good conditions for industrial development although lack of motivation and efficiency of the economic growth. Category four includes the counties of relatively weak economic strength, mainly in Luxi Plain and the north of Yudong Plain, where most are agricultural counties and industrial development is relatively slow with the people's low living standards. Category five includes counties with the weaker economic level, which are mainly in Yudong Plain and are the typical agricultural counties with the lagging behind industrial development and poor investment environment.

5   Conclusions and Discussions

It is very important to analyze the spatial differences and the level of economic strength in a county's economic development to narrowing the regional differences through the 'trickle-down effect' so as to enhance the overall efficiency of regional economic development. In this paper, weighted principal component analysis was used to get the composite score to measure the economic strength. In general, the economic strength of counties along the Lower Yellow River in 2009 shows a U-shaped trend in the southwest - northeast direction and an inverted U-shaped trend in the southeast - northwest direction in the global space with a clear characteristics of spatial concentration. In the local space, the distribution of the economic strength of counties is mainly 'HH' and 'LL' concentration; however, the concentration of counties of the weak economic strength is stronger. Counties with the best economic development such as Xingyang and counties in Dongying and Jinan have formed the 'hot' area, while counties in Yudong plain and Heze became the 'cold' area in the region along the Lower Yellow River. 109 Counties are classified into five economic strength categories of the strongest, stronger, moderate, weak, and weaker by using SOFM network model with three principal component scores as the input and the number of categories derived from hierarchical clustering of gravity method as the number of neurons to optimize parameters of SOFM network. The results could objectively reflect the economic development of counties along the Lower Yellow River. It can be concluded that, to avoid the curse of dimensionality and reduce the impact on the classification of SOFM network caused by different initial parameter settings, it is necessary to take measures to reduce the dimension of the input indicators and predetermine the number of categories to be distinguished.

References

[1] Jiang, G.G.: Empirical analysis of regional circular economy development - study based on Jiangsu, Heilongjiang, Qinghai Province. Energy Procedia 5, 125–129 (2011)

[2] Li, N., Shi, M.J., Wang, F.: Roles of Regional Differences and Linkages on Chinese Regional Policy Effect in CGE Analysis. Systems Engineering - Theory & Practice 29, 35–44 (2009)

[3] Shevtsova, L., Romanenkov, V., Sirotenko, O., Smith, P., Smith, J.U.: Effect of natural and agricultural factors on long-term soil organic matter dynamics in arable soddy-podzolic soils - modeling and observation. Geoderma 1-2, 165–189 (2003)

[4] Chertov, O.G., Komarov, A.S., Crocker, G.: Simulating trends of soil organic carbon in seven long-term experiments using the SOM model of the humus types. Geoderma 81, 121–135 (1997)

[5] Yang, Q., Gao, Q.Q., Chen, M.Y.: Study and integrative evaluation on the development of circular economy of Shaanxi province. Energy Procedia 5, 1568–1578 (2011)

[6] Kim, S.B., Rattakorn, P.: Unsupervised feature selection using weighted principal components. Expert Systems with Applications 38, 5704–5710 (2011)

[7] Getis, A.: Reflections on spatial autocorrelation. Regional Science and Urban Economics 37, 491–496 (2007)

[8] Yang, B.S., Hwang, W.W., Ko, M.H., Lee, S.J.: Cavitation detection of butterfly valve using support vector machines. Journal of Sound and Vibration 287, 25–43 (2005)

[9] Liu, G.Y., Yang, Z.F., Chen, B.: Emergy-based urban health evaluation and development pattern analysis. Ecological Modelling 220, 2291–2301 (2009)

Developing the Software Toolkit on 3DS Max for 3D Modeling of Heritage

Min-Bin Chen, Ya-Ning Yen, Wun-Bin Yang, and Hung-Ming Cheng

Abstract. Monuments and historic buildings, is the cultural legacy of our ancestors, but also social and cultural development for the life experience of technology, customs and heritage. 3D laser scanner technology, not only can do non-contact measurement of high accuracy digital data, but also can save the complete 3D digital data files for future restoration and value-added application. In this study, the digitization process of the original point cloud was obtaining information by reverse engineering software Rapidform XOR, and then converted them to 3DS Max for additional editing and applications. Appropriate Maxscript modeling program is developed to increase the editing efficiency of the designer.

1 Introduction

Historic buildings are cultural heritage left to us by our ancestors, including customs and techniques during social and cultural development. The purpose for preserving historic relics is to avoid improper damage or change to sites which are attractive and can continue their use [3].

Digitalization is an important technique that needs to be improved. The mature 3D laser scanner technique can bring digital technique a big step forward [5]. Extremely high accuracy digital data can be measured through non-contact method. And 3D digital data files can be preserved completely to serve as a basis for future conservation and restoration [9, 12]. In general, functions provided by 3D software are not for any specific object. How to quickly construct 3D models in response to change in quick cycles of reverse engineering is worth discussion [4].

Nowadays computer technology and information broadcasting advance fast. The academic circle and industrial circles also deeply recognize the importance of combining computer technology with other fields. Through digital conservation technology, 3D Studio Max is a powerful application which is generally used by the animation industry. This study adopted 3D Studio Max animation software as

Min-Bin Chen · Ya-Ning Yen · Wun-Bin Yang · Hung-Ming Cheng
China University of Technology, No. 56 Sec. 3 ShingLong Rd., 116 Taipei,
Taiwan(R.O.C.)
e-mail: cmb@cute.edu.tw, alexyen@cute.edu.tw,
    wunbin@cute.edu.tw, hungmc@cute.edu.tw

F.L. Gaol et al. (Eds.): Proc. of the 2011 2nd International Congress on CACS, AISC 144, pp. 503–508.
springerlink.com   © Springer-Verlag Berlin Heidelberg 2012

a platform. And it applied the MaxScript program language provided by 3D Studio Max software. The software possesses all characteristics that general programs have. In addition, it has processing functions such as model construction, animation installation, lighting, etc., to allow users to customize. Moreover, it can work with reverse engineering as a basic process and construct a 3D model required for overall reverse engineering [6]. Hopefully this may correspond to research and application trends of computer technology in other fields and cultivate cross domain talent needed at home and abroad.

Understanding and compilation of reverse engineering knowledge [8, 10], thereby exploring how to transform process into digital computer language which can be executed by computer, will be the basic job for combining both. Maxscript research can serve as a basis for other related research and clarify what role computer should play in cross domains. Moreover, computer aid software should be developed to meet various kinds of needs and achieve the purpose of combining computer technology and design.

This study analyzed historic relic elements often employed in historic relic in New Taipei City during historic relic modeling, preliminarily with pillars and roofs its research subjects. And a suitable Maxscript program was developed for quick modeling and modification by designers.

2   MaxScript Toolkit

Through interviews and records, we may understand methods and process of constructing a 3D model. In the same objects, we may construct it by objection copying and moving. In traditional building elements, buildings of the same attributes, such as temples, Japanese-style buildings, etc., elements which are seen more frequently are roofing tiles, pillars, etc. In the traditional model construction process, drawers often need to repeatedly construct the above elements. As a result, they will spend more time building relevant elements.

In respect of the above model elements which need more time to repeatedly construct, a prototype can be constructed with script so that drawers can create a model more quickly. Maxscript contains all features which general program languages have. Moreover, it can tackle model construction, animation configuration, lighting, etc. This allows users to customize functions. A 3D model can be constructed with the reverse engineering basic process.

2.1   MaxScript

Currently 3D model construction of cultural relics and buildings has many applications, such as building research, virtual guide, cityscape, etc. Some research agencies and firms in Taiwan have also undertaken relevant research. However, a traditional model construction process consumes much time and labor.

In particular, elements constructed with sophisticated curves of traditional relics and buildings are very complicated. A 3D model construction process often needs to repeatedly construct or copy the same elements. If the process can be made more automatic, then much time for constructing a model can be saved.

Currently there are many buildings which have been constructed in 3D. Traditional 2D CAD drawing is time-consuming. Chassagnoux [7] developed an automatic tool which can present 2D Gothic building arches and domes. This study attempted to identify roof and pillar elements which are often used in historic relics. 3D Studio Max is used as a platform, and MaxScript is adopted for 3D automation [1, 2, 11].

2.2 Implementation

This study aimed at roof and pillar elements which are often applied in historic relics, whose interface and program are provided below:

```
 3-11.ms - MAXScript
File  Edit  Search  Debugger  Help
rcMenu mymenu
(
submenu "About Author"
     (
          menuitem myitem1 "陳毓彬 楊文斌"
     )
)
rollout mywin "Untitled" width:162 height:300
(
     button btn1 "產生" pos:[99,59] width:46 height:24 toolTip:""
     label lbl1 "1.**涼亭柱子**" pos:[16,64] width:77 height:18
     label lbl4 "古蹟建築元件輔助工具" pos:[8,8] width:144 height:24
     groupBox grp1 "項目" pos:[8,32] width:144 height:256
     radioButtons rdo1 "形狀" pos:[15,95] width:53 height:46 labels:#("Rectangle", "Circle")
     spinner spn1 "距離" pos:[25,141] width:54 height:16 range:[1,250,50] type:#Integer
     spinner spn2 "數量" pos:[93,141] width:54 height:16 range:[1,6,1] type:#Integer
     label lbl2 "2.**壁瓦**" pos:[18,190] width:77 height:18
     button btn2 "產生" pos:[99,185] width:46 height:24 toolTip:""
     label lbl3 "2008 陳毓彬 楊文斌 著" pos:[25,280] width:120 height:18
     colorpicker mypicker "選顏色" pos:[40,250]
     label lbl5 "═══════════" pos:[15,160] width:120 height:18

     on unnamedRollout open do
(
          )
     on unnamedRollout close do
(
```

Fig. 1 Preview in the source code

If more pillars are needed, choose distance and quantity to add needed quantity and accurate relative distance.

Fig. 2 Rectangular pillar and circular pillar

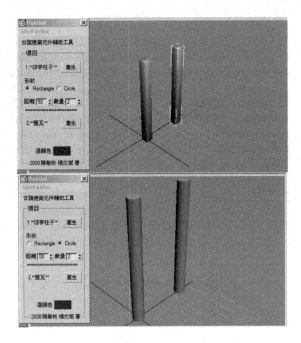

Besides, in respect of roofing tiles, outline of roofing tiles can be automatically drawn through the program. With Extrude, form of roofing tiles (traditional south Fujian roofing tiles) can be generated.

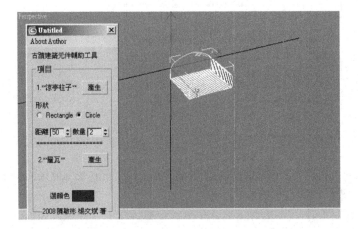

Fig. 3 Roofing tiles function

Fig. 4 The Lin Family Mansion and Garden, Banqiao, Taipei

3 Conclusion

A design computing perspective is provided through customizing with Maxscript in response to RE process in order to speed up 3D model construction efficiency. A 3D laser scanner retrieves point cloud digital file format. After original data is processed, the result can be converted to a linear image file, CAD model, 3D model, or animation (Fig.4). Moreover, through RE software Rapidform XOR, point clouds can be processed fast and a physical 3D model can be constructed. After that, the model can be preserved completely with relevant parameterized software to construct digital data and undertake subsequent editing. Once relevant data of 3D digital historic relics is scanned and a complete 3D digital database for historic buildings is constructed, we may master information and knowledge for promoting digital education. By constructing a 3D model and object, a complete virtual reality is created. Through editing of computer technique, a better digital image can be presented. And the purpose of popularizing sophisticated culture and increasing historic relic preservation is achieved.

References

[1] Bicalho, A., Feltman, S.: Maxscript and the Sdk for 3d Studio Max (2000)
[2] Butterworth-Heinemann, 3ds Max 9 MAXscript Essentials (2006)
[3] Bjork, B.C.: A Case Study of A National Building Industry Strategy for Computer Integrated Construction. International Journal of Construction Information Technology (Autumn 1993)

[4] Puntambekar, N.V., Jablokow, A.G., Sommer, H.J.: Unified review of 3D model generation for reverse engineering. Computer Integrated Manufacturing Systems 7(4), 259–268 (1994)

[5] Cheng, X.J., Jin, W.: Study on Reverse Engineering of Historical Architecture Based on 3D Laser Scanner. In: Proceedings of ISIST 2006, Harbin, 9-12. Conference Series, vol. 48, pp. 843–849 (2006)

[6] Aish, R., Sachs, E., Stoops, D.: 3-DRAW: A three dimensional computer-aided design tool. IEEE Computer Graphics and Applications 11(6), 18–26 (1992)

[7] Chassagnoux, A., Guena, F., Maille, M., Untersteller, L.P.: Modelisation and calculation of gothic or ogive vaults, First european congress on restoration of gothic cathedrals. In: Europa Congress Hall, VITORIA - GASTEIZ (1998)

[8] Liu, J.-Z.: Optimal design for reverse engineering of CAD model reconstruction and machining, Institute of Mechanical Engineering, National Cheng Kung University. Master Thesis (1999)

[9] Lin, C.-Y.: The application of a 3D scanner in the digital information model of historical building– A Taipei Pao-An temple example. Master Thesis, National Taiwan University of Science and Technology (2004)

[10] Zhou, H.-Y.: Introduction to reverse engineering system. Journal of the Mechatronic Industry (141), 130–136 (1999), Master Thesis

[11] Wang, H.: 3ds MAXScript script learning manual. Beijing Hope Electronic Press (2006)

[12] Xu, Z.-Q., Fu, J.-Y., Zhang, B.-F., Zheng, Y.-Z., Ye, S.-H.: Fast forming technique: A study on 3D laser scanning measurement method. State Key Laboratory of Precision Instruments and technology, Tianjin (2004)

Author Index

Ahmed, Moataz   163
Alssir, Fakhreldin T.   163

Bao, Junjiang   341
Baurley, James W.   377

Caarstens, Cobus   21
Cai, Yifan   69
Cai, Zongxi   75
Cao, Lijun   459
Cao, Menglei   9
Chang, Bao Rong   451
Chen, Chi-Ming   451
Chen, Jian-Shiang   43
Chen, Min-Bin   503
Cheng, Hung-Ming   503
Chung, Sung Woo   355

Dai, Bin   75
Declerck, L.   297
Desell, Travis   385
Ding, Linlin   133
Ding, Qingxin   151
Dong, Huijie   471

Edlund, Christopher K.   377
Eziwarman   443

Forbes, G.L.   443

Gatti, Rathishchandra R.   403
Guo, Chao   303
Guo, Lijun   471
Guo, Sy-Jye   215
Guo, Tengfei   9

Hamadicharef, Brahim   101
Hao, Hong-Wei   107, 119
Hao, Lili   61
Haraty, Ramzi A.   181
He, Qi   107, 119
Howard, Ian M.   403, 443
Hsu, Cheng-Chih   43
Hu, Ming   75
Hu, Wen Bin   199
Hu, Zhong   329
Huang, Chan   95
Huang, Chien-Feng   451
Huang, Jun-Bo   483
Huang, Tian   75
Huang, Xuewen   311
Huhtamäki, Jukka   361, 369

Jang, Jiin-Yuh   477, 483
Jiang, Huina   125
Jiang, Linying   53

Kailanto, Meri   361, 369
Kang, Yilan   75
Kasprzak, Andrzej   191
Kawai, Masayuki   29
Kim, Jae Min   355
Koszalka, Leszek   191
Koutaki, Gou   207

Lebègue, P.   297
Lee, Chi-Yung   231
Leng, Jianzhong   75
Li, Chang-Sian   215

Li, Dancheng 53
Li, Duo 303
Li, Lingjuan 113
Li, Suying 341
Li, XiaoBing 341
Li, Xiaochun 151
Liang, Xiao 9
Lin, Cheng-Jian 231
Lin, Peng-chun 349
Lin, Zhong-jie 395
Lin, Zih-Yao 451
Ling, Hangkun 269
Ling, Ning 269
Liu, Chih-Wen 411
Liu, Donghui 39
Liu, Gang 261
Liu, Hanxi 335
Liu, Jingming 53
Liu, Jingyu 329
Liu, Qingyu 421
Liu, Shan 459
Liu, Su 429
Liu, Wenrui 239
Liu, Xiaobing 311
Lu, Guoyu 113
Lu, Hsin-ke 349
Lu, Kaixin 395
Lumia, Ron 1
Luo, Pei 491
Luo, Ting 283

Ma, Xueli 311
Magdon-Ismail, Malik 385
Marttila, Jarno 361, 369
Mendez, M. 297
Miilumäki, Thumas 361, 369
Miksa, Tomasz 191
Mittal, Gauri S. 69
Morimoto, Shoichi 141
Muller, Neil 21

Newberg, Lee A. 385
Nitta, Yoshihiko 223
Nollet, V. Fèvre 297
Nuo, Qun 253

Oumoudian, Ohan R. 181

Pardamean, Bens 377
Peng, Cheng 81
Pozniak-Koszalka, Iwona 191

Qi, Zengying 291
Qiao, Baiyou 133
Qiao, Songyue 395
Qin, Yaochen 497
Qu, Chuanyong 75
Qu, Tsi 253

Ramadhan, Arief 157
Renqian, Duojie 253

Sha, Yun 125
Shao, Jun 335
Shiue, Yu-Jia 231
Silius, Kirsi 361, 369
Sun, Xiaoyun 39
Suroso, Arif Imam 157
Suta Wijaya, I Gede Pasek 207
Szymanski, Boleslaw K. 385

Tebest, Teemo 361, 369
Tervakari, Anne-Maritta 361, 369
Thompson, William 385
Tian, Fei 151
Tong, Qiuli 81
Trashi, Nyima 253
Tsai, Hsiu Fen 451
Tsai, Ying-Chi 477
Tso, Yong 253

Uchimura, Keiichi 207

van der Bijl, Leendert 21

Wang, Honglun 9
Wang, Jingqiu 321
Wang, Shacheng 87
Wang, Shu-ming 349
Wang, Tao 329
Wang, Xiaohua 125
Wang, Xiaolei 321
Wang, Xiaoyang 133
Wang, Xue-shun 395
Wang, Zhenhui 471
Wei, Yong 133
Wu, Chenguang 465
Wu, Chi-Feng 231
Wu, Li-Cheng 411

Xia, Ming   125
Xiao, Min   239
Xiong, Huasheng   303
Xu, Chao   239
Xu, Qifeng   291
Xu, Xin   69
Xue, Guofeng   491

Yamakami, Toshihiko   171
Yan, Taowei   437
Yang, Changming   429
Yang, Fengyan   395
Yang, Huizhen   61
Yang, Junhui   95
Yang, Junlei   335
Yang, Kun   465
Yang, Lili   39
Yang, Simon X.   69
Yang, Ting   395
Yang, Wun-Bin   503
Yang, Xiaoyan   261
Yen, Ya-Ning   503

Yin, Xu-Cheng   107, 119
Yu, Jingyang   465
Yuan, Yixing   465

Zeng, Xue   113
Zhang, Hong   199
Zhang, Hong-yan   247
Zhang, Jian   491
Zhang, Jinping   497
Zhang, Li   277
Zhang, Peichang   471
Zhang, Qian   75
Zhang, Si Di   199
Zhao, Haoyi   29
Zhao, Xunfeng   491
Zheng, Mengze   421
Zhou, Guimei   437
Zhou, Junqing   335
Zhou, Zongfang   239
Zhu, Wenquan   341
Zhu, Zhiliang   53
Zou, Xiao-Ling   283